CONTINUUM THEORY OF INHOMOGENEITIES IN SIMPLE BODIES

A REPRINT OF SIX MEMOIRS

BY

W. NOLL, R. A. TOUPIN, AND C.-C. WANG

WITH A PREFACE BY C. TRUESDELL

SPRINGER-VERLAG NEW YORK INC.

1968

ISBN 978-3-642-85994-6 ISBN 978-3-642-85992-2 (eBook)
DOI 10.1007/978-3-642-85992-2

 Library of Congress Catalog Card Number 68-56948
Softcover reprint of the hardcover 1st edition 1968

Title No. 1531

Preface

The term "dislocation" is used in several different senses in the literature of mechanics. In the classic work of VOLTERRA, WEINGARTEN, and SOMIGLIANA, it refers to particular solutions of the equations of linear elasticity, in which a continuous field of strain does not correspond, globally, to a continuous field of displacement. The configuration of the body so obtained, even when that body is free of all load, is subject to interior stress that does not vanish, and in general no deformation of the body as a whole can bring it into a stress-free configuration. Nevertheless, if any sufficiently small part of the body is considered by itself, a configuration for it in which the stress is everywhere zero may be found at once. In this work *constitutive assumptions* provide the basic data. These consist in prescribed stress-free configurations for each material point and in prescribed elastic moduli governing the response to deformation from the stress-free configuration at each material point. Everything follows from these data, including the dislocations present, if any. In particular, the common boundary-value problems of linear elasticity may be set and solved for the dislocated body.

About twenty years ago solid-state physicists and metallurgists began to use the term "dislocation" in a different way. Unlike VOLTERRA's theory, their formalisms did not rest on any constitutive assumption relating the forces in a continuous body to the deformation giving rise to them. Rather, they envisioned the body as a periodic array of points or tiny balls, which they called an "atomic lattice". Removal of one or more of these balls, or addition of one where there was none before, was named a "dislocation", a "vacancy", an "impurity", or a "stacking fault", and many special kinds of these were described and interpreted. The discrete models, carried heuristically to the limit of infinitely close packing, served to inspire a description of "continuously distributed dislocations" through notions of differential geometry, typically by means of affine connections with non-vanishing Cartan-torsion. Since no constitutive assumptions are laid down explicitly, the literature on these "continuous distributions of dislocations" does not contain the solutions of any specific problems in the sense in which "problem" is used in classical elasticity: deformation resulting from specified loading, or loading required to effect a particular deformation. Indeed, the solid-state physicists seem never to have derived differential equations of motion for the bodies they hypothecated.

In continuum physics it is taken for granted that the physical properties of a body are set forth once and for all by *constitutive relations*, which specify the response produced at each material point as a result of the history of appropriate physical parameters in a neighborhood of that point. Once the constitutive properties of all material points, and the relations among them, have been assumed, everything else is to be discovered and *proved* by mathematics. In the common applications the bodies are assumed homogeneous in the sense of having a global configuration such that the response of each material point in it is the same as any other's. This assumption is not necessary, however, in order that all material points be physically alike. In § 34 of *The Non-Linear Field Theories of Mechanics*[1], NOLL sketched but did not develop the concept of a simple body that is *uniform* without being homogeneous. In this case, there exists for any pair of particles a pair of configurations from which the response is identical, but there is no single such configuration for the whole body. A uniform but inhomogeneous body provides another concept of continuously distributed dislocations: inhomogeneities fixed in a uniform body. It includes VOLTERRA's theory but is far more general, not only in the nature of the dislocations themselves but also in the response of the body, which may be of any kind, mechanical or non-mechanical, and in the material symmetry, which is likewise arbitrary: fluid, solid, or fluid crystal. It is similar to VOLTERRA's theory and different from the lattice theories in that definite constitutive assumptions are laid down once and for all and are made the basis of a concrete, deductive treatment consisting in mathematical definitions and mathematical proofs. No geometric structure is imposed on the body manifold *a priori*; rather, that structure is *determined by precise mathematical process* in terms of the constitutive assumptions laid down. Within this framework NOLL was able to derive the *differential equations of motion* for an arbitrary materially uniform elastic body. He announced them without proof as Equations (44.7) of *The Non-Linear Field Theories of Mechanics.*

The purpose of this volume of reprints is to present the general theory of inhomogeneities and crystalline defects in bodies that are simple with respect to their mechanical, thermal, electrical, or other physical behavior. The emphasis is on mechanics and, more specifically, on the motivation and derivation of NOLL's differential equations.

The first article is a recent survey by NOLL himself, intended to explain the basic ideas in simple words with a minimum of mathematics.

In the second article, TOUPIN develops a different possibility. He defines a *crystal* as a simple body whose material points at similar positions on unit cells are materially isomorphic. In this way he is able to give a precise description of a body with a finite number of defects and to view such a body as consisting of an oriented material.

The detailed mathematical treatment begins with the third article, NOLL's memoir of 1958, which introduces and develops basic concepts of modern continuum mechanics: constitutive functional, simple particle, simple body, objectivity, material isomorphism, isotropy group[2], (simple) fluid, (simple) solid. These concepts and terms have since become the *lingua franca* of continuum mechanics. In this paper, the possibility of a body that is uniform but not homogeneous is first remarked.

Next follows NOLL's recent memoir on the theory of inhomogeneous bodies. The mathematical development is self-contained and requires nothing more of the

---

[1] In FLÜGGE's Handbuch der Physik III/3, Berlin-Heidelberg-New York, 1965.

[2] It should be noticed that the abstract isotropy group as well as the development of materials of higher grade, omitted from the paper of 1958, had been introduced in a report by NOLL issued in June, 1957, by the Mathematics Department of the Carnegie Institute of Technology: "On the foundations of the mechanics of continua."

reader than a good background in modern continuum mechanics and willingness to learn.

The long paper by WANG presents and develops a more general theory, with more liberal possibilities for the global topological structure of the bodies considered. The analysis here rests on the theory of fibre bundles. Again, NOLL's equations of motion result for all uniform elastic bodies.

In the final paper in the volume, WANG obtains the first solutions of NOLL's equations. He constructs the concept of a *laminated body*, which may be visualized as a sandwich made by glueing together infinitely thin slices of a homogeneous body after each slice has been subjected to a deformation symmetric with respect to some preferred direction. Such bodies are uniform but inhomogeneous. WANG considers plane, cylindrical, and spherical lamination, and for each he finds infinitely many universal solutions, provided the material be incompressible and isotropic. In the particular case of a Mooney material, he gets very detailed answers in elementary terms. For example, he shows that by suitable lamination, a circular rod consisting of a given Mooney material and having a given size, may be made to have any desired torsional rigidity within a certain interval.

The papers reprinted in this volume may serve in lieu of a treatise on this new branch of mechanics.

C. TRUESDELL

Contents

Inhomogeneities in Materially Uniform Simple Bodies

By

Walter Noll

Department of Mathematics
Carnegie-Mellon University, Pittsburgh

1. Introduction

Many physical characteristics of continuously distributed matter are *local* in the sense that they pertain to individual material points and their immediate neighborhoods, rather than to a body as a whole. Examples of such local characteristics are: elastic response, viscosity, heat capacity, electrical conductivity, magnetic response, optical refraction, chemical composition, color.

The *theory of simple bodies* deals only with such local characteristics. A given set of physical phenomena is then governed by appropriate physical responses at the various material points of the body. The term "simple" expresses the assumption that deformations whose gradient is the identity at a given material point do not alter the physical response at that point with respect to the set of phenomena under consideration. Roughly speaking, a body is simple if only first spatial gradients occur in the constitutive description of its physical properties. A simple body is called *materially uniform* if the physical response is the same at all points of the body.

I shall show that the theory of the structure of materially uniform simple bodies leads, by the force of logic alone, to the possibility of inhomogeneities. Under sufficient smoothness assumptions, such inhomogeneities can be described locally in terms of a certain third order tensor field, which I call the *inhomogeneity* of the body. This inhomogeneity corresponds to what is called "dislocation density" in the theory of continuous distributions of dislocations as described in other contributions to this Symposium.

A particular type of inhomogeneity is what I call *contorted aeolotropy*, which includes the more familiar curvilinear aeolotropy as a special case. Certain types of laminated and fibrous bodies give intuitive physical examples of bodies with contorted aeolotropy.

If one deals with mechanical material properties one must take into account the principle of balance of forces, which gives rise to Cauchy's familiar equations of balance. The classical form of these equations, however, is unsuited for dealing with inhomogeneous uniform bodies. I shall give a new form for these equations, a form tailored to bodies with a given distribution of inhomogeneities.

The presentation I shall give here will be somewhat informal and I shall omit almost all proofs. A more complete and detailed description of the theory will be found in my paper "Materially Uniform Simple Bodies with Inhomogeneities", Archive for Rational Mechanics and Analysis, Vol. 27, pp. 1—32. That paper contains also the relevant references.

2. The Concept of a Smooth Body. Local Configurations. References

In continuum physics, a *body* $\mathscr{B}$ is described mathematically as a set of *material points* X, Y, . . . The *configurations* of $\mathscr{B}$ in space are mappings

$$\varkappa : \mathscr{B} \to \mathscr{E},$$

where $\mathscr{E}$ is a fixed three-dimensional Euclidean space. The translation space of $\mathscr{E}$, i.e., the space of *spatial vectors*, is denoted by $\mathscr{V}$.

We say that $\mathscr{B}$ is a *smooth* body if all of its configurations can be obtained from one of them by smooth deformations, i.e. by one-to-one mappings having suitable differentiability properties.

Let $f : \mathscr{B} \to \mathscr{R}$ ($\mathscr{R}$ = real line) be a scalar field on $\mathscr{B}$, i.e. a function that assigns a real number to each material point. Given a configuration $\varkappa$ of $\mathscr{B}$, we can consider the composition $f \circ \overset{-1}{\varkappa} : \varkappa(\mathscr{B}) \to \mathscr{R}$, which is a function that assigns a real number to each point in the region $\varkappa(\mathscr{B})$ occupied by $\mathscr{B}$ in the configuration $\varkappa$. We assume that $\varkappa(\mathscr{B})$ is an open set. It may happen that $f \circ \overset{-1}{\varkappa}$ is differentiable. Its gradient $\nabla\big(f \circ \overset{-1}{\varkappa}\big) : \varkappa(\mathscr{B}) \to \mathscr{V}$ is then a function which assigns to each point in the region $\varkappa(\mathscr{B})$ a spatial vector. The composition of this function with $\varkappa$ is denoted by $\nabla_\varkappa f : \mathscr{B} \to \mathscr{V}$, so that

$$\nabla_\varkappa f|_X = \nabla\big(f \circ \overset{-1}{\varkappa}\big)\big|_{\varkappa(X)}, \qquad X \in \mathscr{B}. \tag{2.1}$$

($|_X$ means that the function in question is to be evaluated at X.) The vector field $\nabla_\varkappa f$ on $\mathscr{B}$ defined by (2.1) is called the *gradient of f relative to the configuration $\varkappa$*.

Now let $\boldsymbol{\lambda} : \varkappa(\mathscr{B}) \to \mathscr{E}$ be any smooth deformation. The gradient $\nabla \boldsymbol{\lambda}|_{\varkappa(X)}$ at the point $\varkappa(X)$ in the region $\varkappa(\mathscr{B})$ is an invertible linear transformation of $\mathscr{V}$. The composition $\gamma = \boldsymbol{\lambda} \circ \varkappa : \mathscr{B} \to \mathscr{E}$ is again a configuration. The chain rule for the differentiation of compositions yields

$$\nabla_{\varkappa} f|_X = (\nabla \boldsymbol{\lambda}|_{\varkappa(X)})^T (\nabla_{\gamma} f)|_X, \tag{2.2}$$

where the upper index T denotes transposition.

It is clear from (2.2) that

$$\nabla_{\varkappa} f|_X = \nabla_{\gamma} f|_X \quad \text{if} \quad (\nabla \boldsymbol{\lambda})|_{\varkappa(X)} = \mathbf{1}$$

($\mathbf{1}$ = identity transformation of $\mathscr{V}$). Thus, the gradient $\nabla_{\varkappa} f|_X$ remains unaltered if $\varkappa$ is replaced by a configuration which differs from $\varkappa$ only by a deformation whose gradient at $\varkappa(X)$ is the identity.

Let $X \in \mathscr{B}$ be a specific material point. Two configurations $\varkappa$ and γ are said to be *equivalent at* X if the gradient at $\varkappa(X)$ of the deformation $\boldsymbol{\lambda} = \gamma \circ \overset{-1}{\varkappa}$ relating $\varkappa$ and γ is the identity, i.e. if $(\nabla \boldsymbol{\lambda})|_{\varkappa(X)} = \mathbf{1}$. The equivalence classes for the equivalence relation thus defined are called *local configurations at* X. A function K which associates with each material point $X \in \mathscr{B}$ a local configuration at X is called a *reference* on $\mathscr{B}$. If $\varkappa$ is a configuration, we denote by $\nabla \varkappa$ the reference which associates with each $X \in \mathscr{B}$ the equivalence class (i.e. the local configuration at X) to which $\varkappa$ belongs. References of the form $\nabla \varkappa$ are said to be *homogeneous.*

To visualize a reference K, one has to cut the body into infinitesimal pieces and to view the piece corresponding to the material point X in a configuration that belongs to the equivalence class $K(X)$. The pieces will *not* fit together to form a coherent region in space if K is an inhomogeneous reference. If $K = \nabla \varkappa$ is homogeneous, however, then the pieces can be fit together to form the region $\varkappa(\mathscr{B})$.

The fact that $\nabla_{\varkappa} f|_X = \nabla_{\gamma} f|_X$ holds if $\boldsymbol{\lambda} = \gamma \circ \overset{-1}{\varkappa}$ satisfies $(\nabla \boldsymbol{\lambda})|_{\varkappa(X)} = \mathbf{1}$ can be expressed by saying that the value at X of the gradient relative to $\varkappa$ of the scalar field f depends on $\varkappa$ only through the local configuration at X defined by $\varkappa$. Hence, if K is any reference, it is meaningful to define the *gradient* $\nabla_K f : \mathscr{B} \to \mathscr{V}$ *of* f *relative to the reference* K by

$$\nabla_K f|_X = \nabla_{\varkappa} f|_X \quad \text{for all } \varkappa \in K(X). \tag{2.3}$$

Gradients relative to a reference K can be defined not only for scalar fields, but in an analogous manner also for vector-fields, i.e. functions $\boldsymbol{h} : \mathscr{B} \to \mathscr{V}$, and tensor fields, i.e. functions $\boldsymbol{T} : \mathscr{B} \to \mathscr{L}_1$, where $\mathscr{L}_1$ is the space of all linear transformations of $\mathscr{V}$ into itself. In the same way one can also define the relative gradient $\nabla_K \gamma$ of a configuration γ. Its value $\nabla_K \gamma|_X$ is an invertible member of $\mathscr{L}_1$ which depends only on $K(X)$ and the local configuration $\nabla \gamma(X)$ at X defined by γ. We express this fact by using the notation

$$\nabla_K \gamma|_X = (\nabla \gamma(X))\,(K(X))^{-1}, \qquad \nabla_K \gamma = (\nabla \gamma)\, K^{-1}. \tag{2.4}$$

Conversely, given any invertible $\boldsymbol{L} \in \mathscr{L}_1$, there exists configurations γ such that $\boldsymbol{L} = \nabla_K \gamma|_X$. The class of all these configurations constitute a local configuration at X, which we denote by $\boldsymbol{L}K(X)$. Thus, every local configuration at X can be written in the form $\boldsymbol{L}K(X)$ with a suitable choice of an invertible $\boldsymbol{L} \in \mathscr{L}_1$.

3. The Concept of a Materially Uniform Simple Body

The physical properties of a body $\mathscr{B}$ in a given configuration $\varkappa$ can often be described by functions that assign to every material point $X \in \mathscr{B}$ some quantity with a particular physical meaning. For example, the inertial and gravitational properties can be described by giving the mass density at each $X \in \mathscr{B}$; the forces that hold the body together can be described by giving a stress tensor at each $X \in \mathscr{B}$; the thermal properties can be described by giving the heat capacity and heat conductivity at each $X \in \mathscr{B}$.

In general, given a configuration $\varkappa$ of $\mathscr{B}$, the physical response at $X \in \mathscr{B}$ is described by specifying some member of a set $\mathfrak{R}$. We denote this member of $\mathfrak{R}$ by $\mathfrak{G}_X(\varkappa)$ and call it the *response descriptor* of the material at X. The nature of the set $\mathfrak{R}$ depends on the physical phenomena under consideration. For example, if $\mathscr{B}$ is an elastic body, $\mathfrak{R}$ is the space of all symmetric tensors and $\mathfrak{G}_X(\varkappa) \in \mathfrak{R}$ is the stress tensor at X if $\mathscr{B}$ is in the configuration $\varkappa$. If we consider only the inertial and gravitational behavior of $\mathscr{B}$, then $\mathfrak{R}$ is the set of all positive real numbers and $\mathfrak{G}_X(\varkappa)$ is the mass density at X if $\mathscr{B}$ is in the configuration $\varkappa$.

It may happen that $\mathfrak{G}_X(\varkappa)$ depends on $\varkappa$ only through the local configuration $\nabla\varkappa(X)$ at X defined by $\varkappa$. In this case, we say that $\mathscr{B}$ is a *simple body* and we write $\mathfrak{G}_X(\nabla\varkappa(X))$ instead of $\mathfrak{G}_X(\varkappa)$. We can then regard $\mathfrak{G}_X$ as a mapping from the set of all local configurations at X into the set $\mathfrak{R}$ of all response descriptors.

Now let K be a reference. We have seen in the previous section that every local configuration at X is of the form $\boldsymbol{L}K(X)$, where $\boldsymbol{L}$ is an invertible member of $\mathscr{L}_1$. The response descriptor of the body at X in any configuration belonging to $\boldsymbol{L}K(X)$ is given by $\mathfrak{G}_X(\boldsymbol{L}K(X))$. The physical response to deformations of the body at $X \in \mathscr{B}$, relative to the reference K, is described by the nature of the relation between $\boldsymbol{L}$ and the descriptor $\mathfrak{G}_X(\boldsymbol{L}K(X))$. Therefore, it appears reasonable to say that the response of the material at X is the same as the response of the material at Y if

$$\mathfrak{G}_X(\boldsymbol{L}\,K(X)) = \mathfrak{G}_Y(\boldsymbol{L}\,K(Y)) \tag{3.1}$$

holds for all invertible $\boldsymbol{L} \in \mathscr{L}_1$. The response would be the same everywhere if (3.1) holds identically in $\boldsymbol{L}$ for all material points X and Y.

However, whether or not (3.1) holds for all invertible $\boldsymbol{L}$ and all $X, Y \in \mathscr{B}$ depends on the choice of the reference K. The references, if any, for which (3.1) does hold will be called *uniform references*.

We say that a body is *materially uniform* if it admits uniform references. We say that the body is *homogeneous* if it admits homogeneous uniform references, i.e. uniform references of the form $\nabla \varkappa$, where $\varkappa$ is a (global) configuration. Intuitively, if a body is homogeneous, we can view it in a configuration such that all parts of the body respond in the same way. If the body is inhomogeneous but materially uniform, we must first cut it into infinitesimal pieces that do not fit together before we can make all parts respond in the same way.

A materially uniform body $\mathscr{B}$ always admits infinitely many uniform references. However, from now on we shall fix the attention to one particular uniform reference K. Also, we shall assume that K is smooth in the sense that the invertible tensor field $\nabla_K \gamma$ is smooth for every configuration γ.

4. The Inhomogeneity Field

Let K be a uniform reference and γ an arbitrary configuration. We write

$$\boldsymbol{F} = \nabla_K \gamma = (\nabla \gamma)\, K^{-1} \tag{4.1}$$

for the gradient of γ relative to K. This gradient $\boldsymbol{F}$ is a smooth tensor field on $\mathscr{B}$ with invertible values in $\mathscr{L}_1$. The gradient $\nabla_K \boldsymbol{F}$ relative to K is a field on $\mathscr{B}$ with values in the space $\mathscr{L}_2$ of all linear transformations of $\mathscr{V}$ into $\mathscr{L}_1$. Thus, if $\boldsymbol{u}, \boldsymbol{v} \in \mathscr{V}$ are spatial vectors, then $(\nabla_K \boldsymbol{F})\, \boldsymbol{u} \in \mathscr{L}_1$ and $((\nabla_K \boldsymbol{F})\, \boldsymbol{u})\, \boldsymbol{v} \in \mathscr{V}$.

Consider now the case when $K = \nabla \varkappa$ is a homogeneous uniform reference. In this case, we have $\boldsymbol{F} = \nabla_\varkappa \gamma$ and $\nabla_K \boldsymbol{F} = \nabla_\varkappa \boldsymbol{F} = \nabla_\varkappa \nabla_\varkappa \gamma$. In view of the well-known symmetry property of second gradients, it follows that

$$((\nabla_K \boldsymbol{F}) \boldsymbol{u})\, \boldsymbol{v} - ((\nabla_K \boldsymbol{F})\, \boldsymbol{v})\, \boldsymbol{u} = \boldsymbol{O} \tag{4.2}$$

holds for all $\boldsymbol{u}, \boldsymbol{v} \in \mathscr{V}$ when K is a homogeneous reference. By classical theorems of analysis, the converse is true locally: if (4.2) holds for all $\boldsymbol{u}, \boldsymbol{v} \in \mathscr{V}$, then every material point has a (finite) neighborhood such that K is a homogeneous reference of that neighborhood.

The consideration just shown proves that the field $\boldsymbol{S} : \mathscr{B} \to \mathscr{L}_2$ defined by

$$(\boldsymbol{S}\, \boldsymbol{u})\, \boldsymbol{v} = \boldsymbol{F}^{-1} [((\nabla_K \boldsymbol{F})\, \boldsymbol{u})\, \boldsymbol{v} - ((\nabla_K \boldsymbol{F})\, \boldsymbol{v})\, \boldsymbol{u}] \tag{4.3}$$

vanishes if and only if the uniform reference K is locally homogeneous. Now, it turns out that $\boldsymbol{S}$ depends only on the choice of K and not on the choice of the configuration γ, even though $\boldsymbol{F}$ and $\nabla_K \boldsymbol{F}$ depend on both K and γ. Therefore, the field $\boldsymbol{S}$ defined by (4.3) gives an intrinsic measure of the deviation of the uniform reference K from being homogeneous, at least locally. We call the field $\boldsymbol{S}$ the *inhomogeneity* of the uniform reference K.

5. Contorted Aeolotropy

A uniform reference K may happen to have the following property: There exists a (global) configuration γ such that the gradient

$$\boldsymbol{Q} = \nabla_K \gamma = (\nabla \gamma)\, K^{-1} \tag{5.1}$$

of γ relative to K has only *orthogonal* values. If this is the case, we say that K is a reference of *contorted aeolotropy*. Intuitively, if we view the body in the configuration γ and then cut it into infinitesimal pieces, mere rotations of these pieces will bring them into configurations such that they all respond in the same way.

A real body which can be expected to possess contorted aeolotropy can be manufactured in the following manner: Take very many thin sheets of a homogeneous material and bend them into cylindrical shape in such a way that they can be stacked snugly. Then glue them together with a homogeneous glue (see Figure). One thus obtains a body in a certain configuration γ. If we consider small pieces of the body at material points X and Y, we see that these pieces can be brought into alignment by rotating one of them (see Figure).

For the body we have just described, we can introduce a cylindrical coordinate system and choose the orthogonal tensor field $\boldsymbol{Q}$ in such a way that $\boldsymbol{Q}(X)\,\boldsymbol{e}_i(X)$ is independent of X when the $\boldsymbol{e}_i(X)$, $i = 1, 2, 3$, are unit vectors pointing in the direction of the coordinate lines. When an orthogonal curvilinear coordinate system with the property just mentioned exists, we say that the resulting uniform reference is a reference of *curvilinear aeolotropy*. Not all references of contorted aeolotropy are also of curvilinear aeolotropy. An example is given by taking many very thin fibers of homogeneous material, twisting them together as in a rope, and then glueing them together.

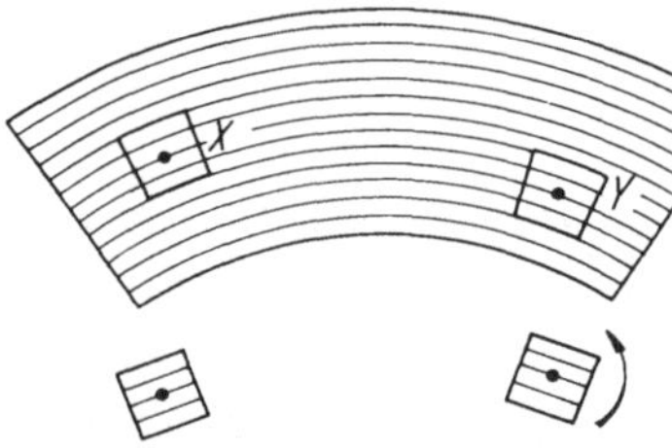

Contorted aeolotropy

The rate at which the orthogonal tensor field $\boldsymbol{Q}$ on $\mathscr{B}$ changes as one proceeds in the direction $\boldsymbol{u}$ is described by

$$\boldsymbol{D}\,\boldsymbol{u} = -\,\boldsymbol{Q}^{\mathsf{T}}\big((\nabla_K \boldsymbol{Q})\,\boldsymbol{u}\big). \tag{5.2}$$

The field $\boldsymbol{D}:\mathscr{B}\to\mathscr{L}_2$ defined by (5.2) is called the *contortion* of the reference of contorted aeolotropy given by (5.1). It is easily seen that this reference is homogeneous if and only if its contortion is zero.

One can prove that the inhomogeneity $\boldsymbol{S}$ and the contortion $\boldsymbol{D}$ determine one another by the following formulas:

$$(\boldsymbol{S}\,\boldsymbol{u})\,\boldsymbol{v}=(\boldsymbol{D}\boldsymbol{u})\,\boldsymbol{v}-(\boldsymbol{D}\boldsymbol{v})\,\boldsymbol{u}, \tag{5.3}$$

$$(\boldsymbol{D}\boldsymbol{u})\,\boldsymbol{v}=\tfrac{1}{2}\{[(\boldsymbol{S}\,\boldsymbol{u})-(\boldsymbol{S}\,\boldsymbol{u})^T]\,\boldsymbol{v}-(\boldsymbol{S}\,\boldsymbol{v})^T\,\boldsymbol{u}\}. \tag{5.4}$$

These formulas are meaningful even if K is not a reference of contorted aeolotropy. Hence, if K is *any* uniform reference, we may define its contortion $\boldsymbol{D}$ by (5.4). If K is a reference of contorted aeolotropy, the field $\overset{*}{\boldsymbol{R}}$ on $\mathscr{B}$ defined by

$$\begin{aligned}\overset{*}{\boldsymbol{R}}(\boldsymbol{u},\boldsymbol{v})=\big((\nabla_K\boldsymbol{D})\,\boldsymbol{v}\big)\,\boldsymbol{u}+(\boldsymbol{D}\boldsymbol{u})\,(\boldsymbol{D}\boldsymbol{v})+\boldsymbol{D}\big((\boldsymbol{D}\boldsymbol{v})\,\boldsymbol{u}\big)-\\-\big[\big((\nabla_K\boldsymbol{D})\,\boldsymbol{u}\big)\,\boldsymbol{v}+(\boldsymbol{D}\boldsymbol{v})\,(\boldsymbol{D}\boldsymbol{u})+\boldsymbol{D}\big((\boldsymbol{D}\boldsymbol{u})\,\boldsymbol{v}\big)\big]\end{aligned}$$

can be shown to vanish. Conversely, if $\overset{*}{\boldsymbol{R}}$ vanishes, then K is a reference of contorted aeolotropy, at least locally. Thus, $\overset{*}{\boldsymbol{R}}$ is an intrinsic measure of the deviation of K from being a reference of contorted aeolotropy, at least locally.

6. Cauchy's Equation of Balance

Let us assume that the body $\mathscr{B}$ is held together by internal contact forces and subject to external body forces. If sufficient smoothness conditions are satisfied and if a configuration $\varkappa$ of $\mathscr{B}$ is given, the internal contact forces are determined by a stress tensor field $\boldsymbol{T}_\varkappa$ and the external body forces by a body force density vector field $\boldsymbol{b}_\varkappa$. The balance of forces is expressed by Cauchy's equation

$$\operatorname{div}_\varkappa \boldsymbol{T}_\varkappa+\boldsymbol{b}_\varkappa=\boldsymbol{O} \tag{6.1}$$

where $\operatorname{div}_\varkappa$ is defined in the obvious way in terms of gradients relative to the configuration $\varkappa$.

Under a change of configuration from $\varkappa$ to $\gamma=\boldsymbol{\lambda}\circ\varkappa$, the stress and the body force density transform according to the formulas

$$\boldsymbol{T}_\gamma=\frac{1}{J}\,\boldsymbol{T}_\varkappa\,\boldsymbol{F}^T,\qquad \boldsymbol{b}_\gamma=\frac{1}{J}\,\boldsymbol{b}_\varkappa, \tag{6.2}$$

where

$$\boldsymbol{F}=\nabla\boldsymbol{\lambda},\qquad J=|\det\boldsymbol{F}|. \tag{6.3}$$

It is clear from (6.2) and (6.3) that $\boldsymbol{T}_\gamma(X)=\boldsymbol{T}_\varkappa(X)$ and $\boldsymbol{b}_\gamma(X)=\boldsymbol{b}_\varkappa(X)$ hold when $\nabla\boldsymbol{\lambda}|_{\varkappa(X)}=\boldsymbol{1}$, i.e., when γ and $\varkappa$ belong to the same local configuration at X. Thus, it is meaningful to define a stress tensor field $\boldsymbol{T}_K$ and a body force vector field $\boldsymbol{b}_K$ relative to a reference K

of $\mathscr{B}$ by the condition that

$$\boldsymbol{T}_K(X) = \boldsymbol{T}_\varkappa(X), \qquad \boldsymbol{b}_K(X) = \boldsymbol{b}_\varkappa(X) \tag{6.4}$$

hold whenever $\varkappa$ belongs to $K(X)$.

The balance equation (6.1) does not remain valid if we replace the configuration $\varkappa$ by a smooth uniform reference K. Rather, the balance of forces is expressed, in terms of K, by

$$\mathrm{div}_K \boldsymbol{T}_K + \boldsymbol{T}_K \mathbf{s} + \boldsymbol{b}_K = \boldsymbol{O}, \tag{6.5}$$

where $\mathbf{s}$ is a vector field that is obtained from the inhomogeneity field $\boldsymbol{S}$ of K by the condition that

$$\mathbf{s} \cdot \boldsymbol{u} = \mathrm{trace}(\boldsymbol{S}\, \boldsymbol{u}) \tag{6.6}$$

hold for all $\boldsymbol{u} \in \mathscr{V}$. When $\mathscr{B}$ is a materially uniform but inhomogeneous simple body, one must use the modified balance equation (6.5) rather than (6.1) in order to solve specific initial value and boundary value problems.

Dislocated and Oriented Media

By

R. A. Toupin

IBM Zürich Research Laboratory
Rüschlikon-ZH

Dislocated and Oriented Media

By

R. A. Toupin

IBM Zürich Research Laboratory
Rüschlikon-ZH

Abstract. A continuum model of perfect or dislocated crystals is considered, and it is shown how one can view such a crystalline medium as an oriented medium, and as a material manifold with an irrotational law of distant parallelism. In dislocated crystals with twist, the Burgers vectors of homologous cycles are not independent of the cycles unless the amount of twist is restricted.

1. Introduction

The concepts of a dislocated material medium and of an oriented material medium had different origins, although both stemmed from special aspects of classical elasticity theory. Dislocated elastic media were first considered by VOLTERRA [1], and the first systematic treatment of a broad class of oriented elastic media appears in the memoir of E. and F. COSSERAT [2]. Both concepts, however, are essentially kinematical and independent of the idea of elastic response and the laws of mechanics which we may suppose govern the motion of either kind of material medium. Oriented media provide a unifying concept for various special theories of elastic rods and shells, of elastic media with microstructure, of liquid crystals, and many other special theories. An oriented medium is nothing more nor less kinematically than a continuous medium of dimension one, two, or three, at each point of which and at each instant of time there is defined a set of vectors $\underset{a}{\boldsymbol{d}}$, $a = 1, 2, \ldots, m$. In many considerations, the physical dimension or any other physical significance of each member of this set of vectors is unimportant. Theories of dislocated media have developed principally as a means to understand the mechanism of initial stress and of the macroscopic inelastic deformations and physical properties of imperfect crystals. In 1958, GÜNTHER [3] called attention to the relevance of the Cosserats theory of oriented media to the then newly developing theory of continuously dislocated material media. More recently, N. FOX [4] has established a connection between these seemingly diverse theories, and W. NOLL [5] has shown how the mathematical structure of a continuously dislocated medium flows naturally from the definition of a materially uniform, yet inhomogeneous simple medium. In this note, we show how one can begin with NOLL's general definition of the isotropy group, define perfect and imperfect crystals, and view a dislocated crystal as an oriented medium.

2. The Isotropy Group of a Continuous Medium

We shall consider a set $\bar{\mathscr{B}}$ of objects $X, Y, \ldots$ called *material points*, and we shall call $\bar{\mathscr{B}}$ a *body*. We shall consider only the cases for which $\bar{\mathscr{B}}$ is a *standard 3-manifold* in the sense defined by WHITNEY [6]. Thus $\bar{\mathscr{B}}$ is a compact, connected topological space $\bar{\mathscr{B}}$, and we are given a closed subset $\partial\bar{\mathscr{B}}$ of $\bar{\mathscr{B}}$ called *boundary points*, and a closed subset $\partial_0\bar{\mathscr{B}}$ of $\partial\bar{\mathscr{B}}$ called *edge* and *corner points*. The set $\mathscr{B} = \bar{\mathscr{B}} - \partial\bar{\mathscr{B}}$ of *interior points* is a smooth 3-dimensional manifold, and the set $\partial\bar{\mathscr{B}} - \partial_0\bar{\mathscr{B}}$ is a finite collection of smooth 2-dimensional manifolds. A standard 3-manifold is an abstraction of a polyhedral region in 3-dimensional Euclidean space, or of a smoothly deformed polyhedral region. Let $\mathscr{T}_X$ denote the *tangent space* of $\mathscr{B}$ at the point X. Each $\mathscr{T}_X$ is a 3-dimensional vector space and we call the elements $\boldsymbol{U}, \boldsymbol{V}, \ldots$ of $\mathscr{T}_X$ *material vectors* at X. An assignment of a material vector in each $\mathscr{T}_X$, $X \in \mathscr{B}$ is a *material vector field over* $\mathscr{B}$. The conjugate space $\mathscr{T}_X^*$ of $\mathscr{T}_X$ is the set of real valued linear functions $A: \mathscr{T}_X \to R$ of material vectors. We call the elements of $\mathscr{T}_X^*$, *material covectors at* X. Material covector fields over $\mathscr{B}$, or, more generally, material tensor fields over $\mathscr{B}$ are defined in the obvious way.

Let $\mathscr{E}$ be 3-dimensional Euclidean space with elements $x, y, \ldots$ we shall call *places* or *positions*, and let $\mathscr{V}$ denote the translation space of $\mathscr{E}$. Then $\mathscr{V}$ is a 3-dimensional vector space and $x = y + \boldsymbol{v}$, ($\boldsymbol{v} \in \mathscr{V}$, $x, y \in \mathscr{E}$) is the point y translated by $\boldsymbol{v}$. Every two points $x, y \in \mathscr{E}$ determine a unique translation $\boldsymbol{v} = x - y$. We call the elements of $\mathscr{V}$, *spatial vectors*.

A configuration $\varkappa$ of $\bar{\mathscr{B}}$ is a mapping

$$\varkappa : \bar{\mathscr{B}} \to \mathscr{E} \tag{2.1}$$

which assigns a position to each point of the body. We write $x_\varkappa = \varkappa(X)$ for the position assigned to X in the configuration $\varkappa$.

A configuration $\varkappa$ of $\bar{\mathscr{B}}$ is faithful if a) $\varkappa$ is one-one. b) $\bar{\mathscr{B}}_\varkappa$ is a standard 3-manifold with boundary points $\partial\bar{\mathscr{B}}_\varkappa = \varkappa(\partial\bar{\mathscr{B}})$, and edge and corner points $\partial_0 \overline{\mathscr{B}}_\varkappa = \varkappa(\partial_0 \overline{\mathscr{B}})$. c) $\varkappa$ is continuous in $\overline{\mathscr{B}}$ and smooth in $\mathscr{B}$.

A *motion* of $\overline{\mathscr{B}}$ is a one-parameter family $\varkappa_t$ of configurations of $\overline{\mathscr{B}}$, one configuration for each value of the *time* t. If each configuration $\varkappa_t$ is faithful, the motion is *regular*. A motion of $\overline{\mathscr{B}}$ can fail to be regular in many ways; e.g., the restriction of $\varkappa$ to $\partial\overline{\mathscr{B}}$ may fail to be one-one.

A *local configuration* of $\overline{\mathscr{B}}$ at a point $X \in \mathscr{B}$ is a one-one linear mapping

$$K_X : \mathscr{T}_X \to \mathscr{V} \tag{2.2}$$

which assigns to each material vector at X a unique spatial vector. A field of local configurations over $\mathscr{B}$ is a *reference*. Every faithful configuration $\varkappa$ of $\overline{\mathscr{B}}$ determines a reference defined by

$$K_X = \nabla_X \varkappa, \tag{2.3}$$

where $\nabla_X \varkappa$ denotes the gradient of $\varkappa$ evaluated at the interior point X of $\overline{\mathscr{B}}$. Not every reference is given by such a gradient.

If $X \neq Y$, the tangent spaces $\mathscr{T}_X$ and $\mathscr{T}_Y$ of $\mathscr{B}$ are distinct 3-dimensional vector spaces, and the addition or subtraction of material vectors at different material points is a meaningless operation. Suppose, however, for each pair of points $(X, Y) \in \mathscr{B} \times \mathscr{B}$ we are given a linear transformation

$$\gamma_{XY} : \mathscr{T}_Y \to \mathscr{T}_X, \tag{2.4}$$

and that this set of transformations satisfies the three conditions

$$\left.\begin{aligned} &1)\ \gamma_{XX} = I_X, \\ &2)\ \gamma_{XY} = \gamma_{YX}^{-1}, \\ &3)\ \gamma_{XY} \circ \gamma_{YZ} = \gamma_{XZ}, \end{aligned}\right\} \tag{2.5}$$

where I_X denotes the identity map of $\mathscr{T}_X$. Every such set of transformations determines an equivalence relation between material vectors at the same or different points of $\mathscr{B}$ defined by

$$\boldsymbol{V}_X \overset{\gamma}{\sim} \boldsymbol{V}_Y \quad \text{iff} \quad \boldsymbol{V}_X = \gamma_{XY} \cdot \boldsymbol{V}_Y. \tag{2.6}$$

We call such a set $\{\gamma_{XY}\}$ a *material parallelism*. Every reference $\{K_X\}$ of $\overline{\mathscr{B}}$ determines a corresponding material parallelism defined by setting

$$\gamma_{XY} = K_X^{-1} \circ K_Y. \tag{2.7}$$

Thus, in particular, every faithful configuration $\varkappa$ of $\overline{\mathscr{B}}$ determines a material parallelism defined by setting $K_X = \nabla_X \varkappa$ in (2.7).

We say that a material covector $\boldsymbol{A}_X$ at $X \in \mathscr{B}$ is *γ-equivalent* to a material covector $\boldsymbol{A}_Y$ at $Y \in \mathscr{B}$ and write $\boldsymbol{A}_X \overset{\gamma}{\sim} \boldsymbol{A}_Y$ if and only if $\boldsymbol{A}_X \cdot \boldsymbol{V}_X = \boldsymbol{A}_Y \cdot \boldsymbol{V}_Y$ whenever $\boldsymbol{V}_X \overset{\gamma}{\sim} \boldsymbol{V}_Y$.

It is not difficult to show that if $\{\gamma_{XY}\}$ is a material parallelism, then there exists a linearly independent set $\{\underset{a}{\boldsymbol{E}}_X;\ a = 1, 2, 3\}$ of γ-equivalent material vector fields such that

$$\gamma_{XY} = \underset{a}{\boldsymbol{E}}_X \otimes \overset{a}{\boldsymbol{E}}_Y, \tag{2.8}$$

where the material covectors $\overset{a}{\boldsymbol{E}}_X$ are *reciprocal* to the material vectors $\underset{a}{\boldsymbol{E}}_X$;

$$\overset{a}{\boldsymbol{E}}_X \cdot \underset{b}{\boldsymbol{E}}_X = \delta^a_b. \tag{2.9}$$

The fields $\{\underset{a}{\boldsymbol{E}}_X\}$ in the representation (2.8) of $\{\gamma_{XY}\}$ are uniquely determined by $\{\gamma_{XY}\}$ up to a non-singular linear transformation $\underset{a}{\boldsymbol{E}}_X \to L^b_a \underset{b}{\boldsymbol{E}}_X$ which is independent of X.

Let $\mathscr{C}$ be a smooth closed curve in $\mathscr{B}$ with differential element $\boldsymbol{dX} \in \mathscr{T}_X$ at $X \in \mathscr{C}$. Choose the point $Y \in \mathscr{B}$ arbitrarily and consider the line integral

$$\boldsymbol{V}_Y(\mathscr{C}) = \oint_{\mathscr{C}} \gamma_{YX} \cdot \boldsymbol{dX}. \tag{2.10}$$

If $\boldsymbol{V}_Y(\mathscr{C}) = 0$ for some $Y \in \mathscr{B}$, then $\boldsymbol{V}_Y(\mathscr{C}) = 0$ for every $Y \in \mathscr{B}$. If $\boldsymbol{V}_Y(\mathscr{C}) = 0$ for every closed curve $\mathscr{C} \subset \mathscr{B}$, we say that the material parallelism $\{\gamma_{XY}\}$ is *torsionless*. Substituting the representation (2.8) of γ_{XY} into (2.10), one concludes that a necessary and sufficient condition for $\{\gamma_{XY}\}$ to be torsionless is that

$$\oint_{\mathscr{C}} \overset{a}{\boldsymbol{E}}_X \cdot \boldsymbol{dX} = 0, \qquad a = 1, 2, 3. \tag{2.11}$$

In other words, the material parallelism $\{\gamma_{XY}\}$ is torsionless if and only if each *exterior differential 1-form* $\overset{a}{E} = \overset{a}{\boldsymbol{E}}_X \cdot \boldsymbol{dX}$ is *exact*. If $\boldsymbol{V}_X(\mathscr{C}) = 0$ whenever $\mathscr{C}$ is the boundary $\partial\mathscr{A}$ of a standard 2-manifold $\mathscr{A} \subset \mathscr{B}$, the material parallelism $\{\gamma_{XY}\}$ is said to be *irrotational*. Then (2.11) must hold for every 1-cycle which is the boundary of an $\mathscr{A} \subset \mathscr{B}$, and the 1-form $\overset{a}{E}$ must be *irrotational*. But not every 1-cycle $\mathscr{C} \subset \mathscr{B}$ may be a boundary $\partial\mathscr{A}$ of a surface $\mathscr{A} \subset \mathscr{B}$, in which case, there exist irrotational material parallelisms which are not torsionless. It is precisely this circumstance which can occur in dislocated crystalline media, as we shall see.

Let $\mathscr{C}_X$ be the set of all local configurations $K_X : \mathscr{T}_X \to \mathscr{V}$ of $\mathscr{B}$ at the point X. Then, following NOLL, we may view the *constitutive relations* of a simple body as a set of mappings

$$\{\mathfrak{G}_X : \mathscr{C}_X \to \mathbb{R};\quad X \in \mathscr{B}\}, \tag{2.12}$$

where the range $\mathbb{R}$ of each $\mathfrak{G}_X$ is the space of *response descriptors*. The nature of the elements of this space is unimportant for the present discussion. Let $\varkappa$ be a faithful configuration of $\overline{\mathscr{B}}$. It need not be a configuration $\varkappa_t$ experienced by $\overline{\mathscr{B}}$ at any time during its motion. It need not be an "equilibrium" configuration of $\overline{\mathscr{B}}$. Define the functions $\mathfrak{G}^{\varkappa}_X$ by

$$\mathfrak{G}^{\varkappa}_X = \mathfrak{G}_X \circ \nabla_X \varkappa. \tag{2.13}$$

Let $\mathscr{L}(\mathscr{V}, \mathscr{V})$ denote the set of all one-one linear transformations $\mathscr{V} \to \mathscr{V}$. Then, by definition, each $\mathfrak{G}^{\varkappa}_X$ is a mapping

$$\mathfrak{G}^{\varkappa}_X : \mathscr{L}(\mathscr{V}, \mathscr{V}) \to \mathbb{R}. \tag{2.14}$$

The functions $\mathfrak{G}^{\varkappa}_X$ and $\varkappa$ determine the $\mathfrak{G}_X$ uniquely. The constitutive relations $\mathfrak{G}^{\varkappa}_X$ depend on $\varkappa$, but the constitutive relations $\mathfrak{G}_X$ are independent of any configuration of $\overline{\mathscr{B}}$.

Every smooth transformation $\chi : \mathscr{E} \to \mathscr{E}$ determines another faithful configuration $\varkappa'$ of $\overline{\mathscr{B}}$ given by $\varkappa' = \chi \circ \varkappa$. Every pair of faithful configurations $(\varkappa, \varkappa')$ of $\overline{\mathscr{B}}$ determines a smooth transformation $\chi : \overline{\mathscr{B}}_{\varkappa} \to \overline{\mathscr{B}}_{\varkappa'}$ given by $\chi = \varkappa' \circ \varkappa^{-1}$. If $\varkappa' = \chi \circ \varkappa$, then

$$\nabla_X \varkappa' = \nabla_{x_\varkappa} \chi \circ \nabla_X \varkappa. \tag{2.15}$$

From this relation, we see that the argument $K^{\varkappa}_X = K_X \circ (\nabla_X \varkappa)^{-1}$ of $\mathfrak{G}^{\varkappa}_X$, for each fixed value of K_X depends on $\varkappa$ as follows

$$K^{\varkappa'}_X = K^{\varkappa}_X \circ (\nabla_{x_\varkappa} \chi)^{-1}. \tag{2.16}$$

From the definition of the functions $\mathfrak{G}^{\varkappa}_X$ we have then

$$\mathfrak{G}^{\varkappa}_X(K^{\varkappa}_X) = \mathfrak{G}^{\varkappa'}_X(K^{\varkappa'}_X) = \mathfrak{G}_X(K_X). \tag{2.17}$$

Let L be any element of $\mathscr{L}(\mathscr{V}, \mathscr{V})$. If there exists an L such that

$$\mathfrak{G}^{\varkappa}_X(S) = \mathfrak{G}^{\varkappa}_Y(S \circ L) \tag{2.18}$$

for all S in the domain $\mathscr{L}(\mathscr{V}, \mathscr{V})$ of $\mathfrak{G}^{\varkappa}_X$ and $\mathfrak{G}^{\varkappa}_Y$, we say that the two material points X and Y are equivalent. While this definition of the equivalence of X and Y might appear to depend on $\varkappa$, the relations (2.16) and (2.17) show that if (2.18) holds for $\varkappa$ and L, then it also holds with $\varkappa$ replaced by $\varkappa'$ and L replaced by $L' = L \circ \nabla(x_\varkappa \chi)^{-1}$, $\chi = \varkappa' \circ \varkappa^{-1}$. Thus the definition of equivalent material points is independent of the configuration.

The *symmetry group* $\mathscr{G}_X$ of the material point $X \in \mathscr{B}$ is the set of all one-one linear transformations

$$P_X : \mathscr{T}_X \to \mathscr{T}_X \tag{2.19}$$

of the tangent space of $\mathscr{B}$ at X such that

$$\mathfrak{G}_X(K_X \circ P_X^{-1}) = \mathfrak{G}_X(K_X) \tag{2.20}$$

for all K_X in the domain of $\mathfrak{G}_X$.

We readily find that (2.20) holds if and only if

$$\mathfrak{G}_X^{\varkappa}(S \circ \overset{-1}{P_X^{\varkappa}}) = \mathfrak{G}_X^{\varkappa}(S). \tag{2.21}$$

For all $S \in \mathscr{L}(\mathscr{V}, \mathscr{V})$, where $P_X^{\varkappa}$ is the linear transformation $\mathscr{V} \to \mathscr{V}$ defined by

$$P_X^{\varkappa} = \nabla_{x_\varkappa} \varkappa \circ P_X \circ (\nabla_{x_\varkappa} \varkappa)^{-1}. \tag{2.22}$$

The set of all $P_X^{\varkappa}$ with $P_X \in \mathscr{G}_X$ comprises a *representation* $\mathscr{G}_X^{\varkappa}$ of the symmetry group $\mathscr{G}_X$ by linear transformations of $\mathscr{V}$. The representations $\mathscr{G}_X^{\varkappa}$ and $\mathscr{G}_X^{\varkappa'}$ are conjugate groups of transformations of $\mathscr{V}$. If X and Y are equivalent material points, the symmetry groups and their representations $\mathscr{G}_X^{\varkappa}$ and $\mathscr{G}_Y^{\varkappa}$ are conjugate.

$$\mathscr{G}_X^{\varkappa} = L \circ \mathscr{G}_X^{\varkappa} \circ L^{-1}. \tag{2.23}$$

If every pair of material points of a body is equivalent, the body is *uniform*. If there exists a faithful configuration $\varkappa$ of a uniform body $\bar{\mathscr{B}}$ such that the representations $\mathscr{G}_X^{\varkappa}$ and $\mathscr{G}_Y^{\varkappa}$ are not only conjugate for every pair of points, but *identical* ($\mathscr{G}_X^{\varkappa} = \mathscr{G}_Y^{\varkappa}$), then the body is *homogeneous*, and $\varkappa$ is an *undistorted* configuration of the homogeneous body. Every configuration $\varkappa'$ related to an undistorted configuration $\varkappa$ of a homogeneous body by an affine transformation of $\mathscr{E}$ is also undistorted. If we think of $\bar{\mathscr{B}}$ as an elastic body, it is then clear that an undistorted configuration need not be an "equilibrium" or stress free, "natural" configuration of $\bar{\mathscr{B}}$. A homogeneous body may never have existed in an undistorted configuration.

Consider a uniform body, so that, in any configuration $\varkappa$, there exists an $L_{XY}^{\varkappa}$ such that

$$\mathfrak{G}_X^{\varkappa}(S) = \mathfrak{G}_Y^{\varkappa}(S \circ L_{YX}^{\varkappa -1}) \tag{2.24}$$

The $L_{XY}^{\varkappa}$ are determined by the constitutive relations $\mathfrak{G}_X^{\varkappa}$ only up to a transformation $L_{XY}^{\varkappa} \to P_X^{\varkappa -1} \circ L_{XY} \circ P_Y^{\varkappa}$, where $P_X^{\varkappa}$ is an element of the representation $\mathscr{G}_X^{\varkappa}$ of the symmetry group $\mathscr{G}_X$. Clearly, we can choose the $L_{XY}^{\varkappa}$ in a uniform body such that the three conditions analogous to (2.5) are satisfied. Now set

$$\gamma_{XY} = (\nabla_X \varkappa)^{-1} \circ P_X^{\varkappa} \circ L_{XY}^{\varkappa} \circ P_Y^{\varkappa -1} \circ \nabla_Y \varkappa = P_X^{-1} \circ (\nabla_X \varkappa)^{-1} \circ L_{XY}^{\varkappa} \circ \nabla_Y \varkappa \circ P_Y. \tag{2.25}$$

We verify that $\{\gamma_{XY}\}$ satisfies the three conditions (2.5) and therefore defines a material parallelism. There is one such material parallelism for each choice of a field $\{P_X\}$ of symmetry transformations.

3. Crystalline Media

The definition of crystalline media rests on the concept of a periodic function and its invariance group. Let $f: \mathscr{E} \to \mathbb{R}$ be a function of points in Euclidean space. It is sufficient for present purposes that the range of f be at least two valued; say, black and white, or 0 and 1. Beyond this, the nature of the set $\mathbb{R}$ is unimportant. Let

$$T_{\boldsymbol{v}}: \mathscr{E} \to \mathscr{E} \tag{3.1}$$

denote the translation of $\mathscr{E}$ by $\boldsymbol{v} \in \mathscr{V}$ so that

$$T_{\boldsymbol{v}}(x) = x + \boldsymbol{v}. \tag{3.2}$$

Say that f is *periodic* in the direction $\boldsymbol{v}$ with period $|v| = \sqrt{\boldsymbol{v} \cdot \boldsymbol{v}}$ if

$$f(T_{n\boldsymbol{v}}(x)) = f(x) \tag{3.3}$$

for every integer n and $x \in \mathscr{E}$. Suppose next that f is periodic in at least three linearly independent directions $\underset{a}{\boldsymbol{D}}$, $a = 1, 2, 3$, and consider the set $\mathscr{D}$ of all such sets of three linearly independent directions in which f is periodic. Put $\operatorname{vol}\{\underset{a}{\boldsymbol{D}}\} = |\underset{1}{\boldsymbol{D}} \cdot (\underset{2}{\boldsymbol{D}} \times \underset{3}{\boldsymbol{D}})|$. Call f a *pattern* if $V_m = \inf\limits_{\{\underset{a}{D}\} \in \mathscr{D}} \operatorname{vol}\{\underset{a}{\boldsymbol{D}}\} > 0$ and if there exists at least one set of directions $\{\underset{a}{D}\} \in \mathscr{D}$ such that $\operatorname{vol}\{\underset{a}{\boldsymbol{D}}\} = V_m$. Any set $\{\underset{a}{\boldsymbol{D}}\} \in \mathscr{D}$ satisfying this condition is a *basis* for the *lattice vectors* of the pattern f. A lattice vector of f is any vectorof the form

$$\boldsymbol{v}_n = \overset{a}{n} \underset{a}{\boldsymbol{D}} \tag{3.4}$$

with integer coefficients $\overset{a}{n}$, $a = 1, 2, 3$. We use the summation convention. The set of all lattice vectors is independent of the basis. If f is periodic in any direction $\boldsymbol{v}$, then $\boldsymbol{v}$ is a lattice vector of f. The point set

$$\Lambda_p = \bigcup_n (p + \boldsymbol{v}_n) = \bigcup_n T_{\boldsymbol{v}_n}(p) \tag{3.5}$$

is a *lattice* in $\mathscr{E}$. Clearly, $\Lambda_{p+\boldsymbol{v}_n} = \Lambda_p$.

Consider next the automorphisms of $\mathscr{E}$ which we call *rigid transformations*. Every rigid transformation $\ell: \mathscr{E} \to \mathscr{E}$ can be represented as a composition

$$\ell = T_{\boldsymbol{v}} \cdot O_p \tag{3.6}$$

of a *rotation* about a point p given by

$$O_p(x) = p + O \cdot (x - p) \tag{3.7}$$

where $O : \mathscr{V} \to \mathscr{V}$ is an orthogonal transformation of the translation space $\mathscr{V}$ of $\mathscr{E}$, followed by a translation $T_{\boldsymbol{v}}$. O_p is a *proper rotation* if $\det O = +1$, and an *improper rotation* if $\det O = -1$. Every l has also a representation

$$l = O_p \cdot T_{\boldsymbol{v}} \tag{3.8}$$

as a translation followed by a rotation. The representations (3.6) and (3.7) are not unique. In fact, the following identities are true.

$$l = T_{\boldsymbol{v}} \cdot O_p = T_{\boldsymbol{v}+O\cdot\boldsymbol{u}-\boldsymbol{u}}\, O_{p+\boldsymbol{u}} = O_p \circ T_{O^{-1}\cdot\boldsymbol{v}} = O_{p+\boldsymbol{u}} \circ T_{\boldsymbol{v}+\boldsymbol{u}-O^{-1}\cdot\boldsymbol{u}}. \tag{3.9}$$

Also,

$$O_p \circ T_{\boldsymbol{v}} \cdot O_p^{-1} = T_{O\cdot\boldsymbol{v}}. \tag{3.10}$$

A *space group* $\mathscr{G}_f$ is the group of all rigid transformations l such that

$$f(l(x)) = f(x), \tag{3.11}$$

for all x, where f is a pattern. In other words, a space group is the invariance group of rigid transformations of a pattern.

Our discussion of dislocations in crystalline media will require some elementary facts about all space groups which will be recorded here without proof since they are all well known. It has been noted already that any *pure translation* $T_{\boldsymbol{v}} \in \mathscr{G}_f$ is of the form $T_{\boldsymbol{v}_n}$, where $\boldsymbol{v}_n$ is a lattice vector of f. The subgroup $\mathscr{G}_f^p$ of $\mathscr{G}_f$ of all pure rotations about the point p such that

$$f(O_p(x)) = f(x) \tag{3.12}$$

is the *group of the point* p. Each $O_p \in \mathscr{G}_f^p$ determines an orthogonal transformation O of $\mathscr{V}$, and the set G_f^p of all these orthogonal transformations, $O_p \in \mathscr{G}_f^p$, is a faithful representation of $\mathscr{G}_f^p$. If $O \in G_f^p$, then it can be shown that the order of O cannot be five nor greater than six. Also, if $O \in G_f^p$, then $O(\boldsymbol{\Lambda}) = \boldsymbol{\Lambda}$, where $\boldsymbol{\Lambda}$ is the set of all lattice vectors of f. Also if $O \neq I$ (the identity transformation) and $O \in G_f^p$, then any proper vector of O must be proportional to a lattice vector. To put it otherwise, if $O\cdot\boldsymbol{v} = \pm\boldsymbol{v}$ and $O \neq I \in G_f^p$, the straight line $\mathscr{L}(O_p) = \{x; x = p + \lambda\boldsymbol{v}\}$ or *axis* of O_p passes through p and at least one other point of the lattice Λ_p. Of course this means it passes through infinitely many points of Λ_p.

Two points p and p' are called *equivalent points* of a pattern f if there exists an element $l \in \mathscr{G}_f$ such that $l(p) = p'$. The groups $\mathscr{G}_f^p$ and $\mathscr{G}_f^{p'}$ of equivalent points p and p' are conjugate subgroups of $\mathscr{G}_f$; i.e., there exists an element $l \in \mathscr{G}_f$ such that $\mathscr{G}_f^{p'} = l\,\mathscr{G}_f^p\, l^{-1}$. Clearly, the latter property of $\mathscr{G}_f^{p'}$ and $\mathscr{G}_f^p$ is not sufficient that p and p' be equivalent points of f.

The set of points τ_p which lie closer to the point p than to any other point p' in the lattice Λ_p is called the *symmetric unit cell about* p. Let $\boldsymbol{\tau}$ denote the set of all vectors in $\mathscr{V}$ such that $|\boldsymbol{u}| < |\boldsymbol{u} - \boldsymbol{v}_n|$ for all lattice vectors $\boldsymbol{v}_n \neq 0$. (Then τ_p is the point set

$$\boldsymbol{\tau}_p = \bigcup_{\boldsymbol{v} \in \boldsymbol{\tau}} (p + \boldsymbol{v}), \qquad (3.13)$$
$$= \bigcup_{\boldsymbol{v} \in \boldsymbol{\tau}} T_{\boldsymbol{v}}(p),$$

and $\bar{\tau}_p$ is a convex polyhedral region in $\mathscr{E}$.

If $l \in \mathscr{G}_f$ has a representation $T_{\boldsymbol{v}} \circ O_p$ then either $\boldsymbol{v}$ is a lattice vector or O_p is not an element of $\mathscr{G}_f^p$. In either case, O_p must commute with every lattice translation $T_{\boldsymbol{v}_n}$ so that $O_p(\Lambda_p) = \Lambda_p$, and the order of O_p must be finite; not five and not greater than six. It follows that for any $l \in \mathscr{G}_f$, $l(\tau_p) = T_{\boldsymbol{v}}(\tau_p)$. That is, l applied to τ_p is equivalent to a translation of τ_p. Every $l \in \mathscr{G}_f$ is the composition of a lattice translation and a transformation of the form $l' = T_{\frac{\boldsymbol{v}_n^o}{r} + \boldsymbol{w}} \circ O_p$, where $O\,\boldsymbol{v}_n^o = \pm \boldsymbol{v}_n^o$, $\boldsymbol{v}_n^o$ is a lattice vector, O has order r, $\boldsymbol{v}_n^o \cdot \boldsymbol{w} = 0$, and $\boldsymbol{w} \in \tau$.

Consider now a simple body $\mathscr{B}$ with constitutive relations $\mathfrak{G}_X$. We say that $\mathscr{B}$ is a *crystalline medium*, or simply a *crystal* if for every point $X \in \mathscr{B}$ there is a neighborhood n_X of X and a faithful configuration $\varkappa$ of $\mathscr{B}$ such that

$$\mathfrak{G}_X^{\varkappa'} = \mathfrak{G}_Y^{\varkappa} \qquad (3.14)$$

whenever $\varkappa' = l \circ \varkappa$, X, $Y \in n_X$, $x_{\varkappa'} = y_\varkappa$, and l is any element of a space group $\mathscr{G}_f$. The constitutive relations $\mathfrak{G}_X^\varkappa$ may be viewed as a set of functions $\mathfrak{G}^\varkappa$ with domain $\mathscr{L}(\mathscr{V}, \mathscr{V}) \times \mathscr{B}_\varkappa$ defined by

$$\mathfrak{G}^\varkappa(S, x_\varkappa) = \mathfrak{G}_X^\varkappa(S). \qquad (3.15)$$

Expressed in terms of the $\mathfrak{G}^\varkappa$, the condition (3.14) reads

$$\mathfrak{G}^\varkappa(S \circ \nabla l^{-1}, l(x_\varkappa)) = \mathfrak{G}^\varkappa(S, x_\varkappa) \qquad (3.16)$$

for all $l \in \mathscr{G}_f$ and $X = \varkappa^{-1}(x_\varkappa)$, $Y = \varkappa^{-1}(y_\varkappa) \in n_X$, $y_\varkappa = l(x_\varkappa)$. It follows from the definitions of § 2, that if X and Y are two material points in n_X with positions $x_\varkappa$ and $y_\varkappa$ in the configuration $\varkappa$ such that $l(x_\varkappa) = y_\varkappa$, $l \in \mathscr{G}_f$, then X and Y are equivalent material points. As a further condition defining a crystal, we shall assume that the set of all transformations l satisfying (3.16) is contained in $\mathscr{G}_f$, and we call $\varkappa$ an *undistorted* configuration of n_X.

If $l \in \mathscr{G}_f$ and l leaves the point $x_\varkappa$ invariant $(l(x_\varkappa) = x_\varkappa)$, then (3.16) requires that

$$\mathfrak{G}^\varkappa(S\, \nabla l^{-1}, x_\varkappa) = \mathfrak{G}^\varkappa(S, x_\varkappa) \qquad (3.17)$$

or, equivalently, that

$$\mathfrak{G}_X^\varkappa(S \circ \nabla l^{-1}) = \mathfrak{G}_X^\varkappa(S). \qquad (3.18)$$

This asserts that ∇l^{-1} is an element $P^{\varkappa}_X$ of the representation $\mathcal{G}^{\varkappa}_X$ of the symmetry group $\mathcal{G}_X$ of X. Thus if $l = T_v \cdot O_{x_\varkappa}$, then $O^{-1} = \nabla l^{-1} \in \mathcal{G}^{\varkappa}_X$.

If $\mathcal{G}_f$ is a space group with elements l, the *point group* $\mathcal{P}_f$ is the factor group $\mathcal{G}_f / \mathcal{G}^T_f$, where $\mathcal{G}^T_f$ are the pure translations. The point group is isomorphic to the group of orthogonal transformations $P_f = \{\nabla l;\ l \in \mathcal{G}_f\}$. In the classical theory of elastic crystals, one consides simple media $\mathcal{B}$ which are uniform and homogeneous for which the common symmetry group $\mathcal{G}^{\varkappa}_X$ of each point in an undistorted configuration $\varkappa$ is a point group representation $P^{\varkappa}_f$. Every such group is a subgroup of the group of all orthogonal transformations which leave some set of lattice vectors invariant; i.e., if $O \in P_f$, then there exists a set of lattice vectors Λ such that $O(\Lambda) = \Lambda$. In general, the crystalline media considered here are neither uniform nor homogeneous. The effects of non-uniformity are unimportant in many applications; however, in others they are of essence.

4. Dislocated Crystalline Media

The definition of a crystalline medium in § 3 is based on local properties of $\mathcal{B}$. The body and its image $\mathcal{B}_\varkappa$ under a faithful configuration $\varkappa$ is connected, but it need not be simply connected. There may not exist a single neighborhood n_X which covers $\mathcal{B}$ and a configuration $\varkappa$ for which the condition (3.13) holds. Let $x_\varkappa$ be a point of $\mathcal{B}_\varkappa$ and suppose that $\varkappa$ is an undistorted configuration of $n_{x_\varkappa} = \varkappa(n_X)$. Consider the lattice $\Lambda_{x_\varkappa}$ and the symmetric unit cells $\bar{\tau}_{y_n}$, $y_n \in \Lambda_{x_\varkappa}$. Call two closed polyhedral n-cells *non-overlapping* if their intersection is of lower dimension. Then $\bar{\tau}_{y_n}$ and $\bar{\tau}_{y_{n'}}$, are non-overlapping if $y_n \neq y_{n'}$, and $\mathcal{E} = \bigcup_{y_n \in \Lambda_{x_\varkappa}} \bar{\tau}_{y_n}$ is a subdivision of space into symmetric unit cells of a lattice. It follows that $n_{x_\varkappa}$ is a union of non-overlapping symmetric unit cells or pieces of cells, and that $n_X = \varkappa^{-1}(n_{x_\varkappa})$ is a union of non-overlapping "*curvilinear*" symmetric unit cells, or pieces of curvilinear cells, $\tilde{\tau}_X = \varkappa^{-1}(\bar{\tau}_{x_\varkappa})$. The whole of $\mathcal{B}$ may be subdivided in this way and for simplicity and definiteness, let us suppose that $\mathcal{B}$ is a union of whole curvilinear symmetric unit cells. Each point $X \in \mathcal{B}$ is contained in a subset $\bar{n}_X$ which can be mapped faithfully by some $\varkappa$ onto a subset of the cells $\bar{\tau}_{y_n}$, $y_n \in \Lambda_x$ in $\mathcal{E}$. If $\mathcal{B}_\varkappa$ is simply connected and for n_X we can choose $\mathcal{B}$, let us say that $\mathcal{B}$ is a *perfect crystal*. If for n_X we can choose $\mathcal{B}$, but $\mathcal{B}$ is *not* simply connected, let us say that $\mathcal{B}$ is a perfect crystal with *vacancies*. If $\mathcal{B}$ is a perfect crystal with vacancies, there exists a faithful configuration $\mathcal{B}_\varkappa$ of $\mathcal{B}$ and a simply connected region $\mathcal{B}^*_\varkappa$ in $\mathcal{E}$ such that

$\overline{\mathscr{B}}_\varkappa^* - \overline{\mathscr{B}}_\varkappa$ is a union of non-overlapping symmetric unit cells τ_{y_n}, $y_n \in \Lambda_{x_\varkappa}$ for some $x_\varkappa \in \overline{\mathscr{B}}_\varkappa$.

A crystalline medium $\overline{\mathscr{B}}$ can fail to be a perfect crystal or a perfect crystal with vacancies if $\overline{\mathscr{B}}$ is not simply connected, in which case we say that it is *dislocated*, or *contains dislocations*. We give an example of a dislocated crystalline medium. For this purpose it is sufficiently general to consider a crystal $\overline{\mathscr{B}}$ of the cubic system for which there exists a

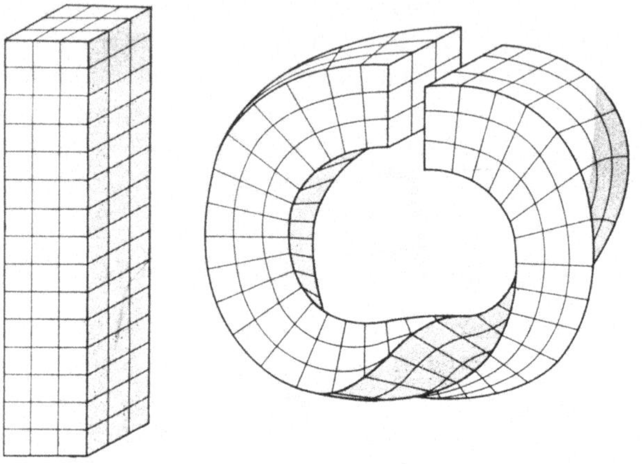

a b

Fig. 1. a) Reference crystal; b) Dislocated crystal with 180° twist.

basis $\{\underset{a}{\boldsymbol{D}}\}$ of mutually orthogonal vectors of equal length. We shall suppose further that there exists a point p which has the full symmetry of the cubic lattice; i.e., for some p, $f(O_p\, x) = f(x)$ for *all* O for which $O(\Lambda) = \Lambda$. Consider a perfect crystal $\overline{\mathscr{C}}$ with this symmetry group $\mathscr{G}_f$ and suppose that $\overline{\mathscr{C}}_\varkappa = \varkappa(\overline{\mathscr{C}})$ is a faithful configuration of $\overline{\mathscr{C}}$. Let $\overline{\mathscr{C}}_\varkappa$ be a prism of square cross section (see Fig. 1). Then its length and two sides are integer multiples of a common length, the side of a symmetric unit cell $\tau_{x_\varkappa}$, which in this case is a perfect cube. Let $\overline{\mathscr{E}}_\varkappa^1$ and $\overline{\mathscr{E}}_\varkappa^2$ denote the ends of the rectangular region $\overline{\mathscr{C}}_\varkappa$. Now let $\lambda: \overline{\mathscr{C}}_\varkappa \to \mathscr{E}$ be a map of $\overline{\mathscr{C}}_\varkappa$ which is smooth in $\mathscr{C}_\varkappa$ and in every $\partial \mathscr{B}_\varkappa - \partial_0 \mathscr{B}_\varkappa$, but fails to be faithful only because a subset $Q_\varkappa^1$ of $\overline{\mathscr{E}}_\varkappa^1$ and a subset $Q_\varkappa^2$ of $\overline{\mathscr{E}}_\varkappa^2$ are mapped by λ onto a common set of points in $\mathscr{E}$; i.e., $\lambda(Q_\varkappa^1) = \lambda(Q_\varkappa^2)$. Let λ_i denote the restriction of λ to $Q_\varkappa^i$. Then $\lambda_2^{-1} \circ \lambda_1,: Q_\varkappa^1 \to Q_\varkappa^2$ and if $\lambda_2^{-1} \circ \lambda_1$, is the restriction of an $l \in \mathscr{G}_f$ to $Q_\varkappa^1$, then equivalent points of $Q_\varkappa^1$ are mapped onto equivalent points of $Q_\varkappa^2$. But if this be true, then the unfaithful configuration $\overline{\mathscr{C}}_{\lambda \circ \varkappa}$ of the perfect crystal $\overline{\mathscr{C}}$ may be viewed as a faithful configuration, or identified with a crystal $\overline{\mathscr{B}}$ which contains a dislocation.

No single faithful configuration $\varkappa'$ of $\bar{\mathscr{C}}_{\lambda\circ\varkappa} = \bar{\mathscr{B}}$ satisfies the conditions of an undistorted configuration of a perfect crystal, or of a perfect crystal with vacancies. Yet, $\bar{\mathscr{C}}_{\lambda\circ\varkappa}$ satisfies all the conditions of a crystalline medium.

5. The Director Fields of a Crystal

Every material covector field $\boldsymbol{A}_X$ in $\mathscr{B}$ determines a corresponding spatial covector field $\boldsymbol{a}_{x_\varkappa}$ in $\mathscr{B}_\varkappa$ such that

$$\int_{\mathscr{C}} \boldsymbol{A}_X \cdot d\boldsymbol{X} = \int_{\mathscr{C}_\varkappa} \boldsymbol{a}_{x_\varkappa} \cdot d\boldsymbol{x}_\varkappa, \tag{5.1}$$

where $\mathscr{C}$ is any smooth curve in $\mathscr{B}$ and $\mathscr{C}_\varkappa = \varkappa(\mathscr{C})$ is its image under $\varkappa$, a smooth regular transformation of $\mathscr{B}$. The field $\boldsymbol{A}_X$ determines the exterior differential 1-form $A = \boldsymbol{A}_X \cdot d\boldsymbol{X}$, and $\boldsymbol{a}_{x_\varkappa}$ the 1-form $a = \boldsymbol{a}_{x_\varkappa} \cdot d\boldsymbol{x}_\varkappa$. In much the same way, material vector fields $\boldsymbol{V}_X$ and spatial vector fields $\boldsymbol{v}_{x_\varkappa}$ are set in correspondence by $\varkappa$ according to the rule

$$\boldsymbol{v}_{x_\varkappa} = \nabla_X \varkappa \cdot \boldsymbol{V}_X. \tag{5.2}$$

If $\bar{\mathscr{B}}$ is a perfect crystal, or a perfect crystal with vacancies, then there exists a faithful configuration $\varkappa$ of $\bar{\mathscr{B}}$ mapping $\bar{\mathscr{B}}$ onto a connected set of non-overlapping symmetric unit cells $\{\bar{\tau}_{y_n}\}$ where each y_n is a point of a space lattice $\Lambda_{x_\varkappa^o}$. Now a symmetric unit cell has a certain number k of pairs of parallel opposite faces separated by a lattice vector $\boldsymbol{d}$ which is perpendicular to these faces. Therefore, with each symmetric unit cell in $\mathscr{B}_\varkappa$, there is associated a finite set $\boldsymbol{\Omega}_{y_n} = \{\pm \underset{1}{\boldsymbol{d}}, \pm \underset{2}{\boldsymbol{d}}, \ldots, \pm \underset{k}{\boldsymbol{d}}\}$ of lattice vectors and $k \geqq 3$. Moreover, one sees that any three linearly independent vectors in $\boldsymbol{\Omega}_{y_n}$ is a basis for the set $\boldsymbol{\Lambda}$ of all lattice vectors. In each cell of $\mathscr{B}_\varkappa$, we may consider any of the *cell vectors* $\boldsymbol{d} \in \boldsymbol{\Omega}_{y_n}$ as a constant vector field in that cell. That is, we define $\boldsymbol{d}_x = \boldsymbol{d}$ for $x \in \bar{\tau}_{y_n}$. Now any symmetry element l of the space group $\mathscr{G}_f$ with gradient O has the property $O(\boldsymbol{\Omega}_{y_n}) = \boldsymbol{\Omega}_{y_n}$. Moreover, if we choose any basis $\{\underset{a}{\boldsymbol{d}}\}$, $a = 1, 2, 3$, from the set $\boldsymbol{\Omega}_{y_n}$, we can construct the reciprocal set $\{\overset{a}{\boldsymbol{d}}\}$ of covectors $(\overset{a}{\boldsymbol{d}} \cdot \underset{b}{\boldsymbol{d}} = \delta^a_b)$ and $O(\{\overset{a}{\boldsymbol{d}}\}) = \{\overset{a}{\boldsymbol{d}}\}$. In a perfect crystal with or without vacancies, there exists at least one way of choosing a basis $\underset{a}{\boldsymbol{d}}_{y_n} \in \boldsymbol{\Omega}_{y_n}$ in each cell of $\mathscr{B}_\varkappa$ such that the fields defined in $\mathscr{B}$ by $\underset{a}{\boldsymbol{d}}_{x_\varkappa} = \underset{a}{\boldsymbol{d}}_{y_n}$, $x \in \bar{\boldsymbol{\tau}}_{y_n}$, $y_n \in \Lambda_{x_\varkappa^o}$, are continuous, and the corresponding covector fields $\overset{a}{\boldsymbol{d}}_{x_\varkappa}$ reciprocal to the $\underset{a}{\boldsymbol{d}}_{x_\varkappa}$ will be continuous, and the 1-forms $\overset{a}{d}_\varkappa = \overset{a}{\boldsymbol{d}}_{x_\varkappa} \cdot d\boldsymbol{x}_\varkappa$ will be continuous throughout $\mathscr{B}_\varkappa$. Thus the material covector fields, the material vector fields, and the material 1-forms, $\overset{a}{\boldsymbol{D}}_X$, $\underset{a}{\boldsymbol{D}}_X$, and $\overset{a}{D}$ set in correspondence with $\overset{a}{\boldsymbol{d}}_{x_\varkappa}$, $\underset{a}{\boldsymbol{d}}_{x_\varkappa}$, and $\overset{a}{d}_\varkappa$ by $\varkappa$ will be smooth throughout $\mathscr{B}$. In an imperfect crystal containing dislocations, we *cannot* make this same construction of smooth fields in $\mathscr{B}$. The same construction can be carried out locally in each n_X and we may cover $\mathscr{B}$ by a finite set of such neighborhoods so as to obtain fields $\overset{a}{\boldsymbol{D}}_X$, $\underset{a}{\boldsymbol{D}}_X$, and forms $\overset{a}{D} = \overset{a}{\boldsymbol{D}} \cdot d\boldsymbol{X}$ defined throughout $\mathscr{B}$; but, in general, there is no way to choose the basis $\underset{a}{\boldsymbol{d}}_{y_n}$ in each cell of a $\varkappa(n_X) = n_{x_\varkappa}$ such that the resulting fields $\underset{a}{\boldsymbol{D}}_X$ are all continuous throughout $\mathscr{B}$. It can be required that the $\underset{a}{\boldsymbol{d}}_{y_n}$ in adjacent cells be equivalent in the sense that $\{\underset{a}{\boldsymbol{d}}_{y_n}\} = O\{\pm \underset{a}{\boldsymbol{d}}_{y_n}\}$ for some $O = \nabla l$, $l \in \mathscr{G}_f$.

To illustrate these ideas consider again the example of the cubic crystal of § 4. If the mapping $\lambda_2^{-1} \circ \lambda_1 : Q_\varkappa^1 \to Q_\varkappa^2$ that we considered in § 3 is a pure translation composed with a rotation of 180° about a symmetry axis in the plane of $Q_\varkappa^1$, then any constant field of cell vectors $\underset{a}{\boldsymbol{d}}_{x_\varkappa}$ in the perfect rectangular crystal is mapped into a continuous field $\underset{a}{\boldsymbol{D}}_{X_{(\lambda\circ\varkappa)^{-1}}} = \underset{a}{\boldsymbol{D}}_X$ in the dislocated crystal. In Addition, $\lambda_2^{-1} \circ \lambda_1$, may contain as a factor any number of complete revolutions about an axis perpendicular to $Q_\varkappa^1$. But if $\lambda_2^{-1} \circ \lambda_1$, contains a factor $l \in \mathscr{G}_f$ which is a rotation about an axis perpendicular to $Q_\varkappa^1$ which is not a complete revolution, then at least one of the fields $\underset{a}{\boldsymbol{D}}_X$ is not continuous in $\mathscr{B}_{\lambda\circ\varkappa}$, but suffers a jump discontinuity across the surface $\lambda(Q_\varkappa^1) = \lambda(Q_\varkappa^2)$ in the dislocated crystal.

Consider, in this same example, a closed curve $C_{\lambda\circ\varkappa}$ in the dislocated crystal which is the image $\lambda(\mathscr{C}_\varkappa)$ of a straight line $\mathscr{C}_\varkappa$ in $\mathscr{B}_\varkappa$ joining points $p_1 \in Q_\varkappa^1$ and $p_2 \in Q_\varkappa^2$ brought into conjunction by λ. Then we always have

$$\oint_{\mathscr{C}_{\lambda\circ\varkappa}} \overset{a}{\boldsymbol{D}}_X \cdot d\boldsymbol{X} = \int_{\mathscr{C}_\varkappa} \overset{a}{\boldsymbol{d}}_{x_\varkappa} \cdot d\boldsymbol{x}_\varkappa . \tag{5.2}$$

In the right-hand integral, the $\overset{a}{\boldsymbol{d}}_{x_\varkappa}$ are constant vector fields and $\mathscr{C}_\varkappa = \int_{\mathscr{C}_\varkappa} d\boldsymbol{x}_\varkappa$ is the vector joining the end points of the straight line $\mathscr{C}_\varkappa$. Let p be a reference point in $Q_\varkappa^1$, and set $p_1 - p = \boldsymbol{w}$. Then $\mathscr{C}_\varkappa$ has the form $\mathscr{C}_\varkappa = \boldsymbol{v}_n + \boldsymbol{w} - O\,\boldsymbol{w}$, and the integral on the right of (5.2) is seen to have the value

$$\overset{a}{\nu}(\mathscr{C}_{\lambda\circ\varkappa}) = \overset{a}{\boldsymbol{d}} \cdot (\boldsymbol{v}_n + \boldsymbol{w} - O \cdot \boldsymbol{w}), \tag{5.3}$$

where $\boldsymbol{v}_n$ is a lattice vector. Thus $\overset{a}{\boldsymbol{d}} \cdot \boldsymbol{v}_n = \overset{a}{n}$ is an integer equal to the number of cells along the length of the rectangle $\mathscr{B}_\varkappa$, and is independent of $\boldsymbol{w} = p_1 - p_0$. The *Burgers vector* for the circuit $\mathscr{C}_{\lambda\circ\varkappa}$ in the dislocated crystal is defined by

$$\boldsymbol{\nu}(\mathscr{C}_{\lambda\circ\varkappa}) = \overset{a}{\nu}(\mathscr{C}_{\lambda\circ\varkappa})\, \underset{a}{\boldsymbol{d}} \tag{5.4}$$

which is independent of $\boldsymbol{w}$ and the same for every circuit $\mathscr{C}_{\lambda\circ\varkappa}$ *if and only if* $O = 1$ in (5.3). When $O = 1$, the forms $\overset{a}{D}$ in the dislocated crystal are smooth and continuous throughout $\mathscr{B}_{\lambda\circ\varkappa}$ the dislocated crystal; moreover, they are irrotational, *rot* $\overset{a}{D}$, the exterior derivative of $\overset{a}{D}$ is zero. The value of $\oint_{\mathscr{C}} \overset{a}{D}$ over all homologous cycles in the dislocated crystal is the same, and the Burgers vector

$$\boldsymbol{\nu}(\mathscr{D}) = \left(\oint_{\mathscr{C}} \overset{a}{D}\right) \underset{a}{\boldsymbol{d}} \tag{5.5}$$

is a function on the homology classes of 1-cycles in the dislocated crystal. But this property of the Burgers vector does not hold generally in a dislocated crystal as defined here and represents a restriction on the kind of dislocations to be considered. If the Burgers vector does not have the same value for every homologous cycle, then the dislocation is known to have *twist* (i.e., $O \neq 1$). On the other hand, one sees from the example that knowledge only of the Burgers vector for every cycle in a dislocated crystal does not permit one to distinguish between the cases where $\lambda\colon \mathscr{C}_\varkappa \to \mathscr{C}_{\lambda\circ\varkappa}$ entails different numbers of whole relative rotations of the ends $Q_\varkappa^1$ and $Q_\varkappa^2$.

The construction of the vector and covector fields $\underset{a}{\boldsymbol{D}}_X$ and $\overset{a}{\boldsymbol{D}}_X$ in a perfect or dislocated crystal provides a corresponding law of material parallelism defined by setting

$$\gamma_{XY} = \underset{a}{\boldsymbol{D}}_X \otimes \overset{a}{\boldsymbol{D}}_Y. \tag{5.6}$$

We have seen that in perfect crystals with or without vacancies, this law of parallelism is irrotational and torsionless. In a dislocated crystal it is, in general, not continuous, but if dislocations without twist only are considered, it is irrotational but not torsionless. In all cases, we see how one may view a crystalline medium in a fairly definite way as an oriented medium with directors $\underset{a}{\boldsymbol{D}}_X$ and reciprocal directors $\overset{a}{\boldsymbol{D}}_X$ interpreted as *"crystal axes"*. The theory of exterior differential forms can then be used to assist in the analysis and description of the dislocated crystalline medium, and a connection can be made with the geometrical idea of distant parallelism. Defining a crystalline medium as we have done here allows also the possibility of a definite subdivision of a crystal into curvilinear ("deformed") symmetrical unit cells and offers the possibility of describing the global topological structure of the crystal using the methods of algebraic topology. Certain aspects of the dynamics of a dislocated crystal can also be inferred from established results in the theory of oriented media.

In summary then, I believe that even the incomplete and cursory results treated here are sufficient indication of how the methods of continuum mechanics and differential geometry can be applied to the problem of crystal structures and crystal dynamics. It is also conceivable that the model of a crystal considered here might be superior to one in which the body $\mathscr{B}$ is a finite or countable set of mass, charge, and spin bearing points or small rigid spheres.

References

[1] Cf. LOVE, A. E. H.: The Mathematical Theory of Elasticity, Fourth Edition, Appendix to Chaps. VIII and IX. New York: Dover Publications 1927.

[2] COSSERAT, E., and F. COSSERAT: Théorie des Corps Déformable. Paris: Hermann 1909.

[3] GÜNTHER, W.: Zur Statik und Kinematik des Cosseratschen Kontinuums. Abh. d. Braunschweigischen Wiss. Ges. **10**, 195 (1958).

[4] FOX, N.: A Continuum Theory of Dislocations for Single Crystals. J. Inst. Maths. Appl. **2**, 285 (1966).

[5] NOLL, W.: Materially Uniform Simple Bodies with Inhomogenities, Report 67-26, Dept. of Math., Carnegie Inst. of Tech. 1967.

[6] WHITNEY, H.: Geometric Integration Theory. Princeton University Press 1957.

A Mathematical Theory of the Mechanical Behavior of Continuous Media

WALTER NOLL

Communicated by C. TRUESDELL

Contents

1. Introduction

Until not long ago continuum mechanics meant to most people the theories of inviscid and linearly viscous fluids and of linearly elastic solids. However, the behavior of only few real materials can be described adequately by these classical theories. Experimental scientists, who had to deal with real materials, developed a science of non-classical materials called rheology. But they did not succeed in fitting their experimental results into a general mathematical framework. Most of the rheological theories are either one-dimensional, and hence appropriate at best for particular experimental situations, or are confined to infinitesimal deformations, in which case they are only of limited use, because large deformations occur easily in the materials these theories are intended to describe.

In the last few years a number of mathematicians have succeeded in devising general three-dimensional theories which are valid for large deformations*. Guided directly by physical experience or by the one-dimensional theories of the rheologists, these authors have proposed various special hypotheses intended to characterize the mechanical behavior of large classes of materials in arbitrary motions. Using the principle of objectivity of material properties (*cf.* Section 11 of this paper) and usually also the assumption of material isotropy** they then derived the general constitutive equation (also called rheological equation of state or stress-strain relation) compatible with the original hypothesis. In many cases they also solved particular problems in order to make possible a comparison with experimental results***.

This paper does not start with a special hypothesis. It starts (Chapter II) with a general principle, called *the principle of determinism for the stress*, which is implicit in all physical experience and applies to any material whatsoever as long as only its mechanical behavior is considered and as long as there are no constraints. From this principle and the principle of objectivity the most general constitutive equation for all materials is derived.

Constitutive equations characterize material properties. But material properties are of a local nature and may change from particle to particle in a body. It turned out to be necessary, therefore, to develop concepts which describe the local behavior of a motion at a particular particle. This is done in Chapter I. Some of the concepts introduced there are similar to those of the theory of "jets" introduced by EHRESMANN [2] into modern differential geometry. The treatment of strain and related notions given in Chapter I is, I believe, more concise and direct than previous treatments. It is based on the well known polar decomposition theorem for linear transformations.

With only one exception**** the ideal materials previously considered in the literature are special cases of what is called a *simple material* in Chapter III of

* A review of the work done up to 1953 was given by TRUESDELL [*12*]. For more recent researches we refer to [*1*], [*4*], [*5*], [*8*], [*9*], [*13*].

** These two things were not clearly distinguished for some time. A clarification was given by the author in [*5*], where the term "principle of isotropy of space" was used for the principle of objectivity.

*** *Cf.* [*3*], [*5*], [*10*], [*12*], [*13*], [*14*].

**** TRUESDELL's theory of Maxwellian fluids (*cf.* [*12*], Chapter V D).

the present paper*. There is reason to believe that most real materials are adequately covered by this category. A detailed discussion of the theory of these simple materials is given. It includes new and very general definitions of isotropic and anisotropic solids and of fluids in terms of the special invariances of the corresponding constitutive equations. Fluids are always isotropic in the sense in which the term is used in this paper. I believe that physically observed anisotropies in fluids are not really intrinsic in these fluids but depend on the past history of the substance.

In Chapter IV it is shown how two large classes of ideal materials fit into the general framework. In the isotropic case, these two classes were introduced by RIVLIN & ERICKSEN [*8*] and COTTER & RIVLIN [*1*]. The general non-isotropic case is discussed here, too.

The theory of simple materials presented in Chapter III generalizes the memory theory of linear visco-elasticity. The theory of materials of the rate type (Section 24) generalizes the differential operator theory of linear visco-elasticity. The materials of the differential type (Section 23) correspond to the special case when only the stress and not the stress rates occur in the basic equation.

Notation and terminology. Physicists customarily use symbols to denote physical quantities. They do not use symbols for the functions that relate these quantities, except sometimes the symbol f, which then stands for any function. Pure mathematicians, on the other hand, distinguish carefully between functions and their values, and they use different symbols for different functions. Both methods have advantages and disadvantages. I have found it necessary here to follow the pure mathematicians. Thus, in this paper, F means something entirely different in nature from $F(t)$. The first is a function, and the second a value of that function and hence a constant.

The terms "function" and "functional" are used here as synonyms for "mapping". A mapping is an operation that assigns to each element of one set, called the *domain* of the mapping, a value in another set. A mapping that has the same values as a given mapping α on a subset of the domain of α but is not defined outside this subset is called the *restriction* of α to the subset in question. Sometimes, if no confusion can arise, we use the same symbol α for such a restriction.

If α and β are two suitable mappings, then $\alpha\circ\beta$ denotes their *composition* defined by $(\alpha\circ\beta)(X)=\alpha(\beta(X))$. The inverse of a one-to-one mapping α is denoted by $\overset{-1}{\alpha}$. If $\mathcal{F}$ and $\mathcal{G}$ are two sets, then $\mathcal{F}\times\mathcal{G}$ means the set of all ordered pairs of elements of $\mathcal{F}$ and $\mathcal{G}$.

In the case of linear transformations F, G of a vector-space into itself we omit the symbol $\circ$ and write simply FG for their composition. The inverse of a linear transformation G is denoted by G^{-1}, its transpose by G^T. The identity transformation is denoted by I. A linear transformation Q is orthogonal if and only if $QQ^T=I$, *i.e.*, $Q^T=Q^{-1}$. The term *tensor* is used as a synonym for linear transformation. Tensors of order higher than two do not occur in this paper.

* More general materials are those "of order n", discussed by the author in [*7*]. They include TRUESDELL's theory of Maxwellian fluids. Simple materials are those of order one.

Points and vectors of a Euclidean space are denoted by boldface letters, except when they are values of mappings. If $\boldsymbol{x}$ and $\boldsymbol{y}$ are two points, then $\boldsymbol{y}-\boldsymbol{x}$ denotes the vector leading from $\boldsymbol{x}$ to $\boldsymbol{y}$. If $\boldsymbol{x}$ is a point and $\boldsymbol{u}$ a vector, then $\boldsymbol{x}+\boldsymbol{u}$ denotes the point $\boldsymbol{y}$ uniquely determined by $\boldsymbol{u}=\boldsymbol{y}-\boldsymbol{x}$.

I. Local kinematics

2. *Basic concepts*

Precise axiomatic definitions of the basic concepts used in this paper are given in [*6*]. Here is a brief summary:

A *body* $\mathscr{B}$ is a smooth manifold of elements $X, Z, \ldots$, called *particles*. The *configurations* $\varphi, \vartheta, \ldots$ of $\mathscr{B}$ are the elements of a set of one-to-one mappings of $\mathscr{B}$ into a three-dimensional Euclidean point space $\mathscr{E}$. The vector space associated with $\mathscr{E}$ will be denoted by $\mathscr{V}$. The *mass-distribution* m of $\mathscr{B}$ is a measure defined on all Borel subsets of $\mathscr{B}$. If φ is a configuration of $\mathscr{B}$, then there is a corresponding *mass density* ϱ_φ such that

$$m(\mathscr{C}) = \int_{\varphi(\mathscr{C})} \varrho_\varphi(\boldsymbol{X})\, dV \tag{2.1}$$

for all Borel subsets $\mathscr{C}$ of $\mathscr{B}$. A *motion* is a one-parameter family $\{\vartheta_\tau\}$, $-\infty<\tau<\infty$, of configurations. The parameter τ is called the *time*. A motion will often simply be denoted by ϑ, which is then regarded as a point-valued function of two variables, a particle and a time. The mass density corresponding to the configuration ϑ_τ is denoted by $\varrho(\tau)$.

A body, its configurations, its mass-distribution, and its possible motions are subject to the axioms given in [*6*]. These axioms insure that the customary notions and operations are meaningful. However, in some of the considerations of this paper, continuity and differentiability conditions stronger than those implied by the axioms of [*6*] have to be assumed. It will be clear from the context when this is the case, and we shall not mention these conditions explicitly.

3. *Deformations and linear transformations*

A smooth homeomorphism δ which maps a neighborhood of the null-vector $\boldsymbol{O}$ of the vector space $\mathscr{V}$ onto another such neighborhood and which maps $\boldsymbol{O}$ into itself,

$$\delta(\boldsymbol{O}) = \boldsymbol{O}, \tag{3.1}$$

shall be called a *local homeomorphism* of $\mathscr{V}$. We define an equivalence relation among all local homeomorphisms by local identity: $\delta\sim\tilde{\delta}$ if and only if $\delta(\boldsymbol{P})=\tilde{\delta}(\boldsymbol{P})$ for all $\boldsymbol{P}$ in some neighborhood of $\boldsymbol{O}$, however small. The resulting equivalence classes Δ will be called *deformations*. The equivalence class of $\delta\circ\hat{\delta}$ depends only on the equivalence classes Δ and $\hat{\Delta}$ of δ and $\hat{\delta}$, and it will therefore be denoted by $\Delta\circ\hat{\Delta}$. With the law of composition thus defined, the deformations form a group denoted by $\mathscr{D}$.

A *linear transformation* G, *i.e.*, a mapping of $\mathscr{V}$ onto itself such that

$$G(\boldsymbol{P}_1+\boldsymbol{P}_2) = G(\boldsymbol{P}_1)+G(\boldsymbol{P}_2), \qquad G(a\boldsymbol{P}) = a\,G(\boldsymbol{P}), \tag{3.2}$$

is a local homeomorphism provided that it is regular, *i.e.*, provided that $G\boldsymbol{P} = \boldsymbol{O}$ is possible only for $\boldsymbol{P} = \boldsymbol{O}$. Two different linear transformations are never equivalent in the sense defined above. Hence a regular linear transformation defines a unique deformation and may thus be regarded as a special deformation. The regular linear transformations thus form a subgroup of the group of deformations $\mathscr{D}$, called the *linear group* and denoted by $\mathscr{L}$.

The gradient at $\boldsymbol{P} = \boldsymbol{O}$ of a local homeomorphism δ is a local property of δ and hence depends only on the equivalence class Δ to which δ belongs. We can, therefore, define the *gradient* $\nabla\Delta$ of a deformation Δ by

$$\nabla\Delta = \nabla\delta(\boldsymbol{O}) \quad \text{if} \quad \delta \in \Delta. \tag{3.3}$$

The gradient $\nabla\Delta$ is a regular linear transformation and hence itself a deformation. Since, by the chain rule,

$$\nabla(\Delta \circ \hat{\Delta}) = (\nabla\Delta)(\nabla\hat{\Delta}), \tag{3.4}$$

it follows that the gradient operation ∇ is an endomorphism of the group $\mathscr{D}$. Its kernel is the group $\mathscr{N}$ of all *null-deformations*, *i.e.*, all deformations whose gradient is the identity I. The quotient group $\mathscr{D}/\mathscr{N}$ consists of the equivalence classes of deformations defined by the equivalence relation

$$\Delta \sim \tilde{\Delta} \quad \text{if and only if} \quad \nabla\Delta = \nabla\tilde{\Delta}. \tag{3.5}$$

The deformations in each class have the same gradient and any representing local homeomorphisms differ only by small terms of order one. The relation (3.5) may be read: Δ and $\tilde{\Delta}$ are equal up to a small term of order one. Each class has a unique representative which is a linear transformation. We have

$$\Delta \sim \nabla\Delta, \tag{3.6}$$

i.e. any deformation differs from its gradient only by a small term of order one. In the case when $\Delta = G$ is a linear transformation, we have

$$\nabla G = G. \tag{3.7}$$

The quotient group $\mathscr{D}/\mathscr{N}$ is isomorphic to the linear group $\mathscr{L}$.

A deformation Δ will be called *isochoric* if there is a volume preserving local homeomorphism δ in the class Δ. The isochoric deformations form a subgroup $\mathscr{I}$ of $\mathscr{D}$. The null-deformations are isochoric, and $\mathscr{N}$ is a subgroup of $\mathscr{I}$. A linear transformation G is isochoric if and only if

$$|\det G| = 1. \tag{3.8}$$

The isochoric linear transformations form a subgroup $\mathscr{U}$ of $\mathscr{L}$, called the *unimodular group*. They are also called *unimodular transformations*. The group of *orthogonal transformations* is a subgroup of $\mathscr{U}$; it will be denoted by $\mathscr{O}$. The space of all *symmetric linear transformations* will be denoted by $\mathscr{S}$. The set of all *positive definite* and symmetric linear transformations will be denoted by $\mathscr{S}_+$. It is a subset of the linear group $\mathscr{L}$.

4. *Local configurations*

Consider a neighborhood $\mathcal{N}(X)$ of a particle X in a body $\mathcal{B}$, *i.e.*, a part of $\mathcal{B}$ containing X. Let ψ be a smooth homeomorphism of $\mathcal{N}(X)$ into the vector space $\mathcal{V}$ mapping X itself into the null-vector,

$$\psi(X) = \boldsymbol{O}. \tag{4.1}$$

We define an equivalence relation among all such homeomorphisms by the condition that $\psi \sim \tilde{\psi}$ if and only if $\psi(Z) = \tilde{\psi}(Z)$ for all Z in some neighborhood of X. The resulting equivalence classes $\boldsymbol{\Phi}$ will be called the *local configurations* of X. The set of all local configurations of X will be denoted by $\mathcal{C}_X$.

Let δ be a local homeomorphism of $\mathcal{V}$. The equivalence class of the mapping $\delta \circ \psi$ depends only on the equivalence class $\boldsymbol{\Phi}$ of ψ and on the deformation Δ of δ. Therefore, it is meaningful to speak about the local configuration $\Delta \circ \boldsymbol{\Phi}$ of X, which is the composite of the local configuration $\boldsymbol{\Phi}$ of X and the deformation Δ. If $\boldsymbol{\Phi}$ and $\hat{\boldsymbol{\Phi}}$ are two local configurations of X, we define

$$\Delta = \hat{\boldsymbol{\Phi}} \circ \overset{-1}{\boldsymbol{\Phi}} \quad \text{if} \quad \hat{\boldsymbol{\Phi}} = \Delta \circ \boldsymbol{\Phi}. \tag{4.2}$$

This Δ will be called the *deformation from the configuration $\boldsymbol{\Phi}$ to the configuration $\hat{\boldsymbol{\Phi}}$.*

Let φ be a configuration of the body $\mathcal{B}$. Then $\psi = \varphi - \varphi(X)$ maps $\mathcal{B}$ into the vector space $\mathcal{V}$ and has the property (4.1). Its equivalence class will be denoted by $\boldsymbol{\Phi}(\varphi, X)$, and it will be called the *localization* at X of φ. Assume that a motion ϑ of $\mathcal{B}$ is given. We then use the notation

$$\boldsymbol{\Theta}(\tau) = \boldsymbol{\Phi}(\vartheta_\tau, X), \tag{4.3}$$

and we call $\boldsymbol{\Theta}$ the *localization* at X of the motion ϑ. We note that $\boldsymbol{\Theta}$ is a function of a real variable whose values are in $\mathcal{C}_X$. Any sufficiently smooth function of this type will be called a *local motion* of X.

Let $\boldsymbol{\Theta}$ be a local motion and $\boldsymbol{\Phi}$ a local configuration of X. We then write

$$\Omega(\tau) = \boldsymbol{\Theta}(\tau) \circ \overset{-1}{\boldsymbol{\Phi}} \tag{4.4}$$

and call Ω the *deformation function* of X in the local motion $\boldsymbol{\Theta}$ relative to the *local reference configuration* $\boldsymbol{\Phi}$. Ω is a function of a real variable with values in $\mathcal{D}$. Any sufficiently smooth function of this type will be called a deformation function.

In the special case when $\boldsymbol{\Phi} = \boldsymbol{\Theta}(t)$ in (4.4), where t is a particular time, we use the notation

$$\Omega_t(\tau) = \boldsymbol{\Theta}(\tau) \circ \overset{-1}{\boldsymbol{\Theta}}(t) \tag{4.5}$$

and call Ω_t the deformation function of X in $\boldsymbol{\Theta}$ relative to the time t.

Let φ be a configuration of the body $\mathcal{B}$ and ϱ_φ the corresponding mass density. The value $\varrho_\varphi(\varphi(X))$ depends only on the localization $\boldsymbol{\Phi}(\varphi, X)$ of φ at X. We can, therefore, define

$$\varrho_{\boldsymbol{\Phi}} = \varrho_\varphi(\varphi(X)), \qquad \boldsymbol{\Phi} = \boldsymbol{\Phi}(\varphi, X), \tag{4.6}$$

and call $\varrho_{\boldsymbol{\Phi}}$ the *mass density* of X in the local configuration $\boldsymbol{\Phi}$.

5. Gradients

Assume that two local configurations Φ and $\hat{\Phi}$ of a particle X are given. The regular linear transformation

$$G = G(\hat{\Phi}, \Phi) = \nabla \Delta, \quad \Delta = \hat{\Phi} \circ \overset{-1}{\Phi} \tag{5.1}$$

will then be called the *deformation gradient* from Φ to $\hat{\Phi}$.

In the case when a local motion Θ and a local reference configuration Φ are given, we use the notation

$$F(\tau) = \nabla \Omega(\tau) = G(\Theta(\tau), \Phi), \tag{5.2}$$

$$F_t(\tau) = \nabla \Omega_t(\tau) = G(\Theta(\tau), \vartheta(t)). \tag{5.3}$$

Assume that $\hat{\Phi}$ is another local reference configuration, and let

$$\hat{F}(\tau) = \nabla \hat{\Omega}(\tau) = G(\Theta(\tau), \hat{\Phi}). \tag{5.4}$$

Then, by the chain rule for gradients, we have

$$F(\tau) = \hat{F}(\tau)\, G, \tag{5.5}$$

where $G = G(\hat{\Phi}, \Phi)$ is defined by (5.1). We note the following formulas:

$$G(\hat{\Phi}, \Phi) = G(\Phi, \hat{\Phi})^{-1}, \tag{5.6}$$

$$F(\tau) = F_t(\tau)\, F(t), \quad F_t(t) = I. \tag{5.7}$$

We define an equivalence relation among all local configurations $\Phi \in \mathscr{C}_X$ by

$$\Phi \sim \tilde{\Phi} \quad \text{if and only if} \quad G(\tilde{\Phi}, \Phi) = I. \tag{5.8}$$

The corresponding equivalence classes M will be called the *configuration gradients* at X. The class of local configurations equivalent to a given $\Phi \in \mathscr{C}_X$ is denoted by $\nabla \Phi$, so that

$$M = \nabla \Phi \quad \text{if and only if} \quad \Phi \in M. \tag{5.9}$$

The equivalence class of the local configuration $\Delta \circ \Phi$, where Δ is a deformation, depends only on the equivalence class of Φ and on the gradient $\nabla \Delta$ of Δ. Therefore, it is meaningful to speak about the configuration gradient GM, which is the product of the configuration gradient M and the regular linear transformation G. This product is characterized by the property

$$\nabla(\Delta \circ \Phi) = (\nabla \Delta)(\nabla \Phi). \tag{5.10}$$

If M and $\hat{M}$ are two configuration gradients, we define

$$G = \hat{M} M^{-1} \quad \text{if} \quad \hat{M} = G M. \tag{5.11}$$

We have

$$G(\hat{\Phi}, \Phi) = \hat{M} M^{-1}, \quad M = \nabla \Phi, \quad \hat{M} = \nabla \hat{\Phi}, \tag{5.12}$$

which shows that $G(\hat{\Phi}, \Phi)$ depends only on the gradients of Φ and $\hat{\Phi}$. In particular, the deformation gradient function F in (5.2) depends only on $\nabla \Phi$, and we may call

$$F(\tau) = (\nabla \Theta(\tau))\, M^{-1} \tag{5.13}$$

the deformation gradient at time τ of the local motion Θ relative to the configuration gradient M as a reference.

If Φ and $\hat{\Phi}$ are two local configuration at X, then the corresponding densities ϱ_Φ and $\varrho_{\hat{\Phi}}$ are related by

$$\varrho_{\hat{\Phi}} = |\det G(\hat{\Phi}, \Phi)|\, \varrho_\Phi. \tag{5.14}$$

It follows from (5.8) that $\varrho_{\hat{\Phi}} = \varrho_\Phi$ if $\hat{\Phi}$ and Φ have the same configuration gradient. Hence we can define

$$\varrho_M = \varrho_\Phi \quad \text{if} \quad M = \nabla \Phi, \tag{5.15}$$

and we call ϱ_M the mass density for the configuration gradient M at X. Equation (5.12) then implies

$$\varrho_{GM} = |\det G|\, \varrho_M. \tag{5.16}$$

It follows that

$$\varrho_M = \varrho_{\hat{M}} \quad \text{if and only if} \quad G = \hat{M} M^{-1} \quad \text{is unimodular.} \tag{5.17}$$

6. *Rotation and strain tensors*

Of fundamental importance for the local kinematics of continuous media is the well known *polar decomposition theorem:*

Let F be any regular linear transformation. Then there are unique decompositions

$$F = RU, \qquad F = VR \tag{6.1}$$

in which R is orthogonal and U and V are symmetric and positive definite, *i.e.*, $R \in \mathcal{O}$, $U, V \in \mathscr{S}_+$. The following relations are valid:

$$U^2 = F^T F, \qquad V^2 = F F^T, \tag{6.2}$$

$$V = R U R^T, \qquad V^2 = R U^2 R^T. \tag{6.3}$$

This theorem, when applied to a deformation gradient $F = G(\hat{\Phi}, \Phi)$, gives rise to the following terminology: We call R the *rotation tensor*, U the *right strain tensor*, and V the *left strain tensor* of the deformation from Φ to $\hat{\Phi}$. The squares

$$C = U^2 = F^T F, \qquad B = V^2 = F F^T = R C R^T \tag{6.4}$$

will be called the *right* and the *left Cauchy-Green tensors* of the deformation from Φ to $\hat{\Phi}$.

The strain tensors U and V describe adequately what is meant physically by strain because their eigenvalues are the principal extension ratios of the deformation from Φ to $\hat{\Phi}$. However, it is often of advantage to use the Cauchy-Green tensors rather than the strain tensors as measures of strain because their components are rational functions of the components of the deformation gradient F, while the components of U and V are complicated irrational functions of the components of F. Both C and B are, of course, symmetric and positive definite. Their eigenvalues are the squares of the principal extension ratios of the deformation.

In the special case when F is the tensor function defined by (5.2) R, U, V, C, and B will also be tensor functions of a real variable. If F is replaced by the tensor function F_t defined in (5.3), we use the analogous notation R_t, U_t, V_t, C_t, and B_t. These functions have the same smoothness properties as F or F_t, respectively. We note that

$$F_t(t) = R_t(t) = U_t(t) = V_t(t) = C_t(t) = B_t(t) = I. \tag{6.5}$$

7. *Histories*

Let $\mathscr{F}$ be an arbitrary set. The class of all functions with values in $\mathscr{F}$ whose domain is the negative real axis will then be denoted by $\mathscr{F}^*$.

Let α be any function of a real variable with values in $\mathscr{F}$. We then define $\alpha^t \in \mathscr{F}^*$ by

$$\alpha^t(s) = \alpha(t+s) \quad \text{for} \quad s \leqq 0, \tag{7.1}$$

and we call it the *history up to time* t of the function α.

For functions $\alpha^* \in \mathscr{F}^*$ we generally use the notation

$$\alpha_0 = \alpha^*(0). \tag{7.2}$$

Then

$$\alpha_0^t = \alpha^t(0) = \alpha(t). \tag{7.3}$$

Let Θ be a local motion of a particle X, so that $\Theta(\tau) \in \mathscr{C}_X$. The corresponding history $\Theta^t \in \mathscr{C}_X^*$, defined by

$$\Theta^t(s) = \Theta(t+s) \quad \text{for} \quad s \leqq 0, \tag{7.4}$$

will then be called the *kinematical history* up to t of X. If, in addition, a local reference configuration Φ is given and if $\Omega = \Theta \circ \overset{-1}{\Phi}$ is the corresponding deformation function, we call $\Omega^t \in \mathscr{D}^*$, defined by

$$\Omega^t(s) = \Omega(t+s) \quad \text{for} \quad s \leqq 0, \tag{7.5}$$

the *deformation history* up to t of X.

8. *Rate of strain and spin*

Consider a local motion Θ of a particle X. Let F_t be the corresponding deformation gradient function defined by (5.3) and F_t^t the history of F_t up to time t defined according to (7.1).

We then call*

$$E(t) = \dot{F}_t(t) = \dot{F}_t^t(0) \tag{8.1}$$

the *velocity gradient* at time t of the local motion Θ. Similarly*

$$E_n(t) = \overset{(n)}{F_t}(t) = \overset{(n)}{F_t^t}(0), \qquad n = 0, 1, 2, \ldots \tag{8.2}$$

---

* A superposed dot denotes the operation of differentiation and a superposed (n) its nth iteration.

is called the nth *acceleration gradient* at time t of Θ. We have, by (6.5) and (8.1),

$$E_0 = I\,, \quad E_1 = E\,. \tag{8.3}$$

If Θ is the localization at X of a global motion ϑ, then $E(t)$ $(E_n(t))$ actually coincides with the gradient with respect to position of the velocity (n^{th} acceleration) at $\boldsymbol{x} = \vartheta_t(X)$ and at time t.

The polar decomposition

$$F_t(\tau) = R_t(\tau)\, U_t(\tau) \tag{8.4}$$

defines the rotation tensor $R_t(\tau)$ and the right strain tensor $U_t(\tau)$ of the deformation from the configuration at time t to the configuration at time τ. The histories R_t^t and U_t^t up to time t of R_t and U_t are defined according to (7.1). The tensor

$$W(t) = \dot{R}_t(t) = \dot{R}_t^t(0) \tag{8.5}$$

is called the *spin* at time t, and the tensor

$$D(t) = \dot{U}_t(t) = \dot{U}_t^t(0) \tag{8.6}$$

the *rate of strain* at time t. Similarly, we define the nth *spin* W_n and the nth *rate of strain* D_n by

$$W_n(t) = \overset{(n)}{R}_t(t) = \overset{(n)}{R_t^t}(0)\,, \tag{8.7}$$

$$D_n(t) = \overset{(n)}{U}_t(t) = \overset{(n)}{U_t^t}(0)\,. \tag{8.8}$$

Replacing U_t in (8.8) by the right Cauchy-Green tensor $C_t = U_t^2$, we get a tensor

$$A_n(t) = \overset{(n)}{C}_t(t) = \overset{(n)}{C_t^t}(0)\,, \tag{8.9}$$

which we shall call the nth *Rivlin-Ericksen tensor**. By (6.5), (8.5), and (8.6) we have

$$R_0 = D_0 = A_0 = I\,, \quad R_1 = R\,, \quad D_1 = D\,. \tag{8.10}$$

One could give definitions similar to (8.8) and (8.9) by replacing the right strain and Cauchy-Green tensors by their left counterparts. But the tensors defined in this way are of little interest. Moreover, for $n = 1$ one would get nothing new, because

$$D(t) = \dot{U}_t(t) = \dot{V}_t(t)\,, \tag{8.11}$$

as is not hard to prove.

9. *Rational expressions for the rates*

It is possible to find explicit expressions for W_n, D_n, and A_n as polynomials in the acceleration gradients E_k and their transpositions E_k^T, $k = 1, 2, \ldots, n$. Differentiating $C_t = F_t^T F_t$ [*cf.* (6.4)] n times, by the product rule we obtain

$$\overset{(n)}{C}_t(\tau) = \sum_{k=0}^{n} \binom{n}{k} \overset{(k)}{F_t^T}(\tau)\, \overset{(n-k)}{F_t}(\tau)\,.$$

* RIVLIN & ERICKSEN [8] recognized the importance of these tensors and used them extensively.

Substituting $\tau=t$ yields, according to (8.2) and (8.9),

$$A_n=\sum_{k=0}^{n}\binom{n}{k}E_k^T E_{n-k}. \tag{9.1}$$

Since $E_0=I$ by (8.3), this may be written in the form

$$A_n=E_n+E_n^T+\sum_{k=1}^{n-1}\binom{n}{k}E_k^T E_{n-k}. \tag{9.2}$$

Differentiating $C_t^{(\tau)}=U_t^2(\tau)$ and then substituting $\tau=t$ we get

$$A_n=2D_n+\sum_{k=1}^{n-1}\binom{n}{k}D_k D_{n-k}, \tag{9.3}$$

and hence

$$D_n=\frac{1}{2}\left\{A_n-\sum_{k=1}^{n-1}\binom{n}{k}D_k D_{n-k}\right\}. \tag{9.4}$$

This is a recursion formula which can be used to find explicit expressions for the D_n as polynomials in the A_k, $k=1, 2, \ldots, n$. Hence, after substitution of (9.2), one would also obtain explicit expressions for the D_n as polynomials in the E_k and E_k^T, $k=1, 2, \ldots, n$.

Differentiating $F_t(\tau)=R_t(\tau)\,U_t(\tau)$ and then putting $\tau=t$, we find

$$E_n=W_n+D_n+\sum_{k=1}^{n-1}\binom{n}{k}W_k D_{n-k}, \tag{9.5}$$

and hence

$$W_n=E_n-D_n-\sum_{k=1}^{n-1}\binom{n}{k}W_k D_{n-k}. \tag{9.6}$$

This is again a recursive formula which permits us to express W_n as a polynomial in the E_k and D_k. Since the D_k are polynomials in the E_k and E_k^T, we can also find expressions for the W_n as polanomials in the E_k and E_k^T, $k=1, 2, \ldots, n$.

In the special case $n=1$ we find

$$D=\tfrac{1}{2}A_1=\tfrac{1}{2}(E+E^T),\qquad W=\tfrac{1}{2}(E-E^T). \tag{9.7}$$

It follows that the spin W is skew ($W^T=-W$). The higher spins W_n, $n>1$ are not necessarily skew. Of course, the D_n and A_n are all symmetric. For $n=2$ we find

$$\begin{aligned} A_2&=E_2+E_2^T+2E^TE,\\ D_2&=\tfrac{1}{4}(A_2-\tfrac{1}{4}A_1^2)=\tfrac{1}{2}[E_2+E_2^T-(E+E^T)^2]+E^TE,\\ W_2&=\tfrac{1}{2}[E_2-E_2^T-E^T(E-E^T)]. \end{aligned} \tag{9.8}$$

II. The general constitutive equation

10. Basic concepts

A *dynamical process* for a body $\mathscr{B}$ is defined as a motion ϑ of $\mathscr{B}$ coupled with a system of forces for $\mathscr{B}$ at each time τ, subject to the principle of linear momentum and the principle of angular momentum (*cf.* [*6*], Section 5). A system of forces can be split into body forces and contact forces. If sufficient continuity assumptions are made, the contact forces are determined, for each time τ, by a field of symmetric *stress tensors* $S(\tau)$. If a motion ϑ and smooth fields of symmetric tensors $S(\tau)$ are arbitrarily prescribed, one can always find a dynamical process such that $S(\tau)$ is the corresponding stress tensor field at time τ. Appropriate body forces can be chosen, for example, by putting the mutual body forces equal to zero and by letting the density $\boldsymbol{b}$ of the external body forces be given by

$$\boldsymbol{b} = \dot{\boldsymbol{v}} - \frac{1}{\varrho} \operatorname{div} S, \tag{10.1}$$

where $\dot{\boldsymbol{v}}$ is the acceleration of the motion ϑ. Then CAUCHY's law of motion ([*6*], (5.25)) holds, and the principles of dynamics are satisfied.

If we subsequently talk about a *process* for a body $\mathscr{B}$ we mean a pair $\{\vartheta, S\}$, where ϑ is an arbitrary motion of $\mathscr{B}$ and S an arbitrary time family of symmetric stress fields. We can always adjoin suitable body forces to a given process so as to make it a dynamical process.

A *local process* for a particle $X \in \mathscr{B}$ is defined as a pair $\{\Theta, S\}$, where Θ is a local motion of X and S a symmetric tensor function of a real variable. The *localization* at X of a process $\{\vartheta, S\}$ is defined by replacing ϑ by its localization Θ at X and S by its values for X. We shall use the same symbol S for these values and write $S(\tau) = S(\tau, X)$. No confusion can arise because, in all local considerations, the particle X will be fixed.

A *constitutive assumption* is a restrictive condition on the possible dynamical processes a body can undergo, and it characterizes the material properties of the body. Restrictive conditions on the possible motions alone are particular constitutive assumptions; they are called *constraints*. Examples of constraints are rigidity (every possible motion is a rigid motion) and incompressibility (every possible motion is isochoric). In this paper, we shall assume that there are no constraints, *i.e.* that all motions ϑ are possible. We shall investigate constitutive assumptions in the form of functional relations between the stress S and the motion ϑ of a process $\{\vartheta, S\}$. Such relations will be called *constitutive equations*.

11. The principle of objectivity of material properties

A *change of frame* is a transformation of space and time specified by a point-valued function $\boldsymbol{c}$ of a real variable, a function Q of a real variable whose values are orthogonal transformations, and a real constant a. It transforms a pair $\{\boldsymbol{x}, \tau\}$ consisting in a point $\boldsymbol{x}$ and a time τ into the pair $\{\boldsymbol{x}', \tau'\}$ given by*

$$\boldsymbol{x}' = \boldsymbol{c}(\tau) + Q(\tau)(\boldsymbol{x} - \boldsymbol{O}), \tag{11.1}$$

$$\tau' = \tau - a, \tag{11.2}$$

where $\boldsymbol{O}$ is an arbitrary point, the same for all transformations. Vectors $\boldsymbol{u} \in \mathscr{V}$ transform according to

$$\boldsymbol{u}' = Q(\tau)\,\boldsymbol{u}. \tag{11.3}$$

* TOUPIN has analyzed these transformations recently [*11*]; he calls them "Euclidean transformations of space-time".

If G is a linear transformation and $\boldsymbol{u} \in \mathscr{V}$, then $G\boldsymbol{u} \in \mathscr{V}$. Hence, by (11.3), $(G\boldsymbol{u})' = Q(\tau)G\boldsymbol{u}$. Defining the transform G' of G by $(G\boldsymbol{u})' = G'\boldsymbol{u}'$, we get $Q(\tau)\,G\boldsymbol{u} = G'\,Q(\tau)\boldsymbol{u}$, and hence the following law of transformation for second order tensors:

$$G' = Q(\tau)\,G\,Q(\tau)^T. \tag{11.4}$$

Two dynamical processes are called *equivalent* if they are related by a change of frame as made precise in Definition 7 of [*6*]. Under such a change of frame the contact forces transform in an objective manner, *i.e.*, according to the law (11.3). Hence the stress tensor must transform according to the law (11.4). The motion ϑ determines the position of the particles and hence transforms according to (11.1). Disregarding the body forces, we can say that *two processes* $\{\vartheta, S\}$ *and* $\{\vartheta', S'\}$ *are equivalent if they are related by a change of frame in the form*

$$\begin{aligned} \vartheta'(Z,\tau') &= \boldsymbol{c}(\tau) + Q(\tau)\,[\vartheta(Z,\tau) - \boldsymbol{O}], \\ S'(Z,\tau') &= Q(\tau)\,S(Z,\tau)\,Q^T(\tau), \\ \tau' &= \tau - a. \end{aligned} \tag{11.5}$$

Constitutive equations are subject to the following invariance requirement:

Principle of objectivity of material properties. *If a process* $\{\vartheta, S\}$ *is compatible with a constitutive equation, then also all processes* $\{\vartheta', S'\}$ *equivalent to it must be compatible with the same constitutive equation.*

This is a special case of the general principle of objectivity stated in [*6*]. Its physical meaning is simply that the material properties of a body should not depend on the observer, no matter how he moves.

12. The principle of determinism for the stress

We ask for guiding principles which will enable us to find the most general form of a constitutive equation. The following two are implied by physical experience:

(*i*) The stress at a particle X should depend only on the physical state of an arbitrarily small neighborhood of X. The state of parts distant from X should have no direct influence on the stress at X. This condition is implicit in the concept of a contact force.

(*ii*) The physical state of a body at a time t should depend only on its past history, *i.e.*, on what happened to it at times $\tau \leq t$, and not on its future, *i.e.*, on what will happen to it at times $\tau > t$. This condition expresses the causality of natural processes.

The physical history of a body consists of several components: its kinematical history, its thermodynamic history, its electromagnetic history, its chemical history, etc. In reality, each of these components may influence the stress. But, in this investigation, we shall disregard all non-mechanical influences.

Principle of determinism. *The stress* $S(t)$ *at a particle* X *and at time* t *is determined by the past history of the motion of an arbitrarily small neighborhood of* X.

In precise terms, this principle states that

(*a*) the stress $S(t)$ at X is a *functional* $\mathfrak{F}_t$ of the motion ϑ,

$$S(t) = \mathfrak{F}_t(\vartheta) \tag{12.1}$$

the domain of $\mathfrak{F}_t$ being the class of all possible motions and its values being symmetric tensors,

(*b*) for any two motions ϑ and $\bar{\vartheta}$ which coincide in some neighborhood of X for times $\tau \leqq t$ the value of the functional $\mathfrak{F}_t$ is the same, *i.e.*,

$$\mathfrak{F}_t(\vartheta) = \mathfrak{F}_t(\bar{\vartheta}) \tag{12.2}$$

whenever

$$\vartheta(Z, \tau) = \bar{\vartheta}(Z, \tau) \tag{12.3}$$

for $\tau \leqq t$ and Z in some neighborhood $\mathcal{N}(X)$ of X, however small.

13. The general constitutive equation

Not any arbitrary functional $\mathfrak{F}_t$ with the Property (*b*) is permissible in (12.1), because (12.1), as a constitutive equation, is subject to the principle of objectivity stated in Section 11. This principle states that (12.1) must hold equally if the process $\{\vartheta, S\}$ is replaced by any equivalent process $\{\vartheta', S'\}$, *i.e.*, that

$$S'(t') = \mathfrak{F}_{t'}(\vartheta') \tag{13.1}$$

must be valid for any process $\{\vartheta', S'\}$ related to $\{\vartheta, S\}$ by a transformation of the form (11.5). Choosing $Q = I$, $a = 0$, and $\boldsymbol{c}(\tau) = \boldsymbol{O} - [\vartheta_\tau(X) - \boldsymbol{O}]$ in (11.5), we get $S' = S$, $\tau' = \tau$, and $\vartheta'_\tau = \boldsymbol{O} + [\vartheta_\tau - \vartheta_\tau(X)]$. It follows that $S(t) = S'(t) = \mathfrak{F}_t(\vartheta')$ holds and hence that $S(t)$ depends only on the vector-valued function ψ defined by $\psi(Z, \tau) = \vartheta(Z, \tau) - \vartheta(X, \tau)$. Moreover, the Property (*b*) of $\mathfrak{F}_t$ implies that $S(t)$ can depend only on the local behavior of ψ in an arbitrarily small neighborhood of X. This means that $S(t)$ is determined by the localization Θ at X of the motion ϑ as defined by (4.3). Thus, (12.1) reduces to the form

$$S(t) = \mathfrak{F}_t(\Theta) \tag{13.2}$$

where $\mathfrak{F}_t$ is a functional with the property that

$$\mathfrak{F}_t(\Theta) = \mathfrak{F}_t(\bar{\Theta}) \tag{13.3}$$

whenever

$$\Theta(\tau) = \bar{\Theta}(\tau) \quad \text{for} \quad \tau \leqq t. \tag{13.4}$$

Now we consider another equivalent process by choosing $\boldsymbol{c}(\tau) = \boldsymbol{O}$, $Q = I$, and $a = t$ in (11.5). We then have

$$\tau' = \tau - t, \quad t' = t - t = 0, \quad S'(t') = S'(0) = S(t),$$

and, for the corresponding local motion Θ,

$$\Theta'(\tau') = \Theta(\tau) = \Theta(t + \tau').$$

It follows from (13.1) that

$$S(t) = S'(0) = \mathfrak{F}_0(\Theta'). \tag{13.5}$$

By (13.4) we have $\mathfrak{F}_0(\Theta') = \mathfrak{F}_0(\overline{\Theta}')$ whenever $\Theta'(\tau') = \overline{\Theta}'(\tau')$ for $\tau' \leqq 0$, which shows that the value $\mathfrak{F}_0(\Theta')$ depends only on the restriction of the function $\Theta'(\tau')$ to $\tau' \leqq 0$. But this restriction is nothing but the kinematical history $\Theta^t \in \mathscr{C}_X^*$ up to time t as defined by (7.4). Omitting the index 0 in (13.5), we see that the constitutive equation (13.2) reduces to the form

$$S(t) = \mathfrak{F}(\Theta^t), \tag{13.6}$$

where Θ^t is the kinematical history of X for the local motion Θ. The form of the functional $\mathfrak{F}$ is independent of t and depends only on the particle X. Its domain is $\mathscr{C}_X^*$.

We finally consider an equivalent process by choosing $\boldsymbol{c}(\tau) = \boldsymbol{O}$ and $a = 0$ in (11.5), leaving the orthogonal tensor function Q arbitrary. The principle of objectivity then implies that the functional $\mathfrak{F}$ of (13.6) must satisfy the relation

$$Q_0 \mathfrak{F}(\Theta^*) Q_0^T = \mathfrak{F}(Q^* \circ \Theta^*), \qquad Q_0 = Q^*(0), \tag{13.7}$$

for all kinematical histories $\Theta^* \in \mathscr{C}_X^*$ and all orthogonal tensor functions $Q^* \in \mathscr{O}^*$.

It is not hard to see that, conversely, the principle of determinism and the principle of objectivity are automatically satisfied for any equation of the form (13.6) provided the functional $\mathfrak{F}$ has the property (13.7). Hence (13.6) with (13.7) is the most general constitutive equation. An equation of this form restricts the class of all possible local processes $\{\Theta, S\}$ for the particle X and characterizes the local material properties of X.

If we subsequently speak about a particle X, we always assume that a functional of the type discussed above is associated with it. We call it the *functional of the particle X*.

14. *Material isomorphisms*

The nature of the domain $\mathscr{C}_X^*$ of the functional $\mathfrak{F}$ of the particle X varies with the particle X. Hence there is no direct way to compare the functionals of different particles. It is desirable to render such a comparison possible, because only then can a precise meaning be given to the statement that two different particles consist of the same material.

A body was defined in [6] as a certain mathematical structure. As in the case with any such structure, it is meaningful to talk about isomorphisms between bodies. An *isomorphism* of a body $\mathscr{B}$ onto a body $\overline{\mathscr{B}}$ is a one-to-one mapping γ of $\mathscr{B}$ onto $\overline{\mathscr{B}}$ such that

(*a*) the configurations $\overline{\varphi}$ of $\overline{\mathscr{B}}$ are of the form

$$\overline{\varphi} = \varphi \circ \overset{-1}{\gamma}, \tag{14.1}$$

(*b*) the mass distributions m and $\overline{m}$ of $\mathscr{B}$ and $\overline{\mathscr{B}}$ are related so that

$$m(\mathscr{C}) = \overline{m}(\gamma(\mathscr{C})) \tag{14.2}$$

for all Borel subsets $\mathscr{C}$ in $\mathscr{B}$.

Assume that X and $\bar{X}$ are two particles and consider the isomorphisms γ, if any, of neighborhoods $\mathcal{N}(X)$ onto neighborhoods $\bar{\mathcal{N}}(\bar{X})$ which map X into $\bar{X}=\gamma(X)$. We define an equivalence relation among all such isomorphisms by the condition that $\gamma\sim\tilde{\gamma}$ if and only if $\gamma(Z)=\tilde{\gamma}(Z)$ for Z in some neighborhood of X, however small. The resulting equivalence classes Γ will be called the *local isomorphisms* of X onto $\bar{X}$. If Γ is such a local isomorphism then the local configurations $\bar{\Phi}$ of $\bar{X}$ are of the form

$$\bar{\Phi}=\Phi\circ\overset{-1}{\Gamma}, \tag{14.3}$$

where Φ is a local configuration of X.

Definition 1. *A local isomorphism Γ of a particle X onto a particle $\bar{X}$ will be called a* material isomorphism *of X onto $\bar{X}$ provided the functionals $\mathfrak{F}$ and $\bar{\mathfrak{F}}$ of X and $\bar{X}$ are related by*

$$\mathfrak{F}(\Theta^*)=\bar{\mathfrak{F}}(\Theta^*\circ\overset{-1}{\Gamma}) \tag{14.4}$$

for all kinematical histories $\Theta^*\in\mathscr{C}_X^*$.

We shall say that two particles consist of the same material if they are materially isomorphic to each other.

15. Constitutive functionals

Assume that a local motion Θ and a local reference configuration Φ of a particle X are given. Let Θ^t and Ω^t be the corresponding kinematical history and deformation history, respectively, as defined in Section 7. By (4.4) they are related by

$$\Theta^t=\Omega^t\circ\Phi. \tag{15.1}$$

Hence the general constitutive equation (13.6) may be rewritten in the form

$$S(t)=\mathfrak{F}(\Omega^t\circ\Phi). \tag{15.2}$$

Definition 2. *A functional $\mathfrak{G}$ whose domain is the set $\mathscr{D}^*$ of all deformation histories and whose values are symmetric tensors is called a* constitutive functional *provided it has the following property: For all deformation histories $\Omega^*\in\mathscr{D}^*$ and all orthogonal tensor functions $Q^*\in\mathcal{O}^*$ the relation*

$$Q_0\,\mathfrak{G}(\Omega^*)\,Q_0^T=\mathfrak{G}(Q^*\circ\Omega^*),\qquad Q_0=Q^*(0), \tag{15.3}$$

holds.

It is clear that the functional $\mathfrak{G}_\Phi$ defined by

$$\mathfrak{G}_\Phi(\Omega^*)=\mathfrak{G}(\Omega^*;\Phi)=\mathfrak{F}(\Omega^*\circ\Phi) \tag{15.4}$$

is a constitutive functional, because the property (15.3) for $\mathfrak{G}=\mathfrak{G}_\Phi$ is equivalent to the property (13.7) for $\mathfrak{F}$. Hence the most general constitutive equation may be formulated in this way:

Theorem 1. *Given a particle X and a local reference configuration Φ of X, there is a constitutive functional $\mathfrak{G}_{\Phi}$ such that, for any local process $\{\Theta, S\}$ of X, the stress S is related to the local motion Θ by*

$$S(t) = \mathfrak{G}_{\Phi}(\Omega^t) = \mathfrak{G}(\Omega^t; \Phi), \tag{15.5}$$

where Ω^t is the deformation history of X defined in Section 7. The constitutive functionals $\mathfrak{G}_{\Phi}$ and $\mathfrak{G}_{\hat{\Phi}}$ corresponding to the local reference configurations Φ and $\hat{\Phi}$ are related, for all deformation histories $\Omega^* \in \mathscr{D}^*$, by

$$\mathfrak{G}(\Omega^* \circ \Delta; \Phi) = \mathfrak{G}(\Omega^*; \hat{\Phi}), \tag{15.6}$$

where $\Delta = \hat{\Phi} \circ \overset{-1}{\Phi}$ is the deformation from Φ to $\hat{\Phi}$.

The relation (15.6) is an immediate consequence of (15.4). We call $\mathfrak{G}_{\Phi}$ the constitutive functional of the particle X relative to the local reference configuration Φ.

Since the nature of the domain $\mathscr{D}^*$ of constitutive functionals is not related to particles, as was the case with the functionals $\mathfrak{F}$ in (13.6), they characterize materials in a manner that is independent of the particular particle. The following theorem is a consequence of Definition 1 and (15.4):

Theorem 2. *Two particles X and $\bar{X}$ are materially isomorphic,* i.e., *consist of the same material, if and only if there is a local configuration Φ of X and a local isomorphism Γ of X onto $\bar{X}$ such that*

$$\mathfrak{G}_{\Phi}(\Omega^*) = \mathfrak{G}_{\bar{\Phi}}(\Omega^*), \quad \bar{\Phi} = \Phi \circ \overset{-1}{\Gamma}, \tag{15.7}$$

for all $\Omega^* \in \mathscr{D}^*$.

16. The local isotropy group

Let Γ be a *material automorphism* of X, *i.e.*, a material isomorphism of X onto itself. By Theorem 2, there is a local configuration Φ such that

$$\mathfrak{G}(\Omega^*; \Phi) = \mathfrak{G}(\Omega^*; \hat{\Phi}), \quad \hat{\Phi} = \Phi \circ \overset{-1}{\Gamma}. \tag{16.1}$$

The material automorphism Γ and the deformation $\Delta = \hat{\Phi} \circ \overset{-1}{\Phi}$ from Φ to $\hat{\Phi}$ are related by

$$\Delta = \Phi \circ \overset{-1}{\Gamma} \circ \overset{-1}{\Phi}, \quad \Gamma = \overset{-1}{\Phi} \circ \overset{-1}{\Delta} \circ \Phi. \tag{16.2}$$

Since Γ, an isomorphism, preserves the mass distribution, it follows that Δ must be an isochoric deformation.

In addition to (16.1) we have the relation (15.6) between the two functionals $\mathfrak{G}_{\Phi}$ and $\mathfrak{G}_{\hat{\Phi}}$. Combining these two relations we obtain

$$\mathfrak{G}_{\Phi}(\Omega^* \circ \Delta) = \mathfrak{G}_{\Phi}(\Omega^*). \tag{16.3}$$

Conversely, if (16.3) holds for an isochoric deformation $\Delta \in \mathscr{I}$, then $\Gamma = \overset{-1}{\Phi} \circ \overset{-1}{\Delta} \circ \Phi$ can easily be seen to be a material automorphism of X.

Definition 3. *Let $\mathfrak{G}$ be a constitutive functional. The group $\mathcal{G}^l$ of all isochoric deformations $\Delta \in \mathcal{I}$ with the property that*

$$\mathfrak{G}(\Omega^* \circ \Delta) = \mathfrak{G}(\Omega^*) \tag{16.4}$$

holds for all $\Omega^* \in \mathcal{D}^*$ is called the local isotropy group *of* $\mathfrak{G}$.

$\mathcal{G}^l$ is a subgroup of the group $\mathcal{I}$ of all isochoric deformations. The material automorphisms of a particle also form a group. The relations (16.2) establish an isomorphism of this group with the local isotropy group $\mathcal{G}^l_\Phi$ of $\mathfrak{G}_\Phi$. The local isotropy groups $\mathcal{G}^l_\Phi$ and $\mathcal{G}^l_{\hat{\Phi}}$ corresponding to different local configurations Φ and $\hat{\Phi}$ will in general be different. But they are all isomorphic to each other because they are isomorphic to the group of material automorphisms of X. In fact, it is not hard to see that $\mathcal{G}^l_{\hat{\Phi}}$ is conjugate to $\mathcal{G}^l_\Phi$:

$$\mathcal{G}^l_{\hat{\Phi}} = \Delta \circ \mathcal{G}^l_\Phi \circ \Delta^{-1}, \qquad \Delta = \hat{\Phi} \circ \overset{-1}{\Phi}. \tag{16.5}$$

III. Simple materials

17. Simple constitutive functionals

It may happen that the values $\mathfrak{G}(\Omega^*)$ of a constitutive functional $\mathfrak{G}$ are not affected if $\Omega^*(s)$, for each $s \leqq 0$, is changed by a small term of order one. By (3.6), $\Omega^*(s)$ differs from $\nabla\Omega^*(s)$ only by such a small term.

Definition 4. *A constitutive functional $\mathfrak{G}$ is said to be* simple *if*

$$\mathfrak{G}(\Omega^*) = \mathfrak{G}(\nabla\Omega^*) \tag{17.1}$$

for all deformation histories $\Omega^* \in \mathcal{D}^*$.

The condition (17.1) is equivalent to the following: If $\nabla\Omega^*(s) = \nabla\tilde{\Omega}^*(s)$ for all $s \leqq 0$, then

$$\mathfrak{G}(\Omega^*) = \mathfrak{G}(\tilde{\Omega}^*). \tag{17.2}$$

The values of a simple constitutive functional $\mathfrak{G}$ are determined for all $\Omega^* \in \mathcal{D}^*$ if they are known for the histories of linear transformations $F^* \in \mathcal{L}^*$. In other words, $\mathfrak{G}$ is determined by its restriction to $\mathcal{L}^*$.

From now on we assume that $\mathfrak{G}$ is simple, and we use the same symbol $\mathfrak{G}$ for its restriction to $\mathcal{L}^*$. By (15.3) $\mathfrak{G}$ satisfies the relation

$$Q_0 \mathfrak{G}(F^*) Q_0^T = \mathfrak{G}(Q^* F^*), \qquad Q_0 = Q^*(0), \tag{17.3}$$

for all $F^* \in \mathcal{L}^*$ and all $Q^* \in \mathcal{O}^*$.

Let $G \in \mathcal{L}$ be an arbitrary regular linear transformation with the polar decomposition

$$G = PT, \qquad P \in \mathcal{O}, \qquad T \in \mathcal{S}_+. \tag{17.4}$$

For any $F^* \in \mathcal{L}^*$ we define

$$\hat{F}^* = F^* G^{-1} = \hat{R}^* \hat{U}^*, \qquad \hat{R}^* \in \mathcal{O}^*, \qquad \hat{U}^* \in \mathcal{S}_+^*, \tag{17.5}$$

where $\hat{R}^*$ and $\hat{U}^*$ are determined by the polar decomposition of $\hat{F}^*$. We then have

$$F^* = \hat{F}^* G = \hat{R}^* \hat{U}^* P\, T = (\hat{R}^* P)(\hat{P}^T U^* P)\, T.$$

Substituting $\hat{R}^* P$ for Q^* and $(P^T \hat{U}^* P)\, T$ for F^* in (17.3), we see that

$$(17.6)\qquad \mathfrak{G}(F^*) = \hat{R}_0 P \mathfrak{G}\big((P^T \hat{U}^* P)\, T\big)\, P^T \hat{R}_0^T, \quad \hat{R}_0 = \hat{R}^*(0).$$

We now define a functional $\mathfrak{K}$ with domain $\mathscr{S}_+^* \times \mathscr{S}_+$ by

$$(17.7)\qquad \mathfrak{G}(U^* T) = \mathfrak{K}(U^*; T)$$

for all $U^* \in \mathscr{S}_+^*$ and $T \in \mathscr{S}_+$. Equation (17.6) then shows that

$$(17.8)\qquad \mathfrak{G}(F^*) = \hat{R}_0 P\, \mathfrak{K}(P^T \hat{U}^* P; T)\, P^T \hat{R}_0^T.$$

The functional $\mathfrak{K}$ has the property that

$$(17.9)\qquad \mathfrak{K}(U^*; T) = \mathfrak{K}(\tilde{U}^*; \tilde{T}) \quad \text{if} \quad U^* T = \tilde{U}^* \tilde{T}.$$

Conversely, if $\mathfrak{K}$ is any functional with domain $\mathscr{S}_+^* \times \mathscr{S}_+$ and with the property (17.9), then (17.8) defines a simple constitutive functional $\mathfrak{G}$. The variables on the right side of (17.8) are defined in terms of F^* and an arbitrary $G \in \mathscr{L}$. For the special choice $G = I$ we get

$$(17.10)\qquad \mathfrak{G}(F^*) = R_0 \mathfrak{K}(U^*; I)\, R_0^T, \quad R_0 = R^*(0),$$

where U^* and R^* are defined by the polar decomposition

$$(17.11)\qquad F^* = R^* U^*, \quad R^* \in \mathscr{O}^*, \quad U^* \in \mathscr{S}_+^*.$$

In the special case when $G = F_0 = F^*(0)$ (17.8) becomes

$$(17.12)\qquad \mathfrak{G}(F^*) = R_0 \mathfrak{K}(R_0^T U_*^* R_0; U_0)\, R_0^T,$$

where $R_0 = R^*(0)$, $U_0 = U^*(0)$, and where $U_*^* \in \mathscr{S}_+^*$ is determined by the polar decomposition

$$(17.13)\qquad F^* F_0^{-1} = R_*^* U_*^*, \quad R_*^* \in \mathscr{O}^*, \quad U_*^* \in \mathscr{S}_+^*.$$

We define a functional $\mathfrak{K}_1$ with domain $\mathscr{S}_+^*$ by

$$(17.14)\qquad \mathfrak{K}_1(U^*) = \mathfrak{K}(U^*; I) \quad \text{for} \quad U^* \in \mathscr{S}_+^*.$$

Then (17.10) takes the form

$$(17.15)\qquad \mathfrak{G}(F^*) = R_0 \mathfrak{K}_1(U^*)\, R_0^T.$$

Conversely, if $\mathfrak{K}_1$ is an arbitrary functional with domain $\mathscr{S}_+^*$, then (17.15) defines a simple constitutive functional.

We also define a functional $\mathfrak{K}_2$ as the restriction of $\mathfrak{K}$ obtained by allowing for its first variable only those functions $U_*^* \in \mathscr{S}_+^*$ whose value for $s = 0$ is the identity,

$$(17.16)\qquad U_*^*(0) = I.$$

We denote the set of all U_*^* with the property (17.16) by $\mathscr{S}_{+*}^*$. The domain of $\mathfrak{K}_2$ is then $\mathscr{S}_{+*}^* \times \mathscr{S}_+$, and it is identical with $\mathfrak{K}$ in this domain. The function

U_*^* defined by (17.13) has the property (17.16), and the same is true for $R_0^T U_*^* R_0$. Hence (17.12) has the form

$$(17.17) \qquad \mathfrak{G}(F^*) = R_0 \mathfrak{K}_2(R_0^T U_*^* R_0; U_0) R_0^T.$$

Conversely, if $\mathfrak{K}_2$ is an arbitrary functional with domain $\mathscr{S}_{+*}^* \times \mathscr{S}_+$, then (17.17) defines a simple constitutive functional $\mathfrak{G}$.

18. Simple materials

We say that the material at a particle X is *simple* or, briefly, that X is simple if the constitutive functional $\mathfrak{G}_\Phi$ of X, for some local configuration Φ of X, is simple. We assume from now on that this is the case. Let $\hat{\Phi}$ be another local configuration of X and let $\Delta = \hat{\Phi} \circ \overset{-1}{\Phi}$ be the deformation from Φ to $\hat{\Phi}$. For every deformation history $\Omega^* \in \mathscr{D}^*$, by (3.7) and the chain rule (3.4), we have

$$(18.1) \qquad \nabla(\Omega^* \circ \Delta) = \nabla[(\nabla\Omega^*) \circ \Delta].$$

It follows from (17.2) that

$$\mathfrak{G}(\Omega^* \circ \Delta; \Phi) = \mathfrak{G}(\nabla\Omega^* \circ \Delta; \Phi),$$

and hence from (15.6) that

$$\mathfrak{G}(\Omega^*; \hat{\Phi}) = \mathfrak{G}(\nabla\Omega^*; \hat{\Phi}),$$

which shows that $\mathfrak{G}_{\hat{\Phi}}$ is also simple. Therefore, if the material at a particle is simple then the constitutive functional $\mathfrak{G}_\Phi$ is simple for all local reference configurations Φ.

It follows from (15.6) that the functionals $\mathfrak{G}_\Phi$ and $\mathfrak{G}_{\hat{\Phi}}$ corresponding to two local configurations Φ and $\hat{\Phi}$ are related by

$$(18.2) \qquad \mathfrak{G}(F^* G; \Phi) = \mathfrak{G}(F^*; \hat{\Phi}),$$

where $G = G(\hat{\Phi}, \Phi)$ is the deformation gradient from Φ to $\hat{\Phi}$ defined by (5.1). It follows from (18.2) that

$$\mathfrak{G}(F^*; \Phi) = \mathfrak{G}(F^*; \hat{\Phi}) \quad \text{if} \quad G(\hat{\Phi}, \Phi) = I.$$

This means, according to (5.8), that $\mathfrak{G}(F^*; \Phi)$ depends only on the gradient $M = \nabla\Phi$ of the local configuration Φ. We can therefore define

$$(18.3) \qquad \mathfrak{G}_M(F^*) = \mathfrak{G}(F^*; M) = \mathfrak{G}(F^*; \Phi) \quad \text{if} \quad M = \nabla\Phi.$$

The following theorem is a corollary of Theorem 1.

Theorem 3. *Let X be a simple particle and let M be a configuration gradient of X. Then there is a simple constitutive functional $\mathfrak{G}_M$ such that, for any local process $\{\Theta, S\}$, the stress is related to the local motion by the constitutive equation*

$$(18.4) \qquad S(t) = \mathfrak{G}_M(F^t) = \mathfrak{G}(F^t; M),$$

where $F^t \in \mathscr{L}^*$ is the history of the deformation gradient of the local motion Θ relative to M as a reference. The functionals $\mathfrak{G}_M$ and $\mathfrak{G}_{\hat{M}}$ corresponding to two configuration *gradients are related by*

$$\mathfrak{G}(F^* G; M) = \mathfrak{G}(F^*; \hat{M}), \qquad G = \hat{M} M^{-1}, \tag{18.5}$$

for all $F^* \in \mathscr{L}^*$.

The results of the previous section enable us to put the constitutive equation into various other forms:

Theorem 4. *The constitutive equation for a simple particle X may be written, with reference to a fixed configuration gradient M of X, in one of the following four forms:*

(a) There is a functional $\mathfrak{K}_1$ *with domain* $\mathscr{S}_+^*$ *such that the stress is given by*

$$S(t) = R(t)\, \mathfrak{K}_1(U^t)\, R(t)^T, \tag{18.6}$$

where $R(t)$ *is the rotation tensor and* U^t *the history of the right strain tensor relative to M as a reference.*

(b) There is a functional $\mathfrak{K}$ *with domain* $\mathscr{S}_+^* \times \mathscr{S}$ *and with the property* (17.9) *such that, for any configuration gradient* $\hat{M}$ *as a reference, the stress is given by*

$$S(t) = \hat{R}(t)\, P\, \mathfrak{K}(P^T \hat{U}^t P; T_r)\, P^T \hat{R}(t)^T, \tag{18.7}$$

where $\hat{R}(t)$ *is the rotation tensor and* $\hat{U}^t$ *the history of the right strain tensor relative to* $\hat{M}$ *as a reference, and where P is the rotation tensor and* T_r *the right strain tensor of the deformation from M to* $\hat{M}$, $\hat{M} M^{-1} = P T$.

(c) In the special case when $\hat{M}$ *is the gradient of the configuration at some time* t_0 *the equation* (18.7) *takes the form*

$$S(t) = R_{t_0}(t)\, R(t_0)\, \mathfrak{K}\big(R(t_0)^T\, U_{t_0}^t\, R(t_0); U(t_0)\big)\, R(t_0)^T R_{t_0}(t)^T, \tag{18.8}$$

where the rotation tensors R and R_{t_0} *and the right strain tensors U and* U_{t_0} *are taken relative to M and the configuration at time* t_0 *as a reference, respectively.*

(d) There is a functional $\mathfrak{K}_2$ *with domain* $\mathscr{S}_{+*}^* \times \mathscr{S}_+$ *such that the stress is given by*

$$S(t) = R(t)\, \mathfrak{K}_2\big(R^T(t)\, U_t^t\, R(t); U(t)\big)\, R(t)^T, \tag{18.9}$$

where R, U, and U_t *are defined as before.*

The forms (18.6) and (18.9) have the advantage that the functionals $\mathfrak{K}_1$ and $\mathfrak{K}_2$ are not subject to any restrictive condition.

19. The isotropy group

If $\mathfrak{G}$ is a simple constitutive functional, then it follows from (17.2) that $\mathfrak{G}(\Omega^*) = \mathfrak{G}(\Omega^* \circ \Delta)$ holds for all null-deformations $\Delta \in \mathscr{N}$, *i.e.*, whenever $\nabla\Delta = I$. Hence, by definition 3, the local isotropy group $\mathscr{G}^l$ of $\mathfrak{G}$ contains the group $\mathscr{N}$ of all null-deformations as a normal subgroup. The quotient group $\mathscr{G}^l/\mathscr{N}$ is isomorphic to a group of unimodular transformations.

Definition 5. *The group* $\mathscr{G}$ *of all unimodular transformations H with the property that*

$$\mathfrak{G}(F^* H) = \mathfrak{G}(F^*) \tag{19.1}$$

holds for all $F^* \in \mathscr{L}^*$ *is called the* isotropy group *of the simple constitutive functional* $\mathfrak{G}$.

$\mathscr{G}$ is a subgroup of the group $\mathscr{U}$ of all unimodular transformations and it is isomorphic to $\mathscr{G}^l/\mathscr{N}$.

Theorem 5. *An orthogonal transformation* Q *is an element of the isotropy group* $\mathscr{G}$ *of a constitutive functional* $\mathfrak{G}$ *if and only if one of the following conditions is satisfied:*

(a) For all $F^* \in \mathscr{L}^*$

$$Q\,\mathfrak{G}(F^*)\,Q^T = \mathfrak{G}(Q F^* Q^T). \tag{19.2}$$

(b) For all $U^* \in \mathscr{S}_+^*$ *and all* $T \in \mathscr{S}_+$

$$Q\,\mathfrak{K}(U^*; T)\,Q^T = \mathfrak{K}(Q\,U^*\,Q^T; Q\,T\,Q^T), \tag{19.3}$$

where $\mathfrak{K}$ *is defined by* (17.7).

(c) For all $U^* \in \mathscr{S}_+^*$

$$Q\,\mathfrak{K}_1(U^*)\,Q^T = \mathfrak{K}_1(Q\,U^*\,Q^T), \tag{19.4}$$

where $\mathfrak{K}_1$ *is defined by* (17.14).

(d) For all $U_*^* \in \mathscr{S}_{+*}^*$ *and all* $T \in \mathscr{S}_+$

$$Q\,\mathfrak{K}_2(U_*^*; T)\,Q^T = \mathfrak{K}_2(Q\,U_*^*\,Q^T; Q\,T\,Q^T) \tag{19.5}$$

where $\mathfrak{K}_2$ *is defined in Section* 17.

Proof. If we substitute the constant tensor Q for the tensor function Q^* in (17.3), and QF^* for F^* and Q for H in (19.1), we see that (19.2) holds if and only if $Q \in \mathscr{G}$. The equivalence of the condition (a) with (b), (c), and (d) follows directly from the definitions of $\mathfrak{K}$, $\mathfrak{K}_1$, and $\mathfrak{K}_2$.

If $\mathscr{G}$ contains the full orthogonal group $\mathscr{O}$, then (19.2), (19.3), (19.4), and (19.5) are valid for all orthogonal Q.

The isotropy group $\mathscr{G}_M$ of the constitutive functional $\mathfrak{G}_M$ of a particle will, in general, depend on the choice of the configuration gradient M. But, as in the case of local isotropy, the groups $\mathscr{G}_M$ and $\mathscr{G}_{\hat{M}}$ corresponding to two configuration gradients M and $\hat{M}$ are conjugate:

$$\mathscr{G}_{\hat{M}} = G\,\mathscr{G}_M\,G^{-1}, \qquad G = \hat{M}\,M^{-1}. \tag{19.6}$$

It follows from (18.5) that $H \in \mathscr{G}_M$ if and only if

$$\mathfrak{G}(F^*; M) = \mathfrak{G}(F^*; HM) \tag{19.7}$$

for all $F^* \in \mathscr{L}^*$.

20. Isotropic and anisotropic solids

We say that a constitutive functional $\mathfrak{G}$ defines a *solid* if its isotropy group is a subgroup of the orthogonal group, *i.e.*, if $\mathscr{G} \subset \mathscr{O}$. A particle X is said to be a solid particle if there is a configuration gradient M of X such that $\mathfrak{G}_M$ defines a solid. A solid is called an *isotropic solid* if the isotropy group of its defining functional is the full orthogonal group, *i.e.*, if $\mathcal{G} = \mathcal{O}$. Let X be an isotropic solid particle. Any configuration gradient M such that $\mathcal{G}_M = \mathcal{O}$ is then called an *undistorted state* of X.

The following theorem follows immediately from Theorems 4 and 5.

Theorem 6. *Let X be an isotropic solid particle. Its constitutive equation may be written, with reference to an undistorted state M of X, in one of the following forms:*

$$S(t) = \mathfrak{K}_1\big(R(t)\, U^t R^T(t)\big), \tag{20.1}$$

$$S(t) = \hat{R}(t)\, \mathfrak{K}(\hat{U}^t; T_l)\, \hat{R}(t)^T, \tag{20.2}$$

$$S(t) = R_{t_0}(t)\, \mathfrak{K}\big(U^t_{t_0}; V(t_0)\big)\, R_{t_0}(t), \tag{20.3}$$

$$S(t) = \mathfrak{K}_2\big(U^t_t; V(t)\big). \tag{20.4}$$

The notation of Theorem 4 applies here. In addition, $V = R\,U\,R^T$ is the left strain tensor relative to M as a reference and T_l is the left strain tensor of the deformation from M to $\hat{M}$. The functionals $\mathfrak{K}$, $\mathfrak{K}_1$ and $\mathfrak{K}_2$ satisfy the conditions (b), (c), and (d) of Theorem 5 for all orthogonal transformations Q.

A solid is called *anisotropic* if the isotropy group of its defining functional is a proper subgroup of the orthogonal group, *i.e.*, if $\mathcal{G} \subset \mathcal{O}$ and $\mathcal{G} \neq \mathcal{O}$. Material symmetries in anisotropic solids, such as *orthotropy*, *transverse isotropy*, and the various types of *crystal symmetry* are defined according to the special nature of the isotropy group $\mathcal{G}$.

21. Fluids

We say that a constitutive functional defines a *fluid* if its isotropy group $\mathcal{G}$ is the full unimodular group $\mathcal{U}$, *i.e.*, if $\mathcal{G} = \mathcal{U}$. A particle X is said to be a fluid particle if, for some configuration gradient M of X, the corresponding constitutive functional defines a fluid, *i.e.*, if $\mathcal{G}_M = \mathcal{U}$. Let $\hat{M}$ be any other configuration gradient. By (19.6) $\mathcal{G}_{\hat{M}}$ is conjugate to $\mathcal{G}_M = \mathcal{U}$. But $\mathcal{U}$ is a normal subgroup of $\mathcal{L}$ and hence coincides with all its conjugates. It follows that, if X is a fluid particle, then $\mathcal{G}_M = \mathcal{U}$ for all configuration gradients M of X.

For a fluid it follows from (19.7) that

$$\mathfrak{G}(F^*; M) = \mathfrak{G}(F^*; \hat{M}) \tag{21.1}$$

whenever $H = \hat{M}\overset{-1}{M}$ is unimodular. But, by (5.17), this is the case if and only if the densities ϱ_M and $\varrho_{\hat{M}}$ coincide. If follows that the value $\mathfrak{G}(F^*; M)$ can depend only on the density ϱ_M. Therefore, we can define a functional $\mathfrak{H}$ with domain $\mathcal{L}^* \times \mathcal{R}_+$ ($\mathcal{R}_+$ = set of positive real numbers) such that

$$\mathfrak{G}(F^*; M) = \mathfrak{H}(F^*; \varrho_M) \tag{21.2}$$

for all $F^* \in \mathcal{L}^*$ and all configuration gradients M of X. Let $F^* = R^* U^*$ be the polar decomposition of F^*. Substituting R^* for Q^* and U^* for F^* in (17.3) and using (21.2) we see that

$$\mathfrak{G}(F^*; M) = R_0\, \mathfrak{H}(U^*; \varrho_M)\, R_0^T, \qquad R_0 = R^*(0), \tag{21.3}$$

which shows that $\mathfrak{H}$ is determined by its restriction to $\mathscr{S}_+^* \times \mathscr{R}_+$. We use the same symbol $\mathfrak{H}$ for this restriction. Since $\mathscr{G}_M = \mathscr{U}$ contains the orthogonal group it follows from (19.2) that $\mathfrak{H}$ satisfies the relation

$$Q\,\mathfrak{H}(U^*; d)\,Q^T = \mathfrak{H}(Q\,U^*\,Q^T; d) \tag{21.4}$$

for all $U^* \in \mathscr{S}_+^*$, all $Q \in \mathscr{O}$, and all $d > 0$.

Theorem 7. *The constitutive equation for a fluid particle X may be written in one of the following forms:*

(a) There is a functional $\mathfrak{H}$ *with domain* $\mathscr{S}_+^* \times \mathscr{R}_+$ *and with the property* (21.4) *such that the stress is given by*

$$S(t) = R(t)\,\mathfrak{H}(U^t; \varrho_M)\,R(t)^T, \tag{21.5}$$

where $R(t)$ *is the rotation tensor and* U^t *the history of the right strain tensor relative to an arbitrary configuration gradient* M *with mass density* ϱ_M.

(b) In the special case when M *is the configuration gradient at some time* t_0, *the equation* (21.5) *takes the form*

$$S(t) = R_{t_0}(t)\,\mathfrak{H}\big(U_{t_0}^t; \varrho(t_0)\big)\,R_{t_0}(t)^T. \tag{21.6}$$

(c) There is a functional $\mathfrak{H}_1$ *with domain* $\mathscr{S}_{+*}^* \times \mathscr{R}_+$ *and with the property that*

$$Q\,\mathfrak{H}_1(U_*^*; d)\,Q^T = \mathfrak{H}_1(Q\,U_*^*\,Q^T; d) \tag{21.7}$$

for all $U_*^* \in \mathscr{S}_{+*}^*$, *all* $Q \in \mathscr{O}$, *and all* $d > 0$, *such that the stress is given by*

$$S(t) = \mathfrak{H}_1\big(U_t^t; \varrho(t)\big). \tag{21.8}$$

Proof. The part (*a*) and its special case (*b*) follow from (21.3) and Theorem 3. The part (*c*) follows from (*b*) by choosing $t_0 = t$ and by defining $\mathfrak{H}_1$ to be the restriction of $\mathfrak{H}$ to $\mathscr{S}_{+*}^* \times \mathscr{R}_+$.

We note that an arbitrary functional $\mathfrak{H}_1$ with domain $\mathscr{S}_{+*}^* \times \mathscr{R}_+$ and with the property (21.7) may define a fluid. The constitutive equation (21.8) is intrinsic in the sense that it does not depend on the choice of a reference configuration. The constitutive equation of a solid cannot be put into such an intrinsic form.

22. Constitutive equations involving the Cauchy-Green tensors

As we pointed out in Section 6, it is often better to use the Cauchy-Green tensors C and B instead of the strain tensors U and V. In order to do so, we define new functionals $\bar{\mathfrak{K}}$, $\bar{\mathfrak{K}}_1$, $\bar{\mathfrak{K}}_2$, $\bar{\mathfrak{H}}$ and $\bar{\mathfrak{H}}_1$, of the same nature as the corresponding functionals without the superposed bars defined in Sections 17 and 20, by the following formulas:

$$\bar{\mathfrak{K}}(U^{*2}; T^2) = U_0\,\mathfrak{K}(U^*; T)\,U_0, \tag{22.1}$$

$$\bar{\mathfrak{K}}_1(U^{*2}) = U_0\,\mathfrak{K}_1(U^*)\,U_0, \tag{22.2}$$

$$\bar{\mathfrak{K}}_2(U_*^{*2}; T^2) = \mathfrak{K}_2(U_*^*; T), \tag{22.3}$$

$$\bar{\mathfrak{H}}(U^{*2}; d) = U_0\,\mathfrak{H}(U^*; d)\,U_0, \tag{22.4}$$

$$\bar{\mathfrak{H}}_1(U_*^{*2}; d) = \mathfrak{H}_1(U_*^*; d), \tag{22.5}$$

valid for all $U^* \in \mathcal{S}_+^*$ $(U_0 = U^*(0))$, all $U_*^* \in \mathcal{S}_{+*}^*$ $(U_*^*(0) = I)$, all $T \in \mathcal{S}_+$, and all $d > 0$.

It is not hard to see that the general constitutive equations (18.6), (18.8), and (18.9) of Theorem 4 for simple materials then take the form

$$F(t)^T S(t) F(t) = \overline{\mathfrak{K}}_1(C^t),\text{*} \tag{22.6}$$

$$F_{t_0}(t)^T S(t) F_{t_0}(t) = R(t_0)\, \overline{\mathfrak{K}}\left(R(t_0)^T C_{t_0}^t R(t_0); C(t_0)\right) R(t_0)^T, \tag{22.7}$$

$$S(t) = R(t)\, \overline{\mathfrak{K}}_2\left(R(t)^T C_t^t R(t); C(t)\right) R(t)^T. \tag{22.8}$$

In the case of isotropic solids, the functionals $\overline{\mathfrak{K}}$, $\overline{\mathfrak{K}}_1$, and $\overline{\mathfrak{K}}_2$ satisfy, for all orthogonal Q, the functional relations obtained from (19.3), (19.4), and (19.5) by superposing bars. Moreover, the simplified constitutive equations (20.3) and (20.4) for isotropic solids take the form

$$F_{t_0}(t)^T S(t) F_{t_0}(t) = \overline{\mathfrak{K}}\left(C_{t_0}^t; B(t_0)\right), \tag{22.9}$$

$$S(t) = \overline{\mathfrak{K}}_2\left(C_t^t; B(t)\right). \tag{22.10}$$

The functionals $\overline{\mathfrak{H}}$ and $\overline{\mathfrak{H}}_1$ have the same properties (21.4) and (21.7) as the corresponding functionals without the superposed bars. The constitutive equations (21.6) and (21.8) for fluids take the form

$$F_{t_0}(t)^T S(t) F_{t_0}(t) = \overline{\mathfrak{H}}\left(C_{t_0}^t; \varrho(t_0)\right), \tag{22.11}$$

$$S(t) = \overline{\mathfrak{H}}_1\left(C_t^t; \varrho(t)\right). \tag{22.12}$$

IV. Special classes of materials

23. Materials of the differential type

The value $\mathfrak{G}(F^*)$ of a simple constitutive functional is determined by the values $F^*(s)$ of the tensor function F^* for $s \leq 0$. It may happen that $\mathfrak{G}(F^*)$ depends only on the values $F^*(s)$ for s very near to zero. If F^* has sufficiently many continuous derivatives then $F^*(s)$ may be approximated, for small values of s, by its Taylor expansion up to some order n. This Taylor expansion is determined by the value of F^* and its derivatives up to the order n at $s = 0$, *i.e.*, by

$$F_0 = F^*(0), \quad \dot{F}_0 = \dot{F}^*(0), \ldots, \overset{(n)}{F_0} = \overset{(n)}{F^*}(0). \tag{23.1}$$

Definition 6. *A simple constitutive functional* $\mathfrak{G}$ *is said to be of the differential type if*

$$\mathfrak{G}(F^*) = \mathfrak{G}(\tilde{F}^*) \tag{23.2}$$

whenever

$$\overset{(k)}{F^*}(0) = \overset{(k)}{\tilde{F}^*}(0), \qquad k = 0, 1, \ldots, n. \tag{23.3}$$

We have seen that every simple constitutive functional $\mathfrak{G}$ has a representation of the form (17.17) in terms of a functional $\mathfrak{K}_2$. Since $\mathfrak{K}_2$ is just a restriction of $\mathfrak{K}$

* This form, in other notation, has been proposed independently by Green & Rivlin [4].

which is defined by (17.7) in terms of $\mathfrak{G}$, we have

$$\mathfrak{K}_2(U_*^*; T) = \mathfrak{G}(U_*^* T) \tag{23.4}$$

for all $U_*^* \in \mathscr{S}_{+*}^*$ and all $T \in \mathscr{S}_+$. If $\mathfrak{G}$ is of the differential type then its value depends only on the values at $s=0$ of its argument and its first n derivatives. Hence, by (23.4), the value $\mathfrak{K}_2(U_*^*; T)$ depends only on the first n derivatives of U_*^* at $s=0$, since $U_*^*(0)=I$. It follows that there is a function $\mathfrak{k}$ of $n+1$ symmetric tensor variables such that

$$\mathfrak{K}_2(U_*^*; T) = \mathfrak{k}(\dot{U}_*^*(0), \ddot{U}_*^*(0), \ldots, \overset{(n)}{U_*^*}(0); T) \tag{23.5}$$

for all $U_*^* \in \mathscr{S}_{+*}^*$ and all $T \in \mathscr{S}_+$. In the case when $U_*^* = U_t^t$ is the history up to time t of the right strain tensor relative to the configuration at time t of a local motion, the derivative $\overset{(k)}{U_t^t}(0)$ coincides by (8.8) with the kth rate of strain $D_k(t)$. For simplicity we use the notation

$$\mathfrak{k}(D_1, D_2, \ldots, D_n; T) = \mathfrak{k}(D_k; T). \tag{23.6}$$

We say that a particle X is of differential type if its constitutive functional G_M, for some configuration gradient M, is of the differential type. It is not hard to see that $G_{\hat{M}}$, for any other configuration gradient $\hat{M}$, is then also of the differential type. The following theorem is a consequence of (18.9) and the remarks made above:

Theorem 8. *The constitutive equation of a particle X of differential type may be written, with reference to a configuration gradient M of X, in the following form:*

There is a symmetric-tensor-valued function $\mathfrak{k}$ of $n+1$ symmetric tensor variables such that the stress is given by

$$S(t) = R(t)\,\mathfrak{k}\big(R^T(t)\,D_k(t)\,R(t);\, U(t)\big)\,R(t)^T, \tag{23.7}$$

where $D_k(t)$ is the kth rate of strain and where the rotation tensor $R(t)$ and the right strain tensor $U(t)$ are taken relative to M as a reference.

In a material of the differential type the stress depends only on the immediate past of the motion and not on its course at times long ago.

Since the rates of strain D_k can be expressed as polynomials in the acceleration gradients E_k and E_k^T as shown in Section 9, it follows that the constitutive equation (23.7) is a relation involving the displacement gradient $G=RU$, the acceleration gradients, and the stress.

In the case of *isotropic solids*, it follows from Theorem 6 and (23.5) that $\mathfrak{k}$ must satisfy the relation

$$Q\,\mathfrak{k}(D_k; T)\,Q^T = \mathfrak{k}(Q\,D_k\,Q^T; Q\,T\,Q^T) \tag{23.8}$$

for all $D_k \in \mathscr{S}$, $k=1, 2, \ldots, n$, all $Q \in \mathfrak{O}$, and all $T \in \mathscr{S}_+$. A function with this property is called an *isotropic tensor function*. By (20.4) the constitutive equation (23.7) reduces for isotropic solids to

$$S(t) = \mathfrak{k}\big(D_k(t); V(t)\big). \tag{23.9}$$

The constitutive equation for a *fluid* of the differential type has the form

$$S(t) = \mathfrak{h}\left(D_k(t); \varrho(t)\right) \tag{23.10}$$

where $\mathfrak{h}$ is an isotropic function of n symmetric tensor variables and one positive scalar variable, *i.e.*, it satisfies the relation

$$Q\,\mathfrak{h}(D_k; d)\,Q^T = \mathfrak{h}(Q\,D_k\,Q^T; d) \tag{23.11}$$

for all $D_k \in \mathscr{S}$, $k = 1, 2, \ldots, n$, all $d > 0$, and all $Q \in \mathcal{O}$. This is an immediate consequence of Theorem 7, (21.8).

If we use the alternate forms (22.8), (22.10), and (22.12) of the general constitutive equations we arrive at the following forms for materials of the differential type: In the general case,

$$S(t) = R(t)\,\bar{\mathfrak{k}}\left(R(t)^T A_k(t)\,R(t); C(t)\right) R(t)^T. \tag{23.12}$$

For isotropic solids*,

$$S(t) = \bar{\mathfrak{k}}\left(A_k(t); B(t)\right). \tag{23.13}$$

For fluids,

$$S(t) = \bar{\mathfrak{h}}\left(A_k(t); \varrho(t)\right). \tag{23.14}$$

In these equations $\bar{\mathfrak{k}}$ and $\bar{\mathfrak{h}}$ are of the same type as $\mathfrak{k}$ and $\mathfrak{h}$; in (23.12) $\bar{\mathfrak{k}}$ may be arbitrary but in (23.13) it must be isotropic. Of course, $\bar{\mathfrak{h}}$ is also isotropic. $A_k(t)$ is the kth Rivlin-Ericksen tensor, defined by (8.9) and related to the acceleration gradients by (9.2).

24. Materials of the rate type

The general constitutive equation of a simple material in the form (18.8) may be rewritten as

$$\bar{S}_{t_0}(t) = \mathfrak{R}\left(\bar{U}^t_{t_0}; U(t_0)\right) \tag{24.1}$$

where

$$\bar{S}_{t_0}(t) = R(t_0)^T R_{t_0}(t)^T S(t)\,R_{t_0}(t)\,R(t_0) \tag{24.2}$$

and

$$\bar{U}^t_{t_0}(s) = \bar{U}_{t_0}(t+s) = R(t_0)^T\,U_{t_0}(t+s)\,R(t_0). \tag{24.3}$$

Keeping t_0 and $U(t_0)$ fixed, we may interpret (24.1) in the following manner: Assuming that the function $\bar{U}_{t_0}$ defined by

$$\bar{U}_{t_0}(t) = R(t_0)\,U_{t_0}(t)\,R(t_0)^T \tag{24.4}$$

is given, the function $\bar{S}_{t_0}$ defined by (24.2) is completely determined. In other words, (24.1) defines an operation on functions $\bar{U}_{t_0}$ with values in $\mathscr{S}_+$ which gives functions $\bar{S}_{t_0}$ with values in $\mathscr{S}$. It may happen that this operation is defined by the process of solution of a differential equation for $\bar{S}_{t_0}$ in the form

$$\mathfrak{f}\left(\bar{S}_{t_0}(t), \dot{\bar{S}}_{t_0}(t), \ldots, \overset{(m)}{\bar{S}}_{t_0}(t); \bar{U}_{t_0}(t), \dot{\bar{U}}_{t_0}(t), \ldots, \overset{(n)}{\bar{U}}_{t_0}(t); U(t_0)\right) = 0, \tag{24.5}$$

---

* This form was first derived by RIVLIN & ERICKSEN [8].

where $\mathfrak{f}$ is a symmetric-tensor-valued function of $m+n+3$ symmetric tensor variables. For simplicity we use a notation similar to (23.6), so that (24.5) becomes

$$\mathfrak{f}\big(\overset{(l)}{\overline{S}}_{t_0}(t)\,;\,\overset{(k)}{\overline{U}}_{t_0}(t)\,;\,U(t_0)\big)=0. \tag{24.6}$$

If the function $\overline{U}_{t_0}$ is given, then (24.6) is a differential equation of order m for the function $\overline{S}_{t_0}$. We assume that the form of $\mathfrak{f}$ is such that there is a unique solution $\overline{S}_{t_0}$ which assumes given initial values $\overset{(l)}{\overline{S}}_{t_0}(t_1)$, $l=0, 1, \ldots, m-1$, no matter how we choose $U(t_0)$, $\overset{(k)}{\overline{U}}_{t_0}$, t_1, and $\overset{(l)}{\overline{S}}_{t_0}(t_1)$.

Since t_0 is arbitrary, in (24.6) we may make the special choice $t_0=t$ obtaining

$$\mathfrak{f}\big(\overset{(l)}{\overline{S}}_{t}(t)\,;\,\overset{(k)}{\overline{U}}_{t}(t)\,;\,U(t)\big)=0. \tag{24.7}$$

By (24.4) and (8.8) we have

$$\overset{(k)}{\overline{U}}_{t}(t)=R^T(t)\,D_k(t)\,R(t). \tag{24.8}$$

The tensor function $\hat{S}_l$ defined by

$$\hat{S}_l(t)=\left.\overset{(l)}{\overline{R_{t_0}(t)^T\,S(t)\,R_{t_0}(t)}}\right|_{t_0=t} \tag{24.9}$$

will be called the lth *invariant stress rate.* If we carry out the differentiation in (24.9) according to the product rule, and if we observe (8.7), we see that

$$\hat{S}_l=\sum_{\substack{p,\,q,\,r=0,\ldots,l\\ p+q+r=l}}\frac{l!}{p!\,q!\,r!}\,W_p^T\,\overset{(q)}{S}\,W_r; \tag{24.10}$$

thus $\hat{S}_l$ can be expressed explicitly in terms of the stress S, its time derivatives $\overset{(q)}{S}$ up to the order l, and the spins W_p up to the order l. We have $\hat{S}_0=S$. For $l=1$, we get the *invariant stress rate**

$$\hat{S}_1=\hat{S}=\dot{S}-W\,S+S\,W. \tag{24.11}$$

Observing (24.2), (24.9), and (24.8), we see that (24.7) has the form

$$\mathfrak{f}\big(R(t)^T\,\hat{S}_l(t)\,R(t)\,;\,R(t)^T D_k(t)\,R(t)\,;\,U(t)\big)=0. \tag{24.12}$$

A material with a constitutive equation of this form will be called a *material of the rate type.*

It must be noted that (24.12) is not really a complete constitutive equation. The stress is not determined by the local motion alone but only when, in addition, initial values $\hat{S}_l(t_1)$, $l=0, 1, \ldots, m-1$, for some initial time t_1, are given. These initial values, on the other hand, should be determined by the history of the local motion up to the time t_1. A constitutive equation of the type (24.12) characterizes not a single material but a family of materials depending on m symmetric tensor parameters.

---

* It was introduced by ZAREMBA [*15*].

In the case of *isotropic solids* it follows from Theorem 6 that the tensor function $\mathfrak{f}$ in (24.12) must be isotropic and that the constitutive equation reduces to

$$\mathfrak{f}\left(\hat{S}_l(t);\, D_k(t);\, V(t)\right) = 0. \tag{24.13}$$

Fluids of the rate type are described by an equation of the form

$$\mathfrak{g}\left(\hat{S}_l(t);\, D_k(t);\, \varrho(t)\right) = 0, \tag{24.14}$$

where $\mathfrak{g}$ is an isotropic tensor function of $m+n+1$ tensor variables and one positive scalar variable. However, not every equation of the form (24.14) defines a class of fluids of the rate type. It may also define a class of isotropic solids, because the stress may depend on the right strain tensor relative to some reference configuration through the initial values.

Starting from the general constitutive equations (22.7), (22.9), and (22.11), one can easily derive alternate forms for the constitutive equations of the rate type:

In the general case we obtain

$$\bar{\mathfrak{f}}\left(R(t)^T \tilde{S}_l(t)\, R(t);\, R(t)^T A_k(t)\, R(t);\, C(t)\right) = 0 \tag{24.15}$$

where $A_k(t)$ is the kth Rivlin-Ericksen tensor (9.2) and where $\tilde{S}_l(t)$ is the lth *stress flux* defined by

$$\tilde{S}_l(t) = \overset{(l)}{\overline{F_{t_0}(t)^T S(t)\, F_{t_0}(t)}}\Big|_{t_0=t}. \tag{24.16}$$

We find that $\tilde{S}_l$ may be expressed explicitly in terms of the stress S, its derivatives up to the order m, and the acceleration gradients E_l up to the order m by the formula

$$\tilde{S}_l = \sum_{\substack{q,\,p,\,r=0,\,\ldots,\,l\\ p+q+r=l}} \frac{l!}{p!\,q!\,r!}\, E_p^T \overset{(q)}{S} E_r. \tag{24.17}$$

For $l=0$ we have $\tilde{S}_0 = S$, and for $l=1$ we get the *stress flux**

$$\tilde{S}_1 = \tilde{S} = \dot{S} + E^T S + S E. \tag{24.18}$$

In the case of isotropic solids $\bar{\mathfrak{f}}$ is isotropic and (24.15) reduces to

$$\bar{\mathfrak{f}}\left(\tilde{S}_l(t);\, A_k(t);\, B(t)\right) = 0^{**}. \tag{24.19}$$

For fluids we get

$$\bar{\mathfrak{h}}\left(\tilde{S}_l(t);\, A_k(t);\, \varrho(t)\right) = 0, \tag{24.20}$$

where $\bar{\mathfrak{h}}$ is an isotropic tensor function.

Acknowledgement. The research leading to this paper was sponsored in part by the National Science Foudation under Grant NSF-G 5250 and in part by the Air Force Office of Scientific Research under Contract AF 18(600)-1138 with Carnegie Institute of Technology.

---

* It differs from the flux introduced by CAUCHY (*cf.* the discussion given by TRUESDELL in [*12*], Section 55 *bis*), in which there are minus signs on the right.

** This form was first derived by COTTER & RIVLIN [*1*].

References

[1] COTTER, B., & R. S. RIVLIN: Tensors associated with time-dependent stress. Quart. Appl. Math. **13**, 177—182 (1955).

[2] EHRESMANN, C.: Introduction à la théorie des structures infinitésimales et des pseudo-groupes de Lie. Colloques internationaux du centre national de la recherche scientifique. Géométrie différentielle, 97—110. Paris 1953.

[3] GREEN, A. E.: Simple extension of a hypo-elastic body of grade zero. J. Rational Mech. Anal. **5**, 637—642 (1956).

[4] GREEN, A. E., & R. S. RIVLIN: The mechanics of non-linear materials with memory. Arch. Rational Mech. Anal. **1**, 1—34 (1957).

[5] NOLL, W.: On the continuity of the solid and fluid states. J. Rational Mech. Anal. **4**, 3—81 (1955).

[6] NOLL, W.: The foundations of classical mechanics in the light of recent advances in continuum mechanics. Proceedings of the Berkeley Symposium on the Axiomatic Method, 1958.

[7] NOLL, W.: On the foundations of the mechanics of continuous media. Carnegie Institute of Technology, Technical Report no. 17, Air Force Office of Scientific Research, 1957.

[8] RIVLIN, R. S., & J. L. ERICKSEN: Stress-deformation relations for isotropic materials. J. Rational Mech. Anal. **4**, 323—425 (1955).

[9] RIVLIN, R. S.: Further remarks on stress-deformation relations for isotropic materials. J. Rational Mech. Anal. **4**, 681—702 (1955).

[10] RIVLIN, R. S.: Solution of some problems in the exact theory of viscoelasticity. J. Rational Mech. Anal. **5**, 179—188 (1956).

[11] TOUPIN, R. A.: World invariant kinematics. Arch. Rational Mech. Anal. **1**, 181—211 (1958).

[12] TRUESDELL, C.: The mechanical foundations of elasticity and fluid dynamics. J. Rational Mech. Anal. **1**, 125—300 (1952) and **2**, 593—616 (1953).

[13] TRUESDELL, C.: Hypo-elasticity. J. Rational Mech. Anal. **4**, 83—133 (1955).

[14] TRUESDELL, C.: Hypo-elastic shear. J. Appl. Physics **27**, 441—447 (1956).

[15] ZAREMBA, S.: Sur une forme perfectionnée de la théorie de la relaxation. Bull. Int. Acad. Sci. Cracovie **1903**, 594—614.

Carnegie Institute of Technology
Pittsburgh, Pennsylvania

(Received July 31, 1958)

Offprint from "Archive for Rational Mechanics and Analysis",
Volume 27, Number 1, 1967, P. 1–32

Springer-Verlag, Berlin · Heidelberg · New York

Materially Uniform Simple Bodies with Inhomogeneities

WALTER NOLL

Contents

1. Introduction

The basic concepts of the theory of simple materials have been introduced in reference [*1*] (see also the exposition in [*2*], Chapter C III). Here I present a detailed study of the structure of bodies that consist of a uniform simple material yet are not necessarily homogeneous.

After assembly of the necessary mathematical tools in Sects. 2–4, the concept of a *simple body* is introduced in Sect. 5. This concept is more inclusive than the one described in [*1*] because it can be appropriate not only to mechanical material properties, but also to thermal, optical, electrical, magnetic, or any other type of material properties. A body may be simple with respect to any particular such material property or to any combination of them. The physical theory relevant to these properties need not be made explicit.

In Sect. 6 a precise definition of a *materially uniform simple body* is given. The nature of the coherence of a uniform body with respect to the local material properties under consideration can be described mathematically in terms of what I call a *material uniformity* or in terms of what I call a *uniform reference*. In general, neither of these is uniquely determined by the simple body structure, but the degree of non-uniqueness can be delimited precisely. There may or may not exist uniform references that are gradients of global configurations. If they do exist, the body is homogeneous, and the theory becomes trivial.

This paper supersedes an unpublished preliminary study written by the author in 1963. Section 34 of reference [*2*] is a summary of that study.

Sections 7–9 contain an exposition of the mathematical prerequisites necessary to describe the local behavior of material uniformities and uniform references that possess a degree of smoothness. In the remainder of the paper, such smoothness is always assumed. A material uniformity is then equivalent to an affine connection, which is defined in Sect. 10 and called a *material connection*. The Cartan torsion of this connection describes locally the deviation from homogeneity and is therefore called, in Sect. 11, the *inhomogeneity* of the given material uniformity.

Associated with each smooth uniform reference is also a Riemannian structure on the body, and the relation of this structure to the material connection is studied in Sect. 12. The difference between the Riemannian connection and the material connection determines what I call the *contortion* of the given uniform reference. Contortion and inhomogeneity determine one another.

Of particular interest is a special type of non-homogeneity called *contorted aeolotropy* in Section 13. It generalizes the more familiar *curvilinear aeolotropy*. In contorted aeolotropy, the deviation from homogeneity is given by a distribution of rotations on a suitable global configuration, and the contortion describes the local behavior of this distribution. The curvature of the Riemannian structure mentioned before describes locally the deviation from contorted aeolotropy.

Section 14 contains a number of results that apply when the response functions of the body have special properties, especially with respect to material symmetry.

The usual version of CAUCHY's equation of balance (*cf.* [*2*], (16.6)) is very useful only when applied to bodies that are homogeneous. For applications to materially uniform but inhomogeneous bodies, a new version of CAUCHY's equation, derived in Sect. 15, is much more suitable than the usual one. This new version gives rise, for example, to a definite differential equation for the theory of inhomogeneous but materially uniform elastic bodies.

Unfortunately, there is no easily accessible exposition of the coordinate-free type of modern differential geometry that is the most appropriate for the applications in this study. The monograph of LANG [*3*], although it explains some of the concepts used here, does not contain sufficient material and emphasizes matters not relevant in the present context. For this reason I develop in this paper all mathematical tools as they come to be needed, tailored to the requirements of the intended applications.

There is a large literature on theories of *continuous distributions of dislocations*, proposed in various forms by KONDO, NYE, BILBY, BULLOUGH, SMITH, SEEGER, KRÖNER, GÜNTHER, and others[1]. Motivated by heuristic considerations, mostly concerning lattice defects in crystals, these authors lay down *a priori* certain geometric structures to describe distributions of dislocations. These geometric structures are formally of the same type as some of those occurring in the present paper. The conceptual status of the theory presented here, however, is very different. I show that once a constitutive assumption defining a materially uniform simple body is laid down, the geometric structures of the body are determined. The geometry is thus the *natural outcome*, not the first assumption, of the theory. Since the underlying constitutive assumption is very general, the real materials to which the theory can be expected to apply need be neither crystalline, nor elastic, nor solid.

[1] For details and references I refer to the expository articles [*4*] and [*5*].

2. Deformations

We shall employ the concept of *absolute physical space*[2], as is customary in classical physics. This space $\mathscr{E}$, whose elements $\boldsymbol{x}, \boldsymbol{y}, \ldots$, we call *spatial points*, has the structure of a three-dimensional Euclidean point space[3]. The translation space of $\mathscr{E}$ is denoted by $\mathscr{V}$; it is a three-dimensional inner product space. The elements $\boldsymbol{u}, \boldsymbol{v}, \ldots$, of $\mathscr{V}$ are called *spatial vectors*. The translation which carries $\boldsymbol{x}\in\mathscr{E}$ to $\boldsymbol{y}\in\mathscr{E}$ is denoted by $\boldsymbol{y}-\boldsymbol{x}\in\mathscr{V}$, and $\boldsymbol{x}+\boldsymbol{u}$ denotes the point into which $\boldsymbol{x}\in\mathscr{E}$ is carried by the translation $\boldsymbol{u}\in\mathscr{V}$. The inner product of two spatial vectors $\boldsymbol{u}, \boldsymbol{v}\in\mathscr{V}$ is denoted by $\boldsymbol{u}\cdot\boldsymbol{v}$. Of course, $\boldsymbol{u}\cdot\boldsymbol{v}\in\mathscr{R}$, where $\mathscr{R}$ is the set of all real numbers.

The set of all linear transformations $\boldsymbol{L}$: $\mathscr{V}\to\mathscr{V}$ of $\mathscr{V}$ into itself is denoted by $\mathscr{L}$. The composition of $\boldsymbol{L}\in\mathscr{L}$ with $\boldsymbol{M}\in\mathscr{L}$ is denoted by $\boldsymbol{ML}\in\mathscr{L}$. The identity transformation on $\mathscr{V}$ is denoted by $\mathbf{1}\in\mathscr{L}$. The transpose of $\boldsymbol{L}\in\mathscr{L}$ is denoted by $\boldsymbol{L}^T$, so that $\boldsymbol{u}\cdot\boldsymbol{L}\boldsymbol{v}=\boldsymbol{L}^T\boldsymbol{u}\cdot\boldsymbol{v}$ holds for all $\boldsymbol{u}, \boldsymbol{v}\in\mathscr{V}$. The trace and determinant of $\boldsymbol{L}\in\mathscr{L}$ are denoted by $\operatorname{tr}\boldsymbol{L}$ and $\det\boldsymbol{L}$, respectively. The set $\mathscr{L}$ of all linear transformations has the natural structure of a nine-dimensional algebra. It is also endowed with a natural inner product, whose values are given by $\boldsymbol{L}\cdot\boldsymbol{M}=\operatorname{tr}(\boldsymbol{L}\boldsymbol{M}^T)$. A transformation $\boldsymbol{L}\in\mathscr{L}$ is said to be *invertible* if it is a bijection (*i.e.*, one-to-one and onto). In this case, there exists an *inverse* $\boldsymbol{L}^{-1}\in\mathscr{L}$ so that $\boldsymbol{L}\boldsymbol{L}^{-1}=\boldsymbol{L}^{-1}\boldsymbol{L}=\mathbf{1}$. The invertible members of $\mathscr{L}$ form a group $\ell\subset\mathscr{L}$ under composition; it is called the *linear group* of $\mathscr{V}$. Important subgroups of ℓ are the *unimodular group*

$$u=\{\boldsymbol{H}\in\ell\,\big|\,|\det\boldsymbol{H}|=1\}$$

and the *orthogonal group*

$$o=\{\boldsymbol{Q}\in\ell\,\big|\,\boldsymbol{Q}\boldsymbol{Q}^T=\mathbf{1}\}.$$

Of course, o is a subgroup of u.

Consider a mapping $\boldsymbol{\varphi}$: $\mathscr{G}\to\mathscr{E}'$ of an open subset $\mathscr{G}\subset\mathscr{E}$ into a point-space or vector-space $\mathscr{E}'$. Let $\mathscr{V}'$ be the translation space of $\mathscr{E}'$ ($\mathscr{V}'=\mathscr{E}'$ if $\mathscr{E}'$ is already a vectorspace) and let $\mathscr{L}(\mathscr{V},\mathscr{V}')$ be the space of all linear transformations of $\mathscr{V}$ into $\mathscr{V}'$. We say that $\boldsymbol{\varphi}$ is *of class* C^1 if there is a continuous mapping $\nabla\boldsymbol{\varphi}$: $\mathscr{G}\to\mathscr{L}(\mathscr{V},\mathscr{V}')$ such that

$$\boldsymbol{\varphi}(\boldsymbol{x}+\boldsymbol{u})=\boldsymbol{\varphi}(\boldsymbol{x})+\big(\nabla\boldsymbol{\varphi}(\boldsymbol{x})\big)\boldsymbol{u}+\boldsymbol{\sigma}(\boldsymbol{x},\boldsymbol{u}),$$

where

$$\lim_{|\boldsymbol{u}|\to 0}\frac{1}{|\boldsymbol{u}|}\,\boldsymbol{\sigma}(\boldsymbol{x},\boldsymbol{u})=\mathbf{0}$$

holds for all $\boldsymbol{x}\in\mathscr{G}$. The mapping $\nabla\boldsymbol{\varphi}$, if it exists, is uniquely determined by $\boldsymbol{\varphi}$ and is called the *gradient* of $\boldsymbol{\varphi}$. If $\nabla\boldsymbol{\varphi}$ exists and is itself of class C^1, we say that $\boldsymbol{\varphi}$ is *of class* C^2. The gradient of $\nabla\boldsymbol{\varphi}$ is denoted by $\nabla^{(2)}\boldsymbol{\varphi}$ and is called the *second gradient* of $\boldsymbol{\varphi}$. Continuing in this manner, we say that $\boldsymbol{\varphi}$ is *of class* C^r, if it is of class C^{r-1} and if its $(r-1)^{\text{st}}$ gradient $\nabla^{(r-1)}\boldsymbol{\varphi}$ is of class C^1. The gradient of $\nabla^{(r-1)}\boldsymbol{\varphi}$ is denoted by $\nabla^{(r)}\boldsymbol{\varphi}$. We say that $\boldsymbol{\varphi}$ is of class C^0 if it is merely continuous. If $\boldsymbol{\varphi}$ is of class C^2, its second gradient has the symmetry property $((\nabla^{(2)}\boldsymbol{\varphi})\,\boldsymbol{u})\,\boldsymbol{v}=((\nabla^{(2)}\boldsymbol{\varphi})\,\boldsymbol{v})\,\boldsymbol{u}$, $\boldsymbol{u}, \boldsymbol{v}\in\mathscr{V}$.

[2] The considerations of this paper can be adapted to the neoclassical space-time explained in [6]. When this is done, absolute space must be replaced by suitably defined "instantaneous spaces".

[3] The exact meaning of this term is explained in [7], Sect. 4.

The modifier "of class C^r" may apply, in particular, to a *scalar field, i.e.* a mapping $f: \mathscr{G} \to \mathscr{R}$, a *vector field, i.e.* a mapping $\boldsymbol{h}: \mathscr{G} \to \mathscr{V}$, or a *tensor field, i.e.*, a mapping $\boldsymbol{T}: \mathscr{G} \to \mathscr{L}$. A one-to-one mapping $\boldsymbol{\lambda}: \mathscr{G} \to \mathscr{E}$ is called a *deformation of class* C^r $(r \geqq 1)$ if it is not only of class C^r but if also the values of its gradient are invertible, *i.e.* if $\nabla \boldsymbol{\lambda}(\boldsymbol{x}) \in \ell$ for all $\boldsymbol{x} \in \mathscr{G}$.

The members of the linear group ℓ are also called *local deformations*, so that a (global) deformation has a gradient whose values are local deformations.

3. Continuous Bodies

A physical object can often be described mathematically by the concept of a *body* $\mathscr{B}$, which is a set whose members $X, Y, \ldots$, are called *material points*, and which is endowed with a structure defined by a class C of mappings $\boldsymbol{\kappa}: \mathscr{B} \to \mathscr{E}$. The mappings $\boldsymbol{\kappa} \in \mathsf{C}$ are called the *configurations* of $\mathscr{B}$ (in the space $\mathscr{E}$). The spatial point $\boldsymbol{\kappa}(X) \in \mathscr{E}$ is called the *place* of the material point $X \in \mathscr{B}$ in the configuration $\boldsymbol{\kappa}$.

We say that $\mathscr{B}$ is a *continuous body of class* C^p $(p \geqq 1)$ if the class C of configurations satisfies the following axioms:

(C 1) Every $\boldsymbol{\kappa} \in \mathsf{C}$ is one-to-one and its range $\boldsymbol{\kappa}(\mathscr{B})$ is an open subset of $\mathscr{E}$, which is called the *region occupied by* $\mathscr{B}$ *in the configuration* $\boldsymbol{\kappa}$.

(C 2) If $\boldsymbol{\gamma}, \boldsymbol{\kappa} \in \mathsf{C}$ then the composite[4] $\boldsymbol{\lambda} = \boldsymbol{\lambda} \circ \overset{-1}{\boldsymbol{\kappa}}: \boldsymbol{\kappa}(\mathscr{B}) \to \boldsymbol{\gamma}(\mathscr{B})$ is a deformation of class C^p, which is called the *deformation of* $\mathscr{B}$ *from the configuration* $\boldsymbol{\kappa}$ *into the configuration* $\boldsymbol{\gamma}$.

(C 3) If $\boldsymbol{\kappa} \in \mathsf{C}$ and if $\boldsymbol{\lambda}: \boldsymbol{\kappa}(\mathscr{B}) \to \mathscr{E}$ is a deformation of class C^p, then $\boldsymbol{\lambda} \circ \boldsymbol{\kappa} \in \mathsf{C}$. The mapping $\boldsymbol{\lambda} \circ \boldsymbol{\kappa}$ is called the *configuration obtained from the configuration* $\boldsymbol{\kappa}$ *by the deformation* $\boldsymbol{\lambda}$.

In the remainder of this paper we shall always assume that $\mathscr{B}$ is a continuous body of class C^p, $p \geqq 1$.

The axioms (C 1)–(C 3) ensure that the class C endows the body $\mathscr{B}$ with the structure of a "C^p-manifold modelled on $\mathscr{E}$" in the sense of Lang ([*3*], Ch. II, § 1). Topologically, it is a very simple manifold because it can be mapped out with a single "chart" ("configuration" in our terminology).

Of central importance for the present paper is the concept of a *local configuration*[5] *at a material point* X. Two (global) configurations $\boldsymbol{\kappa}$ and $\boldsymbol{\gamma}$ are said to be *equivalent at* X, and we write[6]

$$\boldsymbol{\kappa} \sim_X \boldsymbol{\gamma} \quad \text{if} \quad \nabla(\boldsymbol{\kappa} \circ \overset{-1}{\boldsymbol{\gamma}})|_{\boldsymbol{\gamma}(X)} = \mathbf{1}. \tag{3.1}$$

It is an immediate consequence of the chain rule for gradients that $\sim_X$ is an equivalence relation on C. The resulting partition of C is denoted by $\mathscr{C}_X$, and its members $\boldsymbol{K}_X, \boldsymbol{G}_X, \ldots$, *i.e.* the equivalence classes, are called *local configurations* at X. Instead of writing $\boldsymbol{\kappa} \in \boldsymbol{K}_X$ when $\boldsymbol{\kappa}$ is a member of the class $\boldsymbol{K}_X$ we often write

$$\nabla \boldsymbol{\kappa}(X) = \boldsymbol{K}_X \tag{3.2}$$

---

[4] Composition of mappings other than linear mappings is denoted by $\circ$. The inverse of a one-to-one mapping $\boldsymbol{\kappa}$ is denoted by $\overset{-1}{\boldsymbol{\kappa}}$.

[5] The term "configuration gradient" was used and another meaning was assigned to the term "local configuration" in [*1*].

[6] For better reading, we sometimes write $f|_x$ instead of $f(x)$ for the value of f at x.

and say that the local configuration $\boldsymbol{K}_X$ is the *gradient at* X of the (global) configuration $\boldsymbol{\kappa}$.

Let $\boldsymbol{K}_X, \boldsymbol{G}_X \in \mathscr{C}_X$ be two local configurations, and let $\boldsymbol{\kappa} \in \boldsymbol{K}_X$, $\gamma \in \boldsymbol{G}_X$. It is easily seen that the local deformation $\nabla(\gamma \circ \overset{-1}{\boldsymbol{\kappa}})|_{\boldsymbol{\kappa}(X)} \in \ell$ depends only on $\boldsymbol{K}_X$ and $\boldsymbol{G}_X$, and not on the particular choices of $\boldsymbol{\kappa} \in \boldsymbol{K}_X$ and $\gamma \in \boldsymbol{G}_X$. We denote this local deformation by $\boldsymbol{G}_X \boldsymbol{K}_X^{-1}$ and call it the *local deformation from the local configuration* $\boldsymbol{K}_X$ *into the local configuration* $\boldsymbol{G}_X$. Using the notation (3.2), we then have

$$\nabla(\gamma \circ \overset{-1}{\boldsymbol{\kappa}})|_{\boldsymbol{\kappa}(X)} = \nabla\gamma(X)\,[\nabla\boldsymbol{\kappa}(X)]^{-1}. \tag{3.3}$$

If $\boldsymbol{K}_X \in \mathscr{C}_X$ is a local configuration and $\boldsymbol{L} \in \ell$ any local deformation, we can define a new local configuration $\boldsymbol{L}\boldsymbol{K}_X \in \mathscr{C}_X$ by

$$\boldsymbol{L}\boldsymbol{K}_X = \{\boldsymbol{\lambda} \circ \boldsymbol{\kappa} \,|\, \nabla\boldsymbol{\lambda}|_{\boldsymbol{\kappa}(X)} = \boldsymbol{L},\ \nabla\boldsymbol{\kappa}(X) = \boldsymbol{K}_X\}. \tag{3.4}$$

We call $\boldsymbol{L}\boldsymbol{K}_X$ the *local configuration obtained from the local configuration* $\boldsymbol{K}_X$ *by the local deformation* $\boldsymbol{L}$. Clearly, we have the rules

$$(\boldsymbol{G}_X \boldsymbol{K}_X^{-1})\boldsymbol{K}_X = \boldsymbol{G}_X, \qquad (\boldsymbol{L}\boldsymbol{K}_X)\boldsymbol{K}_X^{-1} = \boldsymbol{L}. \tag{3.5}$$

4. Tangent Spaces

Consider pairs $(\boldsymbol{K}_X, \boldsymbol{u})$, where $\boldsymbol{K}_X \in \mathscr{C}_X$ is a local configuration at X and $\boldsymbol{u} \in \mathscr{V}$ a spatial vector. We say that two such pairs $(\boldsymbol{K}_X, \boldsymbol{u})$ and $(\boldsymbol{G}_X, \boldsymbol{v})$ are *equivalent* if

$$(\boldsymbol{K}_X \boldsymbol{G}_X^{-1})\,\boldsymbol{v} = \boldsymbol{u}. \tag{4.1}$$

It follows from the rules (3.5) that (4.1) does indeed define an equivalence relation. The resulting equivalence classes are called *tangent vectors* $\mathfrak{u}_X, \mathfrak{v}_X, \ldots$ at X. The totality of all these tangent vectors is denoted by $\mathscr{T}_X$ and is called the *tangent space* at $X \in \mathscr{B}$. Let $\mathfrak{u}_X \in \mathscr{T}_X$ and $\boldsymbol{K}_X \in \mathscr{C}_X$ be given and let $(\boldsymbol{G}_X, \boldsymbol{v})$ be any pair belonging to the class $\mathfrak{u}_X$. Now, if $(\boldsymbol{K}_X, \boldsymbol{u})$ is to belong to $\mathfrak{u}_X$ then (4.1) must hold. Therefore, we see that $\mathfrak{u}_X \in \mathscr{T}_X$ and $\boldsymbol{K}_X \in \mathscr{C}$ determine a unique spatial vector $\boldsymbol{u} \in \mathscr{V}$ such that $(\boldsymbol{K}_X, \boldsymbol{u}) \in \mathfrak{u}_X$. We can therefore use the notation

$$\boldsymbol{u} = \boldsymbol{K}_X \mathfrak{u}_X, \qquad \mathfrak{u}_X = \boldsymbol{K}_X^{-1}\boldsymbol{u} \quad \text{if} \quad (\boldsymbol{K}_X, \boldsymbol{u}) \in \mathfrak{u}_X, \tag{4.2}$$

and we see that $\boldsymbol{K}_X$ determines a one-to-one mapping of the tangent space $\mathscr{T}_X$ onto the space $\mathscr{V}$ of spatial vectors. The tangent space $\mathscr{T}_X$ has the natural structure of a three-dimensional vector space, with addition defined by

$$\mathfrak{u}_X + \mathfrak{v}_X = \boldsymbol{K}_X^{-1}(\boldsymbol{u} + \boldsymbol{v}) \quad \text{if} \quad \mathfrak{u}_X = \boldsymbol{K}_X^{-1}\boldsymbol{u}, \qquad \mathfrak{v}_X = \boldsymbol{K}_X^{-1}\boldsymbol{v} \tag{4.3}$$

and multiplication with scalars by

$$a\,\mathfrak{u}_X = \boldsymbol{K}_X^{-1}(a\,\boldsymbol{u}) \quad \text{if} \quad \mathfrak{u}_X = \boldsymbol{K}_X^{-1}\boldsymbol{u}, \qquad a \in \mathscr{R}. \tag{4.4}$$

It is immediately seen that these definitions of $\mathfrak{u}_X + \mathfrak{v}_X$ and $a\mathfrak{u}_X$ are legitimate because the results are independent of the choice of the local configuration $\boldsymbol{K}_X$ used to represent $\mathfrak{u}_X$ and $\mathfrak{v}_X$ in $\mathscr{V}$. The local configurations can be identified with the invertible linear transformations of $\mathscr{T}_X$ onto $\mathscr{V}$.

Given a local configuration $\boldsymbol{K}_X \in \mathscr{C}_X$, we can define an inner product $\mathfrak{u}_X * \mathfrak{v}_X$ of $\mathfrak{u}_X, \mathfrak{v}_X, \in \mathscr{T}_X$ by

$$\mathfrak{u}_X * \mathfrak{v}_X = (\boldsymbol{K}_X^{-1} \mathfrak{u}_X) \cdot (\boldsymbol{K}_X^{-1} \mathfrak{v}_X). \tag{4.5}$$

However, we obtain different inner products on $\mathscr{T}_X$ for different choices of $\boldsymbol{K}_X$, and hence $\mathscr{T}_X$ is *not* naturally an inner product space.

5. Simple Bodies, Material Isomorphisms, Intrinsic Isotropy Groups

To describe mathematically the physical characteristics of a body $\mathscr{B}$ we must endow $\mathscr{B}$ with additional structure. Some of these characteristics, such as elasticity, viscosity, heat capacity, and electrical conductivity, are local, *i.e.*, they are attached to the individual material points $X \in \mathscr{B}$ rather than to the body as a whole. Other characteristics, such as mutual gravitation and internal radiative heat transfer, involve more than one material point. We deal here only with local characteristics. The physical response of the body $\mathscr{B}$ at a particular material point $X \in \mathscr{B}$ and a particular time will depend on the configuration $\boldsymbol{\kappa}$ of $\mathscr{B}$ at that time. It may happen that only the local configuration $\nabla \boldsymbol{\kappa}(X)$ at X determined by $\boldsymbol{\kappa}$, and no other properties of $\boldsymbol{\kappa}$, has an influence on the response. If this is the case, we say that the *material at X is simple*. We say that the whole of $\mathscr{B}$ is simple or that $\mathscr{B}$ is a *simple body* if the material at X is simple for all $X \in \mathscr{B}$.

We assume that a possible physical response at a material point is given mathematically by specifying an element from a set R of mathematical objects. The nature of R depends on the particular physical phenomena to be described. For example, in the theory of elasticity R consists of all possible 'stress tensors', *i.e.*, of all symmetric linear transformations of $\mathscr{V}$ into $\mathscr{V}$. In the mechanical theory of simple materials with fading memory, R consists of "memory functionals" that relate relative deformations histories to stresses and are subject to certain smoothness requirements. In theories that include non-mechanical effects R consists of functions or functionals whose independent and dependent variables have interpretations as local temperatures, energy or entropy densities, heat fluxes, electric or magnetic field strengths, polarizations, magnetizations, electric currents, *etc.* For the purpose of the present paper, no specific assumptions about the nature of R need be made.

We can now make our definition of a simple body precise:

Definition 1. *Let* R *be a set, whose elements we call* ***response descriptors****. A continuous body* $\mathscr{B}$ *of class* C^p *will be called a* ***simple body*** *with respect to* R *if it is endowed with a structure by a function* $\mathfrak{G}$ *which assigns to each material point* $X \in \mathscr{B}$ *a mapping*

$$\mathfrak{G}_X: \mathscr{C}_X \to \mathsf{R}. \tag{5.1}$$

The value $\mathfrak{G}_X(\boldsymbol{G}_X)$ *is the response descriptor of the material at X in any configuration* γ *of* $\mathscr{B}$ *such that* $\nabla \gamma(X) = \boldsymbol{G}_X$.

The mappings $\mathfrak{G}_X$ cannot be entirely arbitrary, for they are subject to restrictions imposed by general physical principles such as the principle of frame-indifference and the principle of dissipation. These restrictions need not be made explicit here.

We would like to give now a precise meaning to the statement that the material at one point $X \in \mathscr{B}$ is the same as the material at another point $Y \in \mathscr{B}$. We cannot construe this statement to mean that $\mathfrak{G}_X$ and $\mathfrak{G}_Y$ are the same, for they have different domains and hence cannot be directly compared. However, we can connect the domains $\mathscr{C}_X$ and $\mathscr{C}_Y$ if an isomorphism $\boldsymbol{\Phi}_{XY}\colon \mathscr{T}_Y \to \mathscr{T}_X$ of the tangent space at Y onto the tangent space at X is given. Recalling that a local configuration $\boldsymbol{G}_X \in \mathscr{C}_X$ can be regarded as a mapping $\boldsymbol{G}_X\colon \mathscr{T}_X \to \mathscr{V}$, we can let $\boldsymbol{G}_X \in \mathscr{C}_X$ correspond to the composition $\boldsymbol{G}_X \boldsymbol{\Phi}_{XY} \in \mathscr{C}_Y$. We are thus led to the following definition:

Definition 2. *An invertible linear transformation* $\boldsymbol{\Phi}_{XY}\colon \mathscr{T}_Y \to \mathscr{T}_X$ *is called a* ***material isomorphism*** *from* $\mathscr{T}_Y$ *onto* $\mathscr{T}_X$ *if*

$$\mathfrak{G}_Y(\boldsymbol{G}_Y) = \mathfrak{G}_X(\boldsymbol{G}_Y \boldsymbol{\Phi}_{XY}) \tag{5.2}$$

holds for all $\boldsymbol{G}_X \in \mathscr{C}_X$.

To say that the material at X is the same as the material at Y means that there exists a material isomorphism from $\mathscr{T}_X$ onto $\mathscr{T}_Y$.

It follows immediately from Definition 2 that if $\boldsymbol{\Phi}_{XY}\colon \mathscr{T}_Y \to \mathscr{T}_X$ and $\boldsymbol{\Phi}_{YZ}\colon \mathscr{T}_Z \to \mathscr{T}_Y$ are material isomorphisms, so is their composition $\boldsymbol{\Phi}_{XY} \boldsymbol{\Phi}_{YZ}\colon \mathscr{T}_Z \to \mathscr{T}_X$. Also, if $\boldsymbol{\Phi}_{XY}\colon \mathscr{T}_Y \to \mathscr{T}_X$ is a material isomorphisms, so is its inverse $\boldsymbol{\Phi}_{XY}^{-1}\colon \mathscr{T}_X \to \mathscr{T}_Y$. If we denote the set of all material isomorphisms from $\mathscr{T}_X$ onto $\mathscr{T}_Y$ by g_{XY}, these facts can be expressed by[7]

$$g_{ZY}\, g_{YX} = g_{ZY}, \qquad g_{YX} = g_{XY}^{-1}. \tag{5.3}$$

It is clear that g_{XX}, the set of all material isomorphisms of $\mathscr{T}_X$ onto itself, is a subgroup of the linear group ℓ_X of $\mathscr{T}_X$, which consists of all invertible linear transformations of $\mathscr{T}_X$. We write

$$g_X = g_{XX} \tag{5.4}$$

and call g_X the *intrinsic isotropy group* of the material at X. For any $\boldsymbol{\Phi}_{XY} \in g_{XY}$ one easily establishes the relations

$$g_{XY} = g_X \boldsymbol{\Phi}_{XY}\, g_Y, \qquad g_X = \boldsymbol{\Phi}_{XY}\, g_Y \boldsymbol{\Phi}_{XY}^{-1}. \tag{5.5}$$

It follows from $(5.5)_2$ that if a material isomorphism $\boldsymbol{\Phi}_{XY}\colon \mathscr{T}_Y \to \mathscr{T}_X$ exists, *i.e.* if the material at X is the same as the material at Y, then the intrinsic isotropy groups g_X and g_Y are isomorphic.

6. Material Uniformity, Uniform References, Relative Isotropy Groups

A simple body $\mathscr{B}$ is said to be *materially uniform* if the material at any two of its points is the same, *i.e.* if g_{XY} is never empty. From now on we assume that $\mathscr{B}$ is a materially uniform simple body. We select a member $\boldsymbol{\Phi}'(X, Y)$ from each g_{XY} and thereby define a function $\boldsymbol{\Phi}'$ which assigns to each pair (X, Y) of material points of $\mathscr{B}$ a material isomorphism from $\mathscr{T}_X$ onto $\mathscr{T}_Y$. Choose $X_0 \in \mathscr{B}$ arbitrarily

[7] If g and h are sets of linear transformations of any kind such that the composition $\boldsymbol{LM}$ makes sense whenever $\boldsymbol{L} \in g$, $\boldsymbol{M} \in h$, we write $g h = \{\boldsymbol{LM} \mid \boldsymbol{L} \in g, \boldsymbol{M} \in h\}$. If the $\boldsymbol{L} \in g$ are invertible, we write $g^{-1} = \{\boldsymbol{L}^{-1} \mid \boldsymbol{L} \in g\}$. Also, we write $\boldsymbol{K} g = \{\boldsymbol{KL} \mid \boldsymbol{L} \in g\}$ if $\boldsymbol{KL}$ makes sense for all $\boldsymbol{L} \in g$.

and define Φ by

$$\Phi(X, Y)=\Phi'(X, X_0)\,\Phi'(Y, X_0)^{-1}. \tag{6.1}$$

It follows from (5.3) that $\Phi(X, Y)\in g_{XY}$. Moreover, we have

$$\Phi(Z, Y)\,\Phi(Y, X)=\Phi(Z, X), \qquad \Phi(X, X)=\mathbf{1}_X, \tag{6.2}$$

where $\mathbf{1}_X$ is the identity transformation of $\mathscr{T}_X$.

Definition 3. *A function Φ which assigns to each pair (X, Y) of material points of $\mathscr{B}$ a material isomorphism $\Phi(X, Y)\in g_{XY}$ is called a* ***material uniformity*** *if* (6.2) *holds.*

The construction (6.1) shows that the materially uniform bodies are those that admit material uniformities. It follows from (5.5) and (6.2) that any two material uniformities Φ and $\hat{\Phi}$ are related by

$$\hat{\Phi}(X, Y)=\mathfrak{P}(X)\,\Phi(X, Y)\,\mathfrak{P}(Y)^{-1}, \tag{6.3}$$

where $\mathfrak{P}$ is a function on $\mathscr{B}$ whose values $\mathfrak{P}(X)$ belong to the intrinsic isotropy groups g_X.

Definition 4. *A function $\boldsymbol{K}$ on $\mathscr{B}$ whose values $\boldsymbol{K}(X)\in\mathscr{C}_X$ are local configurations is called a* ***reference*** *for $\mathscr{B}$. If, moreover,*

$$\Phi(X, Y)=\boldsymbol{K}(X)^{-1}\,\boldsymbol{K}(Y) \tag{6.4}$$

is a material isomorphism of $\mathscr{T}_Y$ onto $\mathscr{T}_X$ for any $X, Y\in\mathscr{B}$, then $\boldsymbol{K}$ is called a ***uniform reference*** *for $\mathscr{B}$.*

Actually, (6.2) holds if Φ is defined by (6.4), so that Φ is a material uniformity if $\boldsymbol{K}$ is a uniform reference. Hence, every uniform reference $\boldsymbol{K}$ determines a material uniformity Φ through (6.4). Conversely, if a material uniformity Φ and a local configuration $\boldsymbol{K}_{X_0}\in\mathscr{C}_{X_0}$ for a particular material point $X_0\in\mathscr{B}$ are given, then there exist a unique uniform reference $\boldsymbol{K}$ such that (6.4) and $\boldsymbol{K}(X_0)=\boldsymbol{K}_{X_0}$ hold. In fact, $\boldsymbol{K}$ is given by

$$\boldsymbol{K}(X)=\boldsymbol{K}_{X_0}\,\Phi(X_0, X). \tag{6.5}$$

Therefore, every material uniformity has representations (6.4) in terms of uniform references.

If $\boldsymbol{\kappa}$ is a (global) configuration, then $\nabla\boldsymbol{\kappa}$, which assigns to X the local configuration $\nabla\boldsymbol{\kappa}(X)$ at X, *i.e.*, the equivalence class to which $\boldsymbol{\kappa}$ belongs, is a reference, called the *gradient* of the configuration $\boldsymbol{\kappa}$. We say that a body is *homogeneous* if it admits a gradient as a uniform reference. Of course, not every reference is a gradient, and it may happen that none of the uniform references of a materially uniform body is a gradient.

Let $\boldsymbol{K}$ be a uniform reference. Every local configuration $\boldsymbol{G}_X\in\mathscr{C}_X$ can be characterized by the local deformation $\boldsymbol{F}=\boldsymbol{G}_X\boldsymbol{K}(X)^{-1}\in\ell$ from $\boldsymbol{K}(X)$ into $\boldsymbol{G}_X$, so that

$$\boldsymbol{G}_X=\boldsymbol{F}\,\boldsymbol{K}(X). \tag{6.6}$$

Substituting (6.6) and (6.4) into (5.2) with the choice $\Phi_{XY}=\Phi(X, Y)$, we see that

$$\mathfrak{G}_X(F\,K(X))=\mathfrak{G}_Y(F\,K(Y)) \tag{6.7}$$

must hold for all $F\in\ell$ and all $X, Y\in\mathscr{B}$. Conversely, if (6.7) holds for all $F\in\ell$ and all $X, Y\in\mathscr{B}$, then K is a uniform reference. This result may be formulated as follows:

Theorem 1. *A reference K for $\mathscr{B}$ is uniform if and only if there is a function $\mathfrak{H}_K\colon \ell\to\mathbf{R}$ which satisfies*

$$\mathfrak{H}_K(F)=\mathfrak{G}_X(F\,K(X)) \tag{6.8}$$

for all $X\in\mathscr{B}$ and all $F\in\ell$.

The function $\mathfrak{H}_K$, which assigns to each local deformation a response descriptor, will be called the *response function* of the body *relative to the uniform reference K.*

Let K be uniform reference. If we substitute (6.4) for Φ_{XY} in (5.5), we see that

$$K(X)\,g_X\,K(X)^{-1}=K(Y)\,g_Y\,K(Y)^{-1}, \tag{6.9}$$

i.e. that

$$g_K=K(X)\,g_X\,K(X)^{-1} \tag{6.10}$$

is independent of X. The group g_K is a subgroup of the linear group ℓ. We call g_K *the isotropy group of the body $\mathscr{B}$ relative to the uniform reference K.* In view of (6.10), all the intrinsic isotropy groups g_X, $X\in\mathscr{B}$, are isomorphic to the relative isotropy group g_K. It is easily seen that g_K is given in terms of the response function $\mathfrak{H}_K$ by

$$g_K=\{P\in\ell\,|\,\mathfrak{H}_K(F)=\mathfrak{H}_K(F\,P)\ \text{for all}\ F\in\ell\}. \tag{6.11}$$

The relation between two uniform references and the corresponding response functions and isotropy groups is described by the following theorem:

Theorem 2. *Any two uniform references K and $\hat{K}$ are related by*

$$\hat{K}(X)=L\,P(X)\,K(X), \tag{6.12}$$

where $L\in\ell$ and where P is a function on $\mathscr{B}$ with values in g_K.

The isotropy groups g_K and $g_{\hat{K}}$ relative to K and $\hat{K}$ are conjugate:

$$g_{\hat{K}}=L\,g_K\,L^{-1}. \tag{6.13}$$

The response functions $\mathfrak{H}_{\hat{K}}$ and $\mathfrak{H}_K$ are related by the identity

$$\mathfrak{H}_{\hat{K}}(F)=\mathfrak{H}_K(F\,L) \qquad \textit{for all}\ F\in\ell. \tag{6.14}$$

Proof. The two material uniformities Φ and $\hat{\Phi}$ given by

$$\Phi(X, Y)=K(X)^{-1}K(Y), \qquad \hat{\Phi}(X, Y)=\hat{K}(X)^{-1}K(Y)$$

must be related by (6.3). It follows that

$$\hat{K}(Y)\,\mathfrak{P}(Y)\,K(Y)^{-1}=\hat{K}(X)\,\mathfrak{P}(X)\,K(X)^{-1}=L\in\ell$$

is independent of $X\in\mathscr{B}$. Hence (6.12) holds with the choice

$$P(X)=K(X)\,\mathfrak{P}(X)\,K(X)^{-1}. \tag{6.15}$$

It follows from (6.10) that $\boldsymbol{P}(X)\in g_{\boldsymbol{K}}$ for all $X\in\mathscr{B}$, which proves the first assertion of the theorem. If we write (6.10) with $\boldsymbol{K}$ replaced by $\hat{\boldsymbol{K}}$ and substitute (6.12), we obtain

$$\begin{aligned} g_{\hat{\boldsymbol{K}}} &= \boldsymbol{L}\,\boldsymbol{P}(X)\,\boldsymbol{K}(X)\,g_X\,\boldsymbol{K}(X)^{-1}\,\boldsymbol{P}(X)^{-1}\,\boldsymbol{L}^{-1} \\ &= \boldsymbol{L}\,\boldsymbol{P}(X)\,g_{\boldsymbol{K}}\,\boldsymbol{P}(X)^{-1}\,\boldsymbol{L}^{-1}. \end{aligned}$$

Since $\boldsymbol{P}(X)\in g_{\boldsymbol{K}}$ we have $\boldsymbol{P}(X)\,g_{\boldsymbol{K}}\,\boldsymbol{P}(X)^{-1}=g_{\boldsymbol{K}}$ and hence (6.13). The identity (6.14) is derived by writing (6.8) with $\boldsymbol{K}$ replaced by $\hat{\boldsymbol{K}}$, then substituting (6.12) and observing (6.11). Q.E.D.

The theory of isotropy groups relative to a local reference configuration at a single material point[8] extends without change to isotropy groups relative to a uniform reference $\boldsymbol{K}$ of a whole materially uniform body. In particular, we say that the uniform reference $\boldsymbol{K}$ is *undistorted* if $g_{\boldsymbol{K}}$ is comparable, with respect to inclusion, to the orthogonal group $\mathscr{o}$, *i.e.*, if either $g_{\boldsymbol{K}}\subset\mathscr{o}$ or $\mathscr{o}\subset g_{\boldsymbol{K}}$. If there are uniform references $\boldsymbol{K}$ such that $g_{\boldsymbol{K}}\supset\mathscr{o}$, we say that $\mathscr{B}$ is a *uniform isotropic body*; if there are uniform references $\boldsymbol{K}$ such that $g_{\boldsymbol{K}}\subset\mathscr{o}$, we say that $\mathscr{B}$ is a *uniform solid body*. It is possible that a uniform simple body has no undistorted uniform references at all; such a body would be neither a solid nor isotropic.

7. Vector and Tensor Fields

As before, we assume that $\mathscr{B}$ is a continuous body of class C^p, $p\geqq 1$.

A mapping ψ: $\mathscr{B}\to\mathscr{E}'$ of $\mathscr{B}$ into some point-space or vector-space $\mathscr{E}'$ is said to be *of class* C^r, $0\leqq r\leqq p$, if for every configuration $\boldsymbol{\kappa}\in\mathsf{C}$, the mapping $\psi\circ\overset{-1}{\boldsymbol{\kappa}}$: $\boldsymbol{\kappa}(\mathscr{B})\to\mathscr{E}'$ is of class C^r. In view of the axioms for $\mathscr{B}$ it is clear that $\psi\circ\overset{-1}{\boldsymbol{\kappa}}$ is of class C^r for *every* $\boldsymbol{\kappa}\in\mathsf{C}$ if it is of class C^r for *some* $\boldsymbol{\kappa}\in\mathsf{C}$. These definitions apply, in particular, to *functions (scalar fields)* on $\mathscr{B}$, *i.e.* mappings f: $\mathscr{B}\to\mathscr{R}$, to *vector fields* on $\mathscr{B}$, *i.e.* mappings $\boldsymbol{h}$: $\mathscr{B}\to\mathscr{V}$, and to *tensor fields* on $\mathscr{B}$, *i.e.* mappings $\boldsymbol{T}$: $\mathscr{B}\to\mathscr{L}$.

A mapping $\mathfrak{h}$ which assigns to each material point $X\in\mathscr{B}$ a tangent vector $\mathfrak{h}(X)\in\mathscr{T}_X$ is called a *tangent vector field.* We say that such a tangent vector field $\mathfrak{h}$ is of class C^r, $0\leqq r\leqq p-1$, if the vector field $(\nabla\boldsymbol{\kappa})\,\mathfrak{h}$ on $\mathscr{B}$ defined by

$$(\nabla\boldsymbol{\kappa})\,\mathfrak{h}\,|_X=(\nabla\boldsymbol{\kappa}(X))\,\mathfrak{h}(X) \tag{7.1}$$

is of class C^r for some — and hence every — configurafion $\boldsymbol{\kappa}\in\mathsf{C}$.

The algebra of all linear transformations of the tangent space $\mathscr{T}_X$ into itself will be denoted by $\mathscr{I}_X$. A mapping $\mathfrak{T}$ which assigns to each material point $X\in\mathscr{B}$ a linear transformation $\mathfrak{T}(X)\in\mathscr{I}_X$ is called an *intrinsic tensor field.* We say that $\mathfrak{T}$ is of class C^r, $0\leqq r\leqq p-1$, if the tensor field $(\nabla\boldsymbol{\kappa})\,\mathfrak{T}(\nabla\boldsymbol{\kappa})^{-1}$ on $\mathscr{B}$ defined by

$$(\nabla\boldsymbol{\kappa})\,\mathfrak{T}(\nabla\boldsymbol{\kappa})^{-1}|_X=\nabla\boldsymbol{\kappa}(X)\,\mathfrak{T}(X)\,(\nabla\boldsymbol{\kappa}(X))^{-1} \tag{7.2}$$

is of class C^r for some — and hence every — configuration $\boldsymbol{\kappa}\in\mathsf{C}$.

We shall use the term *field on* $\mathscr{B}$ for any mapping that assigns to every $X\in\mathscr{B}$ an element of some vector space (which may consist of linear or multilinear transformations).

[8] This theory was initiated in [*1*], §§19—21. An exposition is given in [*2*], §§ 31—33.

We shall employ the following scheme of notation:

$\mathscr{F}_{\mathscr{B}}^r$ = set of all functions (scalar fields) of class C^r on $\mathscr{B}$.

$\mathscr{V}_{\mathscr{B}}^r$ = set of all vector fields of class C^r on $\mathscr{B}$.

$\mathscr{T}_{\mathscr{B}}^r$ = set of all tangent vector fields of class C^r on $\mathscr{B}$.

$\mathscr{L}_{\mathscr{B}}^r$ = set of all tensor fields of class C^r on $\mathscr{B}$.

$\mathscr{I}_{\mathscr{B}}^r$ = set of all intrinsic tensor fields of class C^r on $\mathscr{B}$.

The set $\mathscr{F}_{\mathscr{B}}^r$ is a commutative algebra under pointwise addition and multiplication. The sets $\mathscr{V}_{\mathscr{B}}^s$, $\mathscr{T}_{\mathscr{B}}^s$, $\mathscr{L}_{\mathscr{B}}^s$, and $\mathscr{I}_{\mathscr{B}}^s$ can be made modules with respect to any of the algebras $\mathscr{F}_{\mathscr{B}}^r$, $s \leqq r \leqq p-1$, by defining addition and scalar multiplication with functions in $\mathscr{F}_{\mathscr{B}}^r$ pointwise. For example, if $\mathfrak{h}$, $\mathfrak{k} \in \mathscr{T}_{\mathscr{B}}^r$ and $f \in \mathscr{F}_{\mathscr{B}}^r$, we define $\mathfrak{h}+\mathfrak{k} \in \mathscr{T}_{\mathscr{B}}^r$ and $f\,\mathfrak{h} \in \mathscr{T}_{\mathscr{B}}^r$ by

$$(\mathfrak{h}+\mathfrak{k})|_X = \mathfrak{h}(X)+\mathfrak{k}(X), \qquad (f\,\mathfrak{h})|_X = f(X)\,\mathfrak{h}(X), \qquad X \in \mathscr{B}. \tag{7.3}$$

The sets $\mathscr{L}_{\mathscr{B}}^r$ and $\mathscr{I}_{\mathscr{B}}^r$ become associative (but not commutative) algebras over $\mathscr{F}_{\mathscr{B}}^r$ if multiplication is defined pointwise.

It is evident that we have $\mathscr{F}_{\mathscr{B}}^r \subset \mathscr{F}_{\mathscr{B}}^s$ if $s \leqq r$ and similar inclusions for the other sets in the list given above. Actually, $\mathscr{F}_{\mathscr{B}}^r$ is a subalgebra of $\mathscr{F}_{\mathscr{B}}^s$. Also, $\mathscr{V}_{\mathscr{B}}^r$ is not only a $\mathscr{F}_{\mathscr{B}}^r$-module, but also a submodule of $\mathscr{V}_{\mathscr{B}}^s$, regarded as a $\mathscr{F}_{\mathscr{B}}^r$-module. Analogous observations apply to the other modules and algebras of the list above.

If $\boldsymbol{T} \in \mathscr{L}_{\mathscr{B}}^s$ and $\boldsymbol{h} \in \mathscr{V}_{\mathscr{B}}^r$ or $\mathfrak{T} \in \mathscr{I}_{\mathscr{B}}^s$ and $\mathfrak{h} \in \mathscr{T}_{\mathscr{B}}^r$, we define $\boldsymbol{T}\boldsymbol{h}$ or $\mathfrak{T}\mathfrak{h}$ pointwise, *i.e.* by

$$\boldsymbol{T}\boldsymbol{h}|_X = \boldsymbol{T}(X)\,\boldsymbol{h}(X), \qquad \mathfrak{T}\mathfrak{h}|_X = \mathfrak{T}(X)\,\mathfrak{h}(X). \tag{7.4}$$

When $s \leqq r$, one can see that $\boldsymbol{T}\boldsymbol{h} \in \mathscr{V}_{\mathscr{B}}^s$, $\mathfrak{T}\mathfrak{h} \in \mathscr{T}_{\mathscr{B}}^s$. It is evident from $(7.4)_1$ that the rules

$$\boldsymbol{T}(\boldsymbol{h}+\boldsymbol{k}) = \boldsymbol{T}\boldsymbol{h}+\boldsymbol{T}\boldsymbol{k}, \qquad \boldsymbol{T}(f\,\boldsymbol{h}) = f\,\boldsymbol{T}\boldsymbol{h} \tag{7.5}$$

are valid. Hence every $\boldsymbol{T} \in \mathscr{L}_{\mathscr{B}}^s$ gives rise to a mapping

$$\boldsymbol{T}\colon\ \mathscr{V}_{\mathscr{B}}^r \to \mathscr{V}_{\mathscr{B}}^s \tag{7.6}$$

which satisfies the rules (7.5) for $\boldsymbol{h}, \boldsymbol{k} \in \mathscr{V}_{\mathscr{B}}^r$, $f \in \mathscr{F}_{\mathscr{B}}^r$. Mappings of the type (7.6) satisfying the rules (7.5) are homomorphism with respect to the $\mathscr{F}_{\mathscr{B}}^r$-module structures of $\mathscr{V}_{\mathscr{B}}^r$ and $\mathscr{V}_{\mathscr{B}}^s$. We also call them *$\mathscr{F}$-linear mappings*. Thus, every $\boldsymbol{T} \in \mathscr{L}_{\mathscr{B}}^s$ gives rise to an $\mathscr{F}$-linear mapping (7.6). It is remarkable that the converse is also true, *i.e.* that every $\mathscr{F}$-linear mapping of the type (7.6) arises from a tensor field of class C^s on $\mathscr{B}$:

Proposition 1. *If $\boldsymbol{T}\colon \mathscr{V}_{\mathscr{B}}^r \to \mathscr{V}_{\mathscr{B}}^s\ (s \leqq r)$ is $\mathscr{F}$-linear, then there exist a unique tensor field $\bar{\boldsymbol{T}} \in \mathscr{I}_{\mathscr{B}}^s$ such that $\bar{\boldsymbol{T}}\boldsymbol{h} = \boldsymbol{T}\boldsymbol{h}$ holds for all $\boldsymbol{h} \in \mathscr{T}_{\mathscr{B}}^r$.*

Proof. Let $(\boldsymbol{e}_1, \boldsymbol{e}_2, \boldsymbol{e}_3)$ be a basis of $\mathscr{V}$. The vectors $\boldsymbol{e}_i$ can be regarded as constant vector fields on $\mathscr{B}$, so that $\boldsymbol{e}_i \in \mathscr{V}_{\mathscr{B}}^{p-1} \subset \mathscr{V}_{\mathscr{B}}^r$. Every $\boldsymbol{h} \in \mathscr{V}_{\mathscr{B}}^r$ has a unique component representation

$$\boldsymbol{h} = \sum_i h^i \boldsymbol{e}_i, \qquad h^i \in \mathscr{F}_{\mathscr{B}}^r. \tag{7.7}$$

Applying the given $\mathscr{F}$-linear mapping $\boldsymbol{T}$ to (7.7), we obtain

$$\boldsymbol{T}\boldsymbol{h} = \sum_i h^i\, \boldsymbol{T}\boldsymbol{e}_i. \tag{7.8}$$

Now, if there is a tensor field $\bar{T}$ such that $\bar{T}h = Th$ for all $h \in \mathscr{V}_{\mathscr{B}}^r$, we must have, in particular, $Te_i = \bar{T}e_i$, *i.e.*

$$\bar{T}(X)\,e_i = (T\,e_i)|_X \tag{7.9}$$

for all $X \in \mathscr{B}$. But since (e_1, e_2, e_3) is a basis of $\mathscr{V}$, we can find, for each $X \in \mathscr{B}$, exactly one $\bar{T}(X) \in \mathscr{L}$ such that (7.9) holds. Since the vector fields Te_i are of class C^s, it is easily seen that the tensor field $\bar{T}$ obtained in this way is also of class C^s. Moreover, in view of (7.8), (7.9) and the $\mathscr{F}$-linearity of $\bar{T}$ we have

$$\bar{T}h = \sum_i h^i\,\bar{T}e_i = \sum_i h^i\,T\,e_i = T(\sum_i h^i\,e_i) = Th$$

for all h. Q.E.D.

Proposition 1 enables us to identify the set of all $\mathscr{F}$-linear mappings of the type (7.6) with the set $\mathscr{L}_{\mathscr{B}}^s$ of all tensor fields of class C^s on $\mathscr{B}$. Similarly, we can identify the set of all $\mathscr{F}$-linear mappings of the type

$$\mathfrak{T}\colon\ \mathscr{T}_{\mathscr{B}}^r \to \mathscr{T}_{\mathscr{B}}^s \qquad (s \leqq r)$$

with the set $\mathscr{I}_{\mathscr{B}}^s$ of all intrinsic tensor fields of class C^s on $\mathscr{B}$. The proof of this fact follows from Proposition 1 by choosing a configuration κ of $\mathscr{B}$ and letting $\mathfrak{T}$ correspond to $\bar{T} = \nabla\kappa\,\mathfrak{T}(\nabla\kappa)^{-1}\colon \mathscr{V}_{\mathscr{B}}^r \to \mathscr{V}_{\mathscr{B}}^s$. The result just stated is a special case of a general proposition referring to $\mathscr{F}$-multilinear mappings. For later application we state another special case:

Proposition 2. *If*

$$\mathfrak{S}\colon\ \mathscr{T}_{\mathscr{B}}^r \times \mathscr{T}_{\mathscr{B}}^r \to \mathscr{T}_{\mathscr{B}}^s \qquad (or\ \ \mathscr{I}_{\mathscr{B}}^s) \tag{7.10}$$

is $\mathscr{F}$-bilinear (i.e. $\mathscr{F}$-linear in each of the two variables), then there exists a unique field $\bar{\mathfrak{S}}$ on $\mathscr{B}$ whose values $\bar{\mathfrak{S}}(X)$ are bilinear mappings

$$\bar{\mathfrak{S}}(X)\colon\ \mathscr{T}_X \times \mathscr{T}_X \to \mathscr{T}_X \qquad (or\ \ \mathscr{I}_X) \tag{7.11}$$

such that

$$\bar{\mathfrak{S}}(X)\,(\mathfrak{h}(X), \mathfrak{k}(X)) = \mathfrak{S}(\mathfrak{h}, \mathfrak{k})|_X \tag{7.12}$$

holds for all $\mathfrak{h}, \mathfrak{k} \in \mathscr{T}_{\mathscr{B}}^r$ and all $X \in B$. The function $\mathfrak{S}$ is of class C^s (in the obvious sense).

8. Relative Gradients, Brackets

From now on we assume that $\mathscr{B}$ is a continuous body of class C^p with $p \geqq 2$.

Let $\psi\colon \mathscr{B} \to \mathscr{E}'$ be a mapping of class $\mathscr{C}^r$, $1 \leqq r \leqq p$, where $\mathscr{E}'$ is some point space or vector space. Given a configuration κ of $\mathscr{B}$, we can then define

$$\nabla_{\kappa}\psi\colon\ \mathscr{B} \to \mathscr{L}(\mathscr{V}, \mathscr{V}') \tag{8.1}$$

where $\mathscr{V}'$ is the translation space of $\mathscr{E}'$, by $\nabla_{\kappa}\psi = \nabla(\psi \circ \overset{-1}{\kappa}) \circ \kappa$, *i.e.*

$$\nabla_{\kappa}\psi|_X = \nabla(\psi \circ \overset{-1}{\kappa})|_{\kappa(X)}, \qquad X \in \mathscr{B}. \tag{8.2}$$

We call $\nabla_{\kappa}\psi$ the *gradient of ψ relative to the configuration κ*. It is clear that $\nabla_{\kappa}\psi$ is of class C^{r-1}.

Let $\kappa, \gamma \in \mathrm{C}$ be two configurations. Taking the gradient of $\psi \circ \overset{-1}{\kappa} = (\psi \circ \overset{-1}{\gamma}) \circ (\gamma \circ \overset{-1}{\kappa})$ and using the chain rule, we see with the help of (3.3) that the gradients of ψ relative to κ and γ are related by

$$\nabla_{\kappa}\psi|_X=\nabla_{\gamma}\psi|_X\circ[\nabla\gamma(X)(\nabla\kappa(X))^{-1}]. \tag{8.3}$$

Let $\boldsymbol{K}$ and $\boldsymbol{G}$ be two references for $\mathscr{B}$ (see Definition 4). We define $\boldsymbol{K}\boldsymbol{G}^{-1}$ pointwise, *i.e.* by

$$\boldsymbol{K}\boldsymbol{G}^{-1}|_X=\boldsymbol{K}(X)\,\boldsymbol{G}(X)^{-1}. \tag{8.4}$$

Recalling that the configuration gradients $\nabla\kappa$ and $\nabla\gamma$ are references, we see that (8.3) can then be written as

$$\nabla_{\kappa}\psi=\nabla_{\gamma}\psi\circ(\nabla\gamma(\nabla\kappa)^{-1}). \tag{8.5}$$

We say that a reference $\boldsymbol{K}$ is *of class* C^r, $r\leqq p-1$, if for some – and hence every – configuration $\kappa\in\mathsf{C}$ the tensor field $(\nabla\kappa)\boldsymbol{K}^{-1}$ is of class C^r, *i.e.* belongs to $\mathscr{L}^r_{\mathscr{B}}$. It is clear that every gradient reference $\nabla\kappa$ is of class C^{p-1}.

Let a local configuration $\boldsymbol{K}_X\in\mathscr{C}_X$ be given. If κ and γ both belong to the equivalence class that defines $\boldsymbol{K}_X$, which means that $\nabla\kappa(X)=\nabla\gamma(X)=\boldsymbol{K}_X$, we have, by (8.3), $\nabla_{\kappa}\psi(X)=\nabla_{\gamma}\psi(X)$. Hence, $\nabla_{\kappa}\psi(X)$ depends on κ only through the equivalence class $\boldsymbol{K}_X\in\mathscr{C}_X$ to which κ belongs, and it is legitimate to define

$$\nabla_{\boldsymbol{K}_X}\psi(X)=\nabla_{\kappa}\psi(X)\quad\text{if}\quad\kappa\in\boldsymbol{K}_X. \tag{8.6}$$

If $\boldsymbol{K}$ is a reference, we define the *gradient of ψ relative to the reference $\boldsymbol{K}$* by

$$\nabla_{\boldsymbol{K}}\psi|_X=\nabla_{\boldsymbol{K}(X)}\psi(X),\qquad X\in\mathscr{B}. \tag{8.7}$$

If $\boldsymbol{K}$ and $\boldsymbol{G}$ are any two references for $\mathscr{B}$, we see that (8.3) and the definitions (8.6) and (8.7) yield the formula

$$\nabla_{\boldsymbol{K}}\psi=\nabla_{\boldsymbol{G}}\psi\circ(\boldsymbol{G}\boldsymbol{K}^{-1}), \tag{8.8}$$

which generalizes (8.5). By writing (8.8) with $\boldsymbol{G}=\nabla\gamma$, where $\gamma\in\mathsf{C}$, we infer that $\nabla_{\boldsymbol{K}}\psi$ is of class C^{r-1} if ψ is of class C^r and $\boldsymbol{K}$ of class C^{r-1}.

When the range of ψ coincides with the set $\mathscr{R}$ of real numbers, in which case we write f instead of ψ, we can identify $\nabla_{\kappa}f$ with a vector field on $\mathscr{B}$. Thus, if $f\in\mathscr{F}^r_{\mathscr{B}}$ then $\nabla_{\kappa}f\in\mathscr{V}^{r-1}_{\mathscr{B}}$. The formula (8.5) becomes

$$\nabla_{\kappa}f=(\nabla\gamma(\nabla\kappa)^{-1})^T\,\nabla_{\gamma}f. \tag{8.9}$$

Let $\mathfrak{h}\in\mathscr{T}^{r-1}_{\mathscr{B}}$ and $f\in\mathscr{F}^r_{\mathscr{B}}$. The function $\mathfrak{h}(f)$ on $\mathscr{B}$ defined by

$$\mathfrak{h}(f)=\nabla_{\kappa}f\cdot(\nabla\kappa)\mathfrak{h}, \tag{8.10}$$

where the inner product is defined pointwise, does not depend on the choice of the configuration $\kappa\in\mathsf{C}$, as is easily seen with the help of (8.9). Moreover, $\mathfrak{h}(f)$ is of class C^{r-1}. Therefore, every $\mathfrak{h}\in\mathscr{T}^{r-1}_{\mathscr{B}}$ gives rise to a mapping

$$\mathfrak{h}:\ \mathscr{F}^r_{\mathscr{B}}\to\mathscr{F}^{r-1}_{\mathscr{B}}. \tag{8.11}$$

Actually, every $\mathfrak{h}\in\mathscr{T}^{r-1}_{\mathscr{B}}$ can be *identified* with a mapping of the type (8.11), because it is easily seen from (8.10) that $\mathfrak{h}_1(f)=\mathfrak{h}_2(f)$ cannot hold for all $f\in\mathscr{F}^r_{\mathscr{B}}$ unless $\mathfrak{h}_1=\mathfrak{h}_2$. The mapping (8.11) defined by (8.10) has the following basic property, which follows immediately from the chain rule.

Proposition 3. *If $\mathfrak{h}\in\mathcal{T}_{\mathcal{B}}^{r-1}$, if H is a real-valued function of class C^r of any number m of real variables, and if $f_1, f_2, \ldots, f_m \in \mathcal{F}_{\mathcal{B}}^r$, then*

$$\mathfrak{h}(H(f_1, f_2, \ldots, f_m)) = \sum_{k=1}^{m} H_{,k}(f_1, f_2, \ldots, f_m)\,\mathfrak{h}(f_k), \tag{8.12}$$

where $H_{,k}$ denotes the derivative of H with respect to its kth variable.

Actually, the property described in Proposition 3 characterizes the tangent vector fields of class C^{r-1} and hence could have been used for their definition, but we shall neither use nor prove this fact.

Applying (8.12) to the cases when $H(\xi_1, \xi_2)=\xi_1+\xi_2$ and $H(\xi_1, \xi_2)=\xi_1\xi_2$, we obtain

$$\mathfrak{h}(f+g)=\mathfrak{h}(f)+\mathfrak{h}(g), \qquad \mathfrak{h}(f\,g)=f\,\mathfrak{h}(g)+g\,\mathfrak{h}(f). \tag{8.13}$$

Let $\mathfrak{h}, \mathfrak{k}\in\mathcal{T}_{\mathcal{B}}^{r-1}$ with $r\geqq 2$. Since $\mathcal{T}_{\mathcal{B}}^{r-1}\subset\mathcal{T}_{\mathcal{B}}^{r-2}$ and hence also $\mathfrak{h}, \mathfrak{k}\in\mathcal{T}_{\mathcal{B}}^{r-2}$ we can identify $\mathfrak{h}$ and $\mathfrak{k}$ not only with mappings from $\mathcal{F}_{\mathcal{B}}^r$ into $\mathcal{F}_{\mathcal{B}}^{r-1}$, but also with mappings from $\mathcal{F}_{\mathcal{B}}^{r-1}$ into $\mathcal{F}_{\mathcal{B}}^{r-2}$. Therefore, we can form the compositions $\mathfrak{h}\circ\mathfrak{k}$ and $\mathfrak{k}\circ\mathfrak{h}$ as mappings from $\mathcal{F}_{\mathcal{B}}^r$ into $\mathcal{F}_{\mathcal{B}}^{r-2}$. By themselves, these compositions do not correspond to tangent vector fields, but it is remarkable that the difference

$$[\mathfrak{h}, \mathfrak{k}]=\mathfrak{h}\circ\mathfrak{k}-\mathfrak{k}\circ\mathfrak{h}\colon\ \mathcal{F}_{\mathcal{B}}^r\to\mathcal{F}_{\mathcal{B}}^{r-2}, \tag{8.14}$$

called the *bracket* of $\mathfrak{h}$ and $\mathfrak{k}$, has values that belong to $\mathcal{F}_{\mathcal{B}}^{r-1}(\subset\mathcal{F}_{\mathcal{B}}^{r-2})$ and does correspond to a tangent vector field:

Proposition 4. *The bracket of two tangent vector fields $\mathfrak{h}, \mathfrak{k}\in\mathcal{F}_{\mathcal{B}}^{r-1}$ $(r\geqq 2)$ can be identified with the tangent vector field of class C^{r-2} given by*

$$[\mathfrak{h}, \mathfrak{k}]=(\nabla\boldsymbol{\kappa})^{-1}\left[(\nabla_{\boldsymbol{\kappa}}\boldsymbol{k})\,\boldsymbol{h}-(\nabla_{\boldsymbol{\kappa}}\boldsymbol{h})\,\boldsymbol{k}\right], \qquad \boldsymbol{h}=(\nabla\boldsymbol{\kappa})\,\mathfrak{h}, \qquad \boldsymbol{k}=(\nabla\boldsymbol{\kappa})\,\mathfrak{k}, \tag{8.15}$$

where $\boldsymbol{\kappa}$ is an arbitrary configuration.

Proof. We denote the tangent vector field of class C^{r-2} defined by the right-hand side of (8.15) by $\mathfrak{b}$, so that

$$(\nabla\boldsymbol{\kappa})\,\mathfrak{b}=\left[(\nabla_{\boldsymbol{\kappa}}\boldsymbol{k})\,\boldsymbol{h}-(\nabla_{\boldsymbol{\kappa}}\boldsymbol{h})\,\boldsymbol{k}\right]. \tag{8.16}$$

Now let $f\in\mathcal{F}_{\mathcal{B}}^r$. In view of (8.10), it follows from (8.16) that

$$\mathfrak{b}(f)=\nabla_{\boldsymbol{\kappa}}f\cdot\left[(\nabla_{\boldsymbol{\kappa}}\boldsymbol{k})\,\boldsymbol{h}-(\nabla_{\boldsymbol{\kappa}}\boldsymbol{h})\,\boldsymbol{k}\right] \tag{8.17}$$

and from $(8.15)_{2,3}$ that

$$(\mathfrak{h}\circ\mathfrak{k})\,(f)=\mathfrak{h}(\mathfrak{k}(f))=\nabla_{\boldsymbol{\kappa}}(\nabla_{\boldsymbol{\kappa}}f\cdot\boldsymbol{k})\cdot\boldsymbol{h}. \tag{8.18}$$

The rules of ordinary differential calculus yield $\nabla_{\boldsymbol{\kappa}}(\nabla_{\boldsymbol{\kappa}}f\cdot\boldsymbol{k})\cdot\boldsymbol{h}=\boldsymbol{h}\cdot(\nabla_{\boldsymbol{\kappa}}^{(2)}f)\,\boldsymbol{k}+\nabla_{\boldsymbol{\kappa}}f\cdot(\nabla_{\boldsymbol{\kappa}}\boldsymbol{k})\,\boldsymbol{h}$. Hence, since $\nabla_{\boldsymbol{\kappa}}^{(2)}f$ is symmetric, if we write (8.18) with $\mathfrak{h}$ and $\mathfrak{k}$ interchanged, take the difference, and then compare with (8.17), we obtain

$$\mathfrak{b}(f)=(\mathfrak{h}\circ\mathfrak{k})\,(f)-(\mathfrak{k}\circ\mathfrak{h})\,(f)=[\mathfrak{h}, \mathfrak{k}]\,(f)$$

i.e. the desired result $\mathfrak{b}=[\mathfrak{h}, \mathfrak{k}]$. Q.E.D.

The bracket $[\mathfrak{h}, \mathfrak{k}]$ depends linearly (but not $\mathscr{F}$-linearly) on $\mathfrak{h}$ and $\mathfrak{k}$ and satisfies for $\mathfrak{h}, \mathfrak{k}, \mathfrak{l} \in \mathscr{T}_{\mathscr{B}}^{r-1}$, $f \in \mathscr{F}_{\mathscr{B}}^{r}$, $2 \leqq r$, the identities

$$[\mathfrak{h}, \mathfrak{k}] = -[\mathfrak{k}, \mathfrak{l}], \tag{8.19}$$

$$[\mathfrak{h}, f\mathfrak{k}] = f[\mathfrak{h}, \mathfrak{k}] + \mathfrak{h}(f)\,\mathfrak{k}, \tag{8.20}$$

and for $\mathfrak{h}, \mathfrak{k}, \mathfrak{l} \in \mathscr{T}_{\mathscr{B}}^{r}$, $2 \leqq r \leqq p-1$, the Jacobi-identity

$$\sum_{\text{cyclic}} [\mathfrak{h}, [\mathfrak{k}, \mathfrak{l}]] = 0, \tag{8.21}$$

where the sum is taken of all terms obtained from the one written by cyclic permutation of $\mathfrak{h}, \mathfrak{k}, \mathfrak{l}$. The identity (8.19) is obvious from (8.14), and (8.21) is the result of a trivial calculation. The identity (8.20) follows from (8.14) and (8.13).

It would have been possible to define the bracket $[\mathfrak{h}, \mathfrak{k}] \in \mathscr{T}_{\mathscr{B}}^{r-1}$ for $\mathfrak{h}, \mathfrak{k} \in \mathscr{T}_{\mathscr{B}}^{r}$, $1 \leqq r \leqq p-1$, directly by (8.15), for it is easy to see that the right-hand side of (8.15) does not depend on the choice of the configuration $\boldsymbol{\kappa}$.

9. Affine Connections, Torsion, Curvature

From now on we assume that $\mathscr{B}$ is a continuous body of class C^p, $p \geqq 3$.

A mapping

$$\Gamma\colon\ \mathscr{T}_{\mathscr{B}}^{r} \to \mathscr{I}_{\mathscr{B}}^{r-1} \tag{9.1}$$

is called an *affine connection* of class C^{r-1} $(1 \leqq r \leqq p-1)$ on $\mathscr{B}$ if

$$\Gamma(\mathfrak{h} + \mathfrak{k}) = \Gamma\,\mathfrak{h} + \Gamma\,\mathfrak{k} \tag{9.2}$$

holds for all $\mathfrak{h}, \mathfrak{k} \in \mathscr{T}_{\mathscr{B}}^{r}$ and

$$\Gamma(f\,\mathfrak{h})\,\mathfrak{k} = f(\Gamma\,\mathfrak{h})\,\mathfrak{k} + \mathfrak{k}(f)\,\mathfrak{h} \tag{9.3}$$

holds for all $\mathfrak{h} \in \mathscr{T}_{\mathscr{B}}^{r}, f \in \mathscr{F}_{\mathscr{B}}^{r}$ and all $\mathfrak{k} \in \mathscr{T}_{\mathscr{B}}^{r-1}$.

If a is a real constant, then $\mathfrak{k}(a) = 0$ by the definition (8.10). Hence (9.3) reduces to $\Gamma(a\mathfrak{h}) = a\Gamma\mathfrak{h}$ when $a \in \mathscr{R}$, $\mathfrak{h} \in \mathscr{T}_{\mathscr{B}}^{r}$. Thus, Γ is a linear mapping, but it is never $\mathscr{F}$-linear. The rule (9.3) resembles one of the product rules for gradient operators.

A triple $(\mathfrak{e}_1, \mathfrak{e}_2, \mathfrak{e}_3)$ of tangent vector fields of class C^r, $r \leqq p-1$, is called a *frame* of class C^r if the values $\mathfrak{e}_i(X)$ form a basis of the tangent space $\mathscr{T}_X$ for each $X \in \mathscr{B}$. Frames of class C^{p-1} (and hence of class C^r, $r \leqq p-1$) exist. For example, if $(\boldsymbol{e}_1, \boldsymbol{e}_2, \boldsymbol{e}_3)$ is a basis of $\mathscr{V}$ and $\boldsymbol{\kappa}$ a configuration, then $\mathfrak{e}_i = (\nabla\boldsymbol{\kappa})^{-1}\boldsymbol{e}_i$ defines a frame of class C^{p-1}. Every tangent vector field $\mathfrak{h} \in \mathscr{T}_{\mathscr{B}}^{r}$ has a component representation

$$\mathfrak{h} = \sum_i h^i\,\mathfrak{e}_i \tag{9.4}$$

with respect to a given frame $(\mathfrak{e}_1, \mathfrak{e}_2, \mathfrak{e}_3)$ of class C^r such that the component functions h^i belong to $\mathscr{F}_{\mathscr{B}}^{r}$.

Now let Γ be a connection of class C^{r-1}. Substituting (9.4) into $(\Gamma\,\mathfrak{h})\,\mathfrak{e}_j$ and using the rules (9.2) and (9.3), we obtain

$$(\Gamma\,\mathfrak{h})\,\mathfrak{e}_j = \sum_i [h^i(\Gamma\,\mathfrak{e}_i)\,\mathfrak{e}_j + \mathfrak{e}_j(h^i)\,\mathfrak{e}_i]. \tag{9.5}$$

The components Γ_{ij}^k of the three intrinsic tensor fields $\Gamma \mathfrak{e}_i$ with respect to the frame $(\mathfrak{e}_1, \mathfrak{e}_2, \mathfrak{e}_3)$ are defined by

$$(\Gamma \mathfrak{e}_i)\,\mathfrak{e}_j = \sum_k \Gamma_{ij}^k \mathfrak{e}_k. \tag{9.6}$$

These components Γ_{ij}^k belong to $\mathscr{F}_{\mathscr{B}}^{r-1}$ and are called the *components of the connection* Γ with respect to the frame $(\mathfrak{e}_1, \mathfrak{e}_2, \mathfrak{e}_3)$. If we prescribe a frame $(\mathfrak{e}_1, \mathfrak{e}_2, \mathfrak{e}_3)$ of class C^r and 27 functions $\Gamma_{ij}^k \in \mathscr{F}_{\mathscr{B}}^{r-1}$ on $\mathscr{B}$ arbitrarily, then (9.5) and (9.6) determine a unique affine connection of class C^{r-1}.

Let Γ be a connection of class C^{r-1} having components $\Gamma_{ij}^k \in \mathscr{F}_{\mathscr{B}}^{r-1}$ with respect to the frame $(\mathfrak{e}_1, \mathfrak{e}_2, \mathfrak{e}_3)$ of class C^r. When $1 \leqq s \leqq r$, then $(\mathfrak{e}_1, \mathfrak{e}_2, \mathfrak{e}_3)$ is also of class C^s and $\Gamma_{ij}^k \in \mathscr{F}_{\mathscr{B}}^{s-1} \supset \mathscr{F}_{\mathscr{B}}^{r-1}$. Hence, (9.5) and (9.6) define an affine connection of class C^{s-1}. Therefore, the mapping $\Gamma\colon \mathscr{T}_{\mathscr{B}}^{r} \to \mathscr{I}_{\mathscr{B}}^{r-1}$ has a unique extension to $\mathscr{T}_{\mathscr{B}}^{s}$ that is an affine connection of class C^{s-1}. We denote this extension by the same symbol Γ. With this convention, we can say that every affine connection of class C^{r-1} is also of class C^{s-1} when $1 \leqq s \leqq r \leqq p-1$.

Let Γ be a connection of class C^{r-1}, and hence also of class C^{s-1} when $1 \leqq s \leqq r$. Using the notation

$$\Gamma_{\mathfrak{h}} \mathfrak{k} = (\Gamma \mathfrak{k})\,\mathfrak{h}, \tag{9.7}$$

we can identify $\Gamma_{\mathfrak{h}}$ with a mapping

$$\Gamma_{\mathfrak{h}}\colon \mathscr{T}_{\mathscr{B}}^{s} \to \mathscr{T}_{\mathscr{B}}^{s-1} \tag{9.8}$$

for any choice of s, $1 \leqq s \leqq r$, and any choice of $\mathfrak{h} \in \mathscr{T}_{\mathscr{B}}^{s-1}$. In terms of $\Gamma_{\mathfrak{h}}$ the rule (9.3) reads

$$\Gamma_{\mathfrak{h}}(f\mathfrak{k}) = f(\Gamma_{\mathfrak{h}} \mathfrak{k}) + \mathfrak{h}(f)\,\mathfrak{k}. \tag{9.9}$$

Moreover, $\Gamma_{\mathfrak{h}}$ depends $\mathscr{F}$-linearly on $\mathfrak{h}$.

The *Cartan-torsion* (or simply *torsion*) of the connection Γ is the mapping

$$\mathfrak{S}\colon \mathscr{T}_{\mathscr{B}}^{r} \times \mathscr{T}_{\mathscr{B}}^{r} \to \mathscr{T}_{\mathscr{B}}^{r-1} \tag{9.10}$$

defined by

$$\mathfrak{S}(\mathfrak{h}, \mathfrak{k}) = \Gamma_{\mathfrak{h}} \mathfrak{k} - \Gamma_{\mathfrak{k}} \mathfrak{h} - [\mathfrak{h}, \mathfrak{k}]. \tag{9.11}$$

In view of (8.19) it is obvious that $\mathfrak{S}$ is skew in the sense that

$$\mathfrak{S}(\mathfrak{h}, \mathfrak{k}) = -\mathfrak{S}(\mathfrak{k}, \mathfrak{h}). \tag{9.12}$$

It is an almost immediate consequence of (9.9) and the rule (8.20) that $\mathfrak{S}$ is $\mathscr{F}$-bilinear. Hence, by Proposition 2 (Sect. 7), the torsion $\mathfrak{S}$ can be identified with a field on $\mathscr{B}$ of class C^{r-1} whose value $\mathfrak{S}(X)$ at $X \in \mathscr{B}$ is a bilinear mapping from $\mathscr{T}_X \times \mathscr{T}_X$ into $\mathscr{T}_X$. It follows that $\mathfrak{S}(\mathfrak{h}, \mathfrak{k})$ remains meaningful for any, even discontinuous, tangent vector fields $\mathfrak{h}, \mathfrak{k}$, and that $\mathfrak{S}(\mathfrak{h}, \mathfrak{k})|_X = \mathfrak{S}(X)(\mathfrak{h}(X), \mathfrak{k}(X))$ depends on $\mathfrak{h}$ and $\mathfrak{k}$ only through their values at X.

Let $\mathfrak{k}, \mathfrak{h} \in \mathscr{T}_{\mathscr{B}}^{r-1}$, $2 \leqq r \leqq p-1$. In view of (9.8) we can regard $\Gamma_{\mathfrak{h}}$ and $\Gamma_{\mathfrak{k}}$ as mappings from $\mathscr{F}_{\mathscr{B}}^{r}$ into $\mathscr{F}_{\mathscr{B}}^{r-1}$ and also as mappings from $\mathscr{F}_{\mathscr{B}}^{r-1}$ into $\mathscr{F}_{\mathscr{B}}^{r-2}$. Hence we can form the compositions $\Gamma_{\mathfrak{h}} \circ \Gamma_{\mathfrak{k}}$ and $\Gamma_{\mathfrak{k}} \circ \Gamma_{\mathfrak{h}}$ and the bracket

$$[\Gamma_{\mathfrak{h}}, \Gamma_{\mathfrak{k}}] = \Gamma_{\mathfrak{h}} \circ \Gamma_{\mathfrak{k}} - \Gamma_{\mathfrak{k}} \circ \Gamma_{\mathfrak{h}} \tag{9.13}$$

as mappings from $\mathcal{T}_{\mathscr{B}}^{r}$ into $\mathcal{T}_{\mathscr{B}}^{r-2}$. Since $[\mathfrak{h},\mathfrak{k}]\in\mathcal{T}_{\mathscr{B}}^{r-1}$, we can regard $\Gamma_{[\mathfrak{h},\mathfrak{k}]}$ as a mapping from $\mathcal{T}_{\mathscr{B}}^{r}(\subset\mathcal{T}_{\mathscr{B}}^{r-1})$ into $\mathcal{T}_{\mathscr{B}}^{r-2}$. Hence we can define

$$\mathfrak{R}(\mathfrak{h},\mathfrak{k})\colon\ \mathcal{T}_{\mathscr{B}}^{r}\to\mathcal{T}_{\mathscr{B}}^{r-2} \tag{9.14}$$

by

$$\mathfrak{R}(\mathfrak{h},\mathfrak{k})=[\Gamma_{\mathfrak{h}},\Gamma_{\mathfrak{k}}]-\Gamma_{[\mathfrak{h},\mathfrak{k}]}\,. \tag{9.15}$$

An easy calculation, based on the definition (9.13) and the rules (9.9) and (8.20), shows that the mapping (9.15) is $\mathscr{F}$-linear. Hence, by an analogue of Proposition 1 (Sect. 7) for intrinsic tensor fields, $\mathfrak{R}(\mathfrak{h},\mathfrak{k})$ can be identified with an element of $\mathscr{I}_{\mathscr{B}}^{r-2}$ and $\mathfrak{R}$ can be regarded as a mapping

$$\mathfrak{R}\colon\ \mathcal{T}_{\mathscr{B}}^{r-1}\times\mathcal{T}_{\mathscr{B}}^{r-1}\to\mathscr{I}_{\mathscr{B}}^{r-2}, \tag{9.16}$$

which is called the *Riemann-curvature* (or simply *curvature*) of the connection Γ. It is obvious that $\mathfrak{R}$ is skew in the sense that

$$\mathfrak{R}(\mathfrak{h},\mathfrak{k})=-\mathfrak{R}(\mathfrak{k},\mathfrak{h})\,. \tag{9.17}$$

A short calculation shows that the mapping (9.16) is $\mathscr{F}$-bilinear. Therefore, by Proposition 2 (Sect. 7), we can identify the curvature $\mathfrak{R}$ with a field on $\mathscr{B}$ of class C^{r-2} whose value $\mathfrak{R}(X)$ at $X\in\mathscr{B}$ is a bilinear transformation from $\mathcal{T}_X\times\mathcal{T}_X$ into $\mathscr{I}_X$. We have $\mathfrak{R}(\mathfrak{h},\mathfrak{k})|_X=\mathfrak{R}(X)(\mathfrak{h}(X),\mathfrak{k}(X))$, which shows that $\mathfrak{R}(\mathfrak{h},\mathfrak{k})$ is meaningful for any tangent vector fields $\mathfrak{h}$, $\mathfrak{k}$.

There is an important relation between the torsion and the curvature of an affine connection:

Proposition 5. *Let Γ be an affine connection of class C^{r-1} on $\mathscr{B}$, $2\leqq r\leqq p-1$. The torsion $\mathfrak{S}$ and the curvature $\mathfrak{R}$ of Γ satisfy*

$$\sum_{\text{cyclic}}\{\Gamma_{\mathfrak{l}}(\mathfrak{S}(\mathfrak{h},\mathfrak{k}))+\mathfrak{S}(\mathfrak{k},[\mathfrak{l},\mathfrak{h}])-\mathfrak{R}(\mathfrak{l},\mathfrak{h})\,\mathfrak{k}\}=0 \tag{9.18}$$

for all $\mathfrak{h},\mathfrak{k},\mathfrak{l}\in\mathcal{T}_{\mathscr{B}}^{r}$. The sum is to be taken over all terms obtained from the one written by cyclic permutation of $\mathfrak{h},\mathfrak{k},\mathfrak{l}$.

The identity (9.18) is often called the *First Bianchi Identity*.

Proof. Operating with $\Gamma_{\mathfrak{l}}$ on (9.11) gives

$$\Gamma_{\mathfrak{l}}(\mathfrak{S}(\mathfrak{h},\mathfrak{k}))-(\Gamma_{\mathfrak{l}}\circ\Gamma_{\mathfrak{h}})\,\mathfrak{k}+(\Gamma_{\mathfrak{l}}\circ\Gamma_{\mathfrak{k}})\,\mathfrak{h}+\Gamma_{\mathfrak{l}}[\mathfrak{h},\mathfrak{k}]=0\,.$$

The cyclic sum of the left side of this equation remains unchanged if the third term is changed by one and the fourth by two cyclic permutations of $\mathfrak{h}$, $\mathfrak{k}$, $\mathfrak{l}$. Hence we have

$$\sum_{\text{cyclic}}\{\Gamma_{\mathfrak{l}}(\mathfrak{S}(\mathfrak{h},\mathfrak{k}))-(\Gamma_{\mathfrak{l}}\circ\Gamma_{\mathfrak{h}})\,\mathfrak{k}+(\Gamma_{\mathfrak{h}}\circ\Gamma_{\mathfrak{l}})\,\mathfrak{k}+\Gamma_{\mathfrak{k}}[\mathfrak{l},\mathfrak{h}]\}$$
$$=\sum_{\text{cyclic}}\{\Gamma_{\mathfrak{l}}(\mathfrak{S}(\mathfrak{h},\mathfrak{k}))-[\Gamma_{\mathfrak{l}},\Gamma_{\mathfrak{h}}]\,\mathfrak{k}+\Gamma_{\mathfrak{k}}[\mathfrak{l},\mathfrak{h}]\}=0\,.$$

Using the definitions (9.15) and (9.11), we obtain

$$\sum_{\text{cyclic}}\{\Gamma_{\mathfrak{l}}(\mathfrak{S}(\mathfrak{h},\mathfrak{k}))-\mathfrak{R}(\mathfrak{l},\mathfrak{h})\,\mathfrak{k}+\mathfrak{S}(\mathfrak{k},[\mathfrak{l},\mathfrak{h}])+[\mathfrak{k},[\mathfrak{l},\mathfrak{h}]]\}=0\,.$$

In view of the Jacobi identity (8.21) the last term gives no contribution and (9.18) results. Q.E.D.

Let Γ and $\overset{*}{\Gamma}$ be two connections of class C^{r-1} on $\mathscr{B}$. Using the notation (9.7) and observing the rule (9.9), we see that, for each s, $1 \leqq s \leqq r$, and each $\mathfrak{h} \in \mathscr{T}_{\mathscr{B}}^{s-1}$, the difference

$$\mathfrak{D}_{\mathfrak{h}} = \Gamma_{\mathfrak{h}} - \overset{*}{\Gamma}_{\mathfrak{h}}: \mathscr{T}_{\mathscr{B}}^{s} \to \mathscr{T}_{\mathscr{B}}^{s-1} \tag{9.19}$$

is actually $\mathscr{F}$-linear. Hence $\mathfrak{D}_{\mathfrak{h}}$ can be identified with an intrinsic tensor field of class C^{s-1}, *i.e.* $\mathfrak{D}_{\mathfrak{h}} \in \mathscr{I}_{\mathscr{B}}^{s-1}$. Since $\mathfrak{D}_{\mathfrak{h}}$ depends $\mathscr{F}$-linearly on $\mathfrak{h}$, $\mathfrak{D}$ can be regarded as an $\mathscr{F}$-linear mapping

$$\mathfrak{D}: \mathscr{T}_{\mathscr{B}}^{s-1} \to \mathscr{I}_{\mathscr{B}}^{s-1}, \quad 1 \leqq s \leqq r, \tag{9.20}$$

and hence can be identified with a field of class C^{r-1} whose values $\mathfrak{D}(X)$ are linear transformations from $\mathscr{T}_X$ into $\mathscr{I}_X$. The possibility of identifying $\mathfrak{D}_{\mathfrak{h}}$ and $\mathfrak{D}$ with fields on $\mathscr{B}$ follows from analogues of Proposition 1 (Sect. 7).

Let $\mathfrak{S}$ and $\overset{*}{\mathfrak{S}}$ denote the torsions and $\mathfrak{R}$ and $\overset{*}{\mathfrak{R}}$ the curvatures of Γ and $\overset{*}{\Gamma}$, respectively. If we write the definition (9.11) of the torsion for both Γ and $\overset{*}{\Gamma}$ and take the difference, we obtain

$$\mathfrak{S}(\mathfrak{h}, \mathfrak{k}) - \overset{*}{\mathfrak{S}}(\mathfrak{h}, \mathfrak{k}) = \mathfrak{D}_{\mathfrak{h}} \mathfrak{k} - \mathfrak{D}_{\mathfrak{k}} \mathfrak{h}. \tag{9.21}$$

If we write the definition (9.15) of the curvature for $\overset{*}{\Gamma}$ and substitute $\overset{*}{\Gamma}_{\mathfrak{h}} = \Gamma_{\mathfrak{h}} - \mathfrak{D}_{\mathfrak{h}}$, we find

$$\overset{*}{\mathfrak{R}}(\mathfrak{h}, \mathfrak{k}) = \mathfrak{R}(\mathfrak{h}, \mathfrak{k}) - [\Gamma_{\mathfrak{h}}, \mathfrak{D}_{\mathfrak{k}}] - [\mathfrak{D}_{\mathfrak{h}}, \Gamma_{\mathfrak{k}}] + [\mathfrak{D}_{\mathfrak{h}}, \mathfrak{D}_{\mathfrak{k}}] + \mathfrak{D}_{[\mathfrak{h}, \mathfrak{k}]}. \tag{9.22}$$

10. Material Connections

Let Φ be a material uniformity for the simple body $\mathscr{B}$ of class C^p. (See Definition 3, Sect. 6.) We say that a tangent vector field $\mathfrak{c}$ is *materially constant* if

$$\mathfrak{c}(X) = \Phi(X, Y)\, \mathfrak{c}(Y) \tag{10.1}$$

holds for all $X, Y \in \mathscr{B}$. If $X_0 \in \mathscr{B}$ is fixed and $\mathfrak{u}_{X_0} \in \mathscr{T}_{X_0}$ is prescribed arbitrarily, then

$$\mathfrak{c}(X) = \Phi(X, X_0)\, \mathfrak{u}_{X_0} \tag{10.2}$$

is easily seen to define a materially constant field $\mathfrak{c}$ such that $\mathfrak{c}(X_0) = \mathfrak{u}_{X_0}$. Moreover, every materially constant field $\mathfrak{c}$ can be obtained in this fashion. Thus (10.2) describes a one-to-one correspondence between $\mathscr{T}_{X_0}$ and the set $\mathscr{T}_{\Phi}$ of all materially constant vector fields. This correspondence is actually a vector-space isomorphism, showing that $\mathscr{T}_{\Phi}$ is a three-dimensional vector-space when addition and multiplication with scalars in $\mathscr{T}_{\Phi}$ are defined pointwise.

Let $\boldsymbol{K}$ be a uniform reference (see Definition 4, Sect. 6) and $\mathfrak{c} \in \mathscr{T}_{\Phi}$. Then it follows from (10.1) and (6.4) that $\boldsymbol{K}(X)\, \mathfrak{c}(X) = \boldsymbol{K}(Y)\, \mathfrak{c}(Y)$ for all $X, Y \in \mathscr{B}$, *i.e.* that

$$\boldsymbol{K} \mathfrak{c} = \boldsymbol{c} = \text{constant}. \tag{10.3}$$

Conversely, if $c \in \mathscr{V}$, then $\mathfrak{c} = \boldsymbol{K}^{-1} c$ is easily seen to be materially constant. Thus $\mathscr{T}_\Phi$ is exactly the set of all tangent vector fields $\mathfrak{c}$ with the property (10.3).

A material uniformity Φ is said to be *of class* C^r, $r \leqq p-1$, if $\mathscr{T}_\Phi \subset \mathscr{T}_{\mathscr{B}}^r$, *i.e.*, if all tangent vector fields materially constant with respect to Φ are of class C^r. For the remainder of this paper we lay down the following:

Smoothness assumption[9]: *$\mathscr{B}$ is a materially uniform continuous body of class C^p, $p \geqq 3$*[10], *which admits a material uniformity Φ of class C^{p-1}.*

Let Φ be a material uniformity of class C^{p-1} and let $\boldsymbol{K}$ be a uniform reference such that (6.4) holds. Then for $c \in \mathscr{V}$, $\kappa \in \mathsf{C}$, and $\mathfrak{c} = \boldsymbol{K}^{-1} c \in \mathscr{T}_\Phi$, the vector field $((\nabla \kappa) \boldsymbol{K}^{-1})\, c = (\nabla \kappa)\, \mathfrak{c}$ is of class C^{p-1} because $\mathfrak{c}$ is of class C^{p-1}. This is possible for all $c \in \mathscr{V}$ only if $(\nabla \kappa) \boldsymbol{K}$ and hence $\boldsymbol{K}$ is of class C^{p-1}. Thus, if $\boldsymbol{K}$ is a uniform reference such that

$$\Phi(X, Y) = \boldsymbol{K}(X)^{-1} \boldsymbol{K}(Y), \qquad X, Y \in \mathscr{B}, \tag{10.4}$$

then $\boldsymbol{K}$ is of class C^{p-1}.

Theorem 3. *Given a material uniformity Φ of class C^{p-1}, there is a unique affine connection Γ such that $\Gamma \mathfrak{c} = 0$ holds for all material constant tangent vector fields $\mathfrak{c} \in \mathscr{T}_\Phi$. In terms of any uniform reference $\boldsymbol{K}$ satisfying* (10.4), *Γ is given by*

$$\Gamma \mathfrak{h} = \boldsymbol{K}^{-1} \nabla_{\boldsymbol{K}}(\boldsymbol{K} \mathfrak{h}) \boldsymbol{K}, \qquad \mathfrak{h} \in \mathscr{T}_{\mathscr{B}}^1. \tag{10.5}$$

Also, Γ is of class C^{p-2}.

Proof. To prove the uniqueness, assume that Γ and $\bar{\Gamma}$ are connections such that $\Gamma \mathfrak{c} = \bar{\Gamma} \mathfrak{c}$ for all $\mathfrak{c} \in \mathscr{T}_\Phi$. Putting $\mathfrak{D}_\mathfrak{h} = \Gamma_\mathfrak{h} - \bar{\Gamma}_\mathfrak{h}$, we then have $\mathfrak{D}_\mathfrak{h} \mathfrak{c} = 0$ for all $\mathfrak{c} \in \mathscr{T}_\Phi$. We have seen at the end of the previous section that $\mathfrak{D}_\mathfrak{h}$ can be identified with an intrinsic tensor field in $\mathscr{I}_{\mathscr{B}}^0$ when $\mathfrak{h} \in \mathscr{T}_{\mathscr{B}}^0$. Hence $\mathfrak{D}_\mathfrak{h}(X_0) \mathfrak{c}(X_0) = 0$ for all $X_0 \in \mathscr{B}$ and all $\mathfrak{c} \in \mathscr{T}_\Phi$. Since for any prescribed $\mathfrak{u}_{X_0} \in \mathscr{T}_{X_0}$ the $\mathfrak{c} \in \mathscr{T}_\Phi$ given by (10.2) has the property $\mathfrak{c}(X_0) = \mathfrak{u}_{X_0}$, it follows that $\mathfrak{D}_\mathfrak{h}(X_0)\, \mathfrak{u}_{X_0} = 0$ for all $\mathfrak{u}_{X_0} \in \mathscr{T}_{X_0}$, *i.e.*, that $\mathfrak{D}_\mathfrak{h}(X_0) = 0$. Since $X_0 \in \mathscr{B}$ is arbitrary, we infer that $\mathfrak{D}_\mathfrak{h} = 0$, *i.e.* that $\Gamma = \bar{\Gamma}$.

To prove the existence of Γ we choose a uniform reference $\boldsymbol{K}$ with the property (10.4), define Γ by (10.5), and show that it has all the necessary properties. It is clear that $\Gamma \mathfrak{c} = 0$ when $\mathfrak{c} \in \mathscr{T}_\Phi$ because, by (10.3), $\nabla_{\boldsymbol{K}}(\boldsymbol{K} \mathfrak{c}) = 0$ when $\mathfrak{c} \in \mathscr{T}_\Phi$. Since $\boldsymbol{K}$ is of class C^{p-1} it follows that $\Gamma \mathfrak{h}$ is of class C^{p-2} when $\mathfrak{h} \in \mathscr{T}_{\mathscr{B}}^{p-1}$. The validity of the rules (9.2) and (9.3) follows from the validity of the analogous rules for the relative gradient $\nabla_{\boldsymbol{K}}$. Hence Γ is indeed an affine connection of class C^{p-2}. Q.E.D.

Definition 5. *The affine connection (of class C^{p-2}) with the property $\Gamma \mathfrak{c} = 0$ for all $\mathfrak{c} \in \mathscr{T}_\Phi$ is called the* ***material connection*** *for the material uniformity Φ (of class C^{p-1}).*

Theorem 4. *Material connections have zero Riemann-curvature.*

Proof. Let $X \in \mathscr{B}$ and $\mathfrak{u}_X \in \mathscr{T}_X$ be given. We can determine $\mathfrak{c} \in \mathscr{T}_\Phi \subset \mathscr{T}_{\mathscr{B}}^{p-1}$ such that $\mathfrak{c}(X) = \mathfrak{u}_X$. If Γ is the material connection for Φ we have $\Gamma_\mathfrak{h} \mathfrak{c} = 0$ for all $\mathfrak{h} \in \mathscr{T}_{\mathscr{B}}^0$. Hence the definition (9.15) shows that $\mathfrak{R}(\mathfrak{h}, \mathfrak{k})\, \mathfrak{c} = 0$ for all $\mathfrak{h}, \mathfrak{k} \in \mathscr{T}_{\mathscr{B}}^{p-1}$. Since $\mathfrak{R}(\mathfrak{h}, \mathfrak{k})$

[9] C. C. Wang [8] has recently shown that the theory given here can be extended to the case when each point has a neighborhood that admits a smooth materially uniformity. This can happen even when all material uniformities for the whole body are discontinuous.

[10] For all considerations not referring to curvature, $p \geqq 2$ is actually sufficient.

can be identified with an intrinsic tensor field it follows that

$$\mathfrak{R}(\mathfrak{h},\mathfrak{k})\,\mathfrak{c}|_X=\mathfrak{R}(\mathfrak{h},\mathfrak{k})|_X\,\mathfrak{c}(X)=\mathfrak{R}(\mathfrak{h},\mathfrak{k})|_X\,\mathfrak{u}_X=0\,.$$

This can be valid for all $X\in\mathscr{B}$ and all $\mathfrak{u}_X\in\mathscr{T}_X$ only if $\mathfrak{R}(\mathfrak{h},\mathfrak{k})=0$. Hence, since $\mathfrak{h},\mathfrak{k}\in\mathscr{T}_{\mathscr{B}}^{p-1}$ are arbitrary, we must have $\mathfrak{R}=0$. Q.E.D.

11. Inhomogeneity

Let Φ be a material uniformity of class C^{p-1}, let Γ be the associated material connection (of class C^{p-2}) with torsion $\mathfrak{S}$, and let $\boldsymbol{K}$ be a uniform reference (of class C^{p-1}) such that (10.4) holds. We can define a field $\boldsymbol{S}$ of class C^{p-2} with values $\boldsymbol{S}(X)$: $\mathscr{V}\to\mathscr{L}$ by the condition

$$(\boldsymbol{S}\,\boldsymbol{u})\,\boldsymbol{v}=\boldsymbol{K}\,\mathfrak{S}(\boldsymbol{K}^{-1}\boldsymbol{u},\boldsymbol{K}^{-1}\boldsymbol{v}) \tag{11.1}$$

for all $\boldsymbol{u}$, $\boldsymbol{v}\in\mathscr{V}$. In view of the linearity of the values $\boldsymbol{S}(X)$, $\mathfrak{S}(X)$, and $\boldsymbol{K}(X)$, (11.1) continues to hold if the fixed vectors $\boldsymbol{u}$ and $\boldsymbol{v}$ in (11.1) are replaced by vector fields $\boldsymbol{h}$ and $\boldsymbol{k}$. The following theorem shows how $\boldsymbol{S}$ and hence $\mathfrak{S}$ can be expressed directly in terms of $\boldsymbol{K}$:

Theorem 5. *Let γ be an arbitrary configuration of $\mathscr{B}$ and*

$$\boldsymbol{F}=(\nabla\gamma)\,\boldsymbol{K}^{-1}\in\mathscr{L}_{\mathscr{B}}^{p-1}\,. \tag{11.2}$$

Then $\boldsymbol{S}$ is given by

$$(\boldsymbol{S}\,\boldsymbol{u})\,\boldsymbol{v}=\boldsymbol{F}^{-1}[((\nabla_{\boldsymbol{K}}\boldsymbol{F})\,\boldsymbol{v})\,\boldsymbol{u}-((\nabla_{\boldsymbol{K}}\boldsymbol{F})\,\boldsymbol{u})\,\boldsymbol{v}]\,, \tag{11.3}$$

$$\boldsymbol{S}\,\boldsymbol{u}=\boldsymbol{F}^{-1}[\nabla_{\boldsymbol{K}}(\boldsymbol{F}\,\boldsymbol{u})-(\nabla_{\boldsymbol{K}}\boldsymbol{F})\,\boldsymbol{u}]\,, \tag{11.4}$$

or

$$(\boldsymbol{S}\,\boldsymbol{u})\,\boldsymbol{v}=((\nabla_{\gamma}\boldsymbol{F}^{-1})\,\boldsymbol{h})\,\boldsymbol{k}-((\nabla_{\gamma}\boldsymbol{F}^{-1})\,\boldsymbol{k})\,\boldsymbol{h}\,, \tag{11.5}$$

where $\boldsymbol{u}$, $\boldsymbol{v}\in\mathscr{V}$ and $\boldsymbol{h}=\boldsymbol{F}\,\boldsymbol{u}$, $\boldsymbol{k}=\boldsymbol{F}\,\boldsymbol{v}$.

Proof. The tangent vector fields $\boldsymbol{K}^{-1}\boldsymbol{u}$ and $\boldsymbol{K}^{-1}\boldsymbol{v}$ are materially constant and hence are annihilated by Γ. Hence, the definitions (11.1) and (9.11) of $\boldsymbol{S}$ and $\mathfrak{S}$ yield

$$(\boldsymbol{S}\,\boldsymbol{u})\,\boldsymbol{v}=-\boldsymbol{K}\,[\boldsymbol{K}^{-1}\boldsymbol{u},\boldsymbol{K}^{-1}\boldsymbol{v}]=-\boldsymbol{K}\,[(\nabla\gamma)^{-1}\boldsymbol{h},(\nabla\gamma)^{-1}\boldsymbol{k}]\,. \tag{11.6}$$

Using Proposition 4, (8.15), we find

$$(\boldsymbol{S}\,\boldsymbol{u})\,\boldsymbol{v}=\boldsymbol{F}^{-1}[(\nabla_{\gamma}\boldsymbol{h})\,\boldsymbol{k}-(\nabla_{\gamma}\boldsymbol{k})\,\boldsymbol{h}]\,. \tag{11.7}$$

The formula (8.8), with the choices $\psi=\boldsymbol{h}$ and $\boldsymbol{G}=\nabla\gamma$, gives

$$(\nabla_{\gamma}\boldsymbol{h})\,\boldsymbol{k}=(\nabla_{\gamma}\boldsymbol{h})\,\boldsymbol{F}\,\boldsymbol{v}=(\nabla_{\boldsymbol{K}}\boldsymbol{h})\,\boldsymbol{v}=\nabla_{\boldsymbol{K}}(\boldsymbol{F}\,\boldsymbol{u})\,\boldsymbol{v}=((\nabla_{\boldsymbol{K}}\boldsymbol{F})\,\boldsymbol{v})\,\boldsymbol{u}\,. \tag{11.8}$$

Substituting (11.8) and the formula obtained from (11.8) by interchanging $\boldsymbol{h}$ and $\boldsymbol{k}$ into (11.7), we obtain (11.3) and (11.4).

To prove (11.5) we note that $\nabla_{\gamma}\boldsymbol{u}=\boldsymbol{0}$ for constant $\boldsymbol{u}\in\mathscr{V}$. Using one of the product rules for gradient operators we find

$$\boldsymbol{0}=(\nabla_{\gamma}\boldsymbol{u})\,\boldsymbol{k}=\nabla_{\gamma}(\boldsymbol{F}^{-1}\boldsymbol{h})\,\boldsymbol{k}=((\nabla_{\gamma}\boldsymbol{F}^{-1})\,\boldsymbol{k})\,\boldsymbol{h}+\boldsymbol{F}^{-1}(\nabla_{\gamma}\boldsymbol{h})\,\boldsymbol{k}\,. \tag{11.9}$$

Of course, (11.9) remains valid if we interchange $\boldsymbol{h}$ and $\boldsymbol{k}$. The formula (11.5) follows from (11.7) and (11.9). Q.E.D.

Recall that the body $\mathscr{B}$ is homogeneous if it admits a configuration gradient $\boldsymbol{K}=\nabla\boldsymbol{\kappa}$ as a uniform reference. We can then choose $\gamma=\boldsymbol{\kappa}$ in (11.2), obtaining $\boldsymbol{F}=\mathbf{1}$, which is constant and hence has gradient zero. Thus, Theorem 5 shows that $\boldsymbol{S}=\mathbf{0}$ and hence $\mathfrak{S}=0$ for suitable uniform references if the body is homogeneous. The converse of this result is not true, but it becomes true if "homogeneous" is replaced by "locally homogeneous" in a sense we shall now make precise.

Let $\mathscr{N}$ be an open subset of a simple body $\mathscr{B}$ of class C^p. We can give $\mathscr{N}$ the structure of a continuous body of class C^p by letting $\gamma\colon\mathscr{N}\to\mathscr{E}$ be a configuration of $\mathscr{N}$ if $\boldsymbol{\kappa}\circ\overset{-1}{\gamma}:\gamma(\mathscr{N})\to\mathscr{E}$ is of class C^p for all configurations $\boldsymbol{\kappa}\in\mathsf{C}$ of $\mathscr{B}$. We denote the set of all configurations of $\mathscr{N}$ by $\mathsf{C}_{\mathscr{N}}$. If $\boldsymbol{\kappa}\in\mathsf{C}$, then the restriction of $\boldsymbol{\kappa}$ to $\mathscr{N}$ belongs to $\mathsf{C}_{\mathscr{N}}$. However, not all configurations $\gamma\in\mathsf{C}_{\mathscr{N}}$ of $\mathscr{N}$ can be obtained in this manner. Still, given any $X\in\mathscr{N}$ and any configuration γ of $\mathscr{N}$, one can easily construct a configuration $\boldsymbol{\kappa}$ of $\mathscr{B}$ such that (3.1) holds. Therefore, an equivalence class $\boldsymbol{K}_X$ which defines a local configuration at X relative to $\mathscr{N}$ can be made to correspond to the non-empty set $\{\boldsymbol{\kappa}\in\mathsf{C}\,|\,\nabla(\boldsymbol{\kappa}\circ\overset{-1}{\gamma})|_{\gamma(X)}=\mathbf{1}$ for all $\gamma\in\boldsymbol{K}_X\}$, which is a local configuration at X relative to $\mathscr{B}$. This correspondence is one-to-one and can be used to identify local configurations at X relative to $\mathscr{N}$ with local configurations at X relative to $\mathscr{B}$. Using this identification, we can endow $\mathscr{N}$ with the structure of a simple body by using the restriction to $\mathscr{N}$ of the function $\mathfrak{G}$ which defines the simple body structure on $\mathscr{B}$ according to Definition 1 (Sect. 5). Thus, every open subset $\mathscr{N}$ of $\mathscr{B}$ has a natural structure of a simple body of class C^p, *i.e.*, every open subset $\mathscr{N}$ of $\mathscr{B}$ can be regarded as a simple body of class C^p. Such a subset is called a *neighborhood* of a material point if it contains that point.

A simple body $\mathscr{B}$ is called *locally homogeneous* if every $X\in\mathscr{B}$ has a neighborhood $\mathscr{N}$ that is homogeneous. A body can be locally homogeneous without being homogeneous, even if it is simply connected.

Definition 6. *The Cartan torsion $\mathfrak{S}$ of the material connection Γ associated with a material uniformity Φ of class C^{p-1} is called the* ***inhomogeneity*** *of Φ.*

The field $\boldsymbol{S}$ defined by (11.1) *is called the* ***inhomogeneity***[11] *relative to the reference $\boldsymbol{K}$.*

This definition finds its motivation in the result already mentioned:

Theorem 6[12]. *If $\mathscr{B}$ is homogeneous, then it admits a material uniformity of class C^{p-1} with zero inhomogeneity. If $\mathscr{B}$ admits a material uniformity of class C^{p-1} with zero inhomogeneity, then it is locally homogeneous.*

Proof. Only the second part of the theorem remains to be proved. Assume, therefore, that Φ is a material uniformity of class C^{p-1} with zero inhomogeneity. Using the same notation as before, we then have $\mathfrak{S}=0$ and hence $\boldsymbol{S}=\mathbf{0}$. Theorem 5, (11.5), shows that if $\gamma\in\mathsf{C}$ is arbitrary and $\boldsymbol{F}$ defined by (11.2), $\nabla_\gamma\boldsymbol{F}^{-1}$ has the symmetry property

$$((\nabla_\gamma\boldsymbol{F}^{-1})\,\boldsymbol{u})\,\boldsymbol{v}=((\nabla_\gamma\boldsymbol{F}^{-1})\,\boldsymbol{v})\,\boldsymbol{u} \tag{11.10}$$

[11] It corresponds to what is called "dislocation density" in the theory of continuous distributions of dislocations (*cf.* [4]).

[12] The theorem stated in the middle of p. 90 in reference [2] is incorrect and should be replaced by Theorem 6.

for all $\boldsymbol{u}, \boldsymbol{v} \in \mathscr{V}$. Let $X \in \mathscr{B}$ be given and let $\mathscr{N}'$ be a simply connected neighborhood of $\mathscr{B}$. By a classical theorem of analysis, the symmetry (11.10) implies the existence of a mapping $\boldsymbol{\lambda}: \gamma(\mathscr{N}') \to \mathscr{E}$ such that $\boldsymbol{F}^{-1} = (\nabla \boldsymbol{\lambda}) \circ \overset{-1}{\gamma}$ holds in $\mathscr{N}'$. Moreover, $\boldsymbol{K}$, ∇_γ and hence $\boldsymbol{F}$ and $\boldsymbol{F}^{-1}$ being of class C^{p-1}, $\boldsymbol{\lambda}$ is of class C^p. Since $\nabla \boldsymbol{\lambda} = \boldsymbol{F}^{-1} \circ \gamma$ is invertible, it follows by the inverse function theorem that $\boldsymbol{\lambda}$ is locally (but not necessarily globally) invertible, *i.e.* that X has a neighborhood $\mathscr{N} \subset \mathscr{N}'$ on which $\boldsymbol{\lambda}$ is invertible. The mapping $\boldsymbol{\kappa} = \boldsymbol{\lambda} \circ \gamma$, when restricted to $\mathscr{N}$, is therefore a configuration of $\mathscr{N}$ with gradient $\nabla \boldsymbol{\kappa} = ((\nabla \boldsymbol{\lambda}) \circ \overset{-1}{\gamma}) = \boldsymbol{F}^{-1} \, \nabla \gamma = \boldsymbol{K}$ on $\mathscr{N}$. Hence the uniform reference $\boldsymbol{K}$ on $\mathscr{N}$ is the gradient of the configuration $\boldsymbol{\kappa}$ of $\mathscr{N}$, *i.e.*, $\mathscr{N}$ is homogeneous. Q.E.D.

The first Bianchi identity gives rise to the following identity for the relative inhomogeneity $\boldsymbol{S}$ and its gradient $\nabla_{\boldsymbol{K}} \boldsymbol{S}$ relative to $\boldsymbol{K}$:

$$\sum [(((\nabla_{\boldsymbol{K}} \boldsymbol{S}) \boldsymbol{u}) \boldsymbol{v}) \boldsymbol{w}) - (\boldsymbol{S} \boldsymbol{u})(\boldsymbol{S} \boldsymbol{v}) \boldsymbol{w}] = \boldsymbol{0}. \tag{11.11}$$

To prove (11.11), substitute $\mathfrak{l} = \boldsymbol{K}^{-1} \boldsymbol{u}$, $\mathfrak{k} = \boldsymbol{K}^{-1} \boldsymbol{w}$, $\mathfrak{h} = \boldsymbol{K}^{-1} \boldsymbol{v}$ into (9.18), observe that $\mathfrak{R} = 0$ (Theorem 4, Sect. 10), and make use of $(11.6)_1$.

12. Relative Riemannian Structures, Contortion

Let $\boldsymbol{K}$ be a uniform reference of class C^{p-1}. If we choose $\boldsymbol{K}_X = \boldsymbol{K}(X)$ in (4.5), then this equation defines an inner product $*$ on each of the tangent spaces $\mathscr{T}_X$, $X \in \mathscr{B}$. The structure on $\mathscr{B}$ defined by these inner products will be called the *Riemannian structure of $\mathscr{B}$ relative to the uniform reference $\boldsymbol{K}$*. If $\mathfrak{h}$ and $\mathfrak{k}$ are tangent vector fields, we define $\mathfrak{h} * \mathfrak{k}$ pointwise. For such fields, (4.5) then yields

$$\mathfrak{h} * \mathfrak{k} = (\boldsymbol{K} \mathfrak{h}) \cdot (\boldsymbol{K} \mathfrak{k}). \tag{12.1}$$

It is clear that $\mathfrak{h} * \mathfrak{k} \in \mathscr{F}_{\mathscr{B}}^r$ if $\mathfrak{h}, \mathfrak{k} \in \mathscr{T}_{\mathscr{B}}^r$ for $0 \leqq r \leqq p-1$. This fact is expressed by saying that the Riemannian structure relative to $\boldsymbol{K}$ is *of class C^{p-1}*.

Although the following proposition is one of the basic facts of Riemannian geometry, we shall give an independent proof:

Proposition 6. *There is a unique affine connection $\overset{*}{\Gamma}$ of class C^{p-2} with the following properties:*

(a) *The torsion $\overset{*}{\mathfrak{S}}$ of $\overset{*}{\Gamma}$ vanishes.*

(b) *For any $\mathfrak{h}, \mathfrak{k}, \mathfrak{l} \in \mathscr{T}_{\mathscr{B}}^{r-1}$ the relation*

$$\mathfrak{h}(\mathfrak{k} * \mathfrak{l}) = \mathfrak{k} * \overset{*}{\Gamma}_{\mathfrak{h}} \mathfrak{l} + \mathfrak{l} * \overset{*}{\Gamma}_{\mathfrak{h}} \mathfrak{k} \tag{12.2}$$

is valid.

Proof. First we assume the existence of $\overset{*}{\Gamma}$. Let Γ be the material connection associated with $\boldsymbol{K}$ and consider the difference

$$\mathfrak{D}_{\mathfrak{h}} = \Gamma_{\mathfrak{h}} - \overset{*}{\Gamma}_{\mathfrak{h}}, \quad \mathfrak{h} \in \mathscr{T}_{\mathscr{B}}^{p-1}. \tag{12.3}$$

According to the results given at the end of Sect. 9, $\mathfrak{D}$ can be identified with a field of class C^{p-2} on $\mathscr{B}$ whose values are linear transformations from $\mathscr{T}_X$ into $\mathscr{I}_X$.

Therefore, we can define a field $\boldsymbol{D}$ on $\mathscr{B}$ of class C^{p-2} with values $\boldsymbol{D}(X): \mathscr{V} \to \mathscr{L}$ by the condition

$$\boldsymbol{D}\,\boldsymbol{u} = \boldsymbol{K}\,\mathfrak{D}_{\boldsymbol{K}^{-1}\boldsymbol{u}}\,\boldsymbol{K}^{-1} \tag{12.4}$$

for all $\boldsymbol{u} \in \mathscr{V}$. Since $\overset{*}{\mathfrak{S}} = 0$, the relation (9.21) and the definitions (12.4) and (11.1) give

$$(\boldsymbol{S}\,\boldsymbol{u})\,\boldsymbol{v} = (\boldsymbol{D}\,\boldsymbol{u})\,\boldsymbol{v} - (\boldsymbol{D}\,\boldsymbol{v})\,\boldsymbol{u}, \qquad \boldsymbol{u}, \boldsymbol{v} \in \mathscr{V}. \tag{12.5}$$

By the Definition 5 (Sect. 10), we have $\Gamma_{\mathfrak{h}}\,\mathfrak{c} = 0$ whenever $\mathfrak{c}$ is materially constant. Hence, since $\mathfrak{c} = \boldsymbol{K}^{-1}\boldsymbol{c}$ is materially constant when $\boldsymbol{c} \in \mathscr{V}$, we infer from (12.3) and (12.4) that

$$\boldsymbol{K}\,\overset{*}{\Gamma}_{\mathfrak{h}}(\boldsymbol{K}^{-1}\,\boldsymbol{c}) = -(\boldsymbol{D}\,\boldsymbol{K}\,\mathfrak{h})\,\boldsymbol{c} \tag{12.6}$$

when $\boldsymbol{c} \in \mathscr{V}$. Now let $\boldsymbol{u}, \boldsymbol{v}, \boldsymbol{w} \in \mathscr{V}$. If we substitute $\mathfrak{h} = \boldsymbol{K}^{-1}\boldsymbol{u}$, $\mathfrak{k} = \boldsymbol{K}^{-1}\boldsymbol{v}$, and $\mathfrak{l} = \boldsymbol{K}^{-1}\boldsymbol{w}$ into (12.2) and observe (12.6) and (12.1) we obtain

$$\mathfrak{h}(\boldsymbol{v} \cdot \boldsymbol{w}) = -\boldsymbol{v} \cdot (\boldsymbol{D}\,\boldsymbol{u})\,\boldsymbol{w} + \boldsymbol{w} \cdot (\boldsymbol{D}\,\boldsymbol{u})\,\boldsymbol{v}. \tag{12.7}$$

Since $\boldsymbol{v} \cdot \boldsymbol{w}$ is constant, we have $\mathfrak{h}(\boldsymbol{v} \cdot \boldsymbol{w}) = 0$ (see the definition (8.10)). Therefore (12.7) states that $\boldsymbol{D}\,\boldsymbol{u} \in \mathscr{L}$ is skew for all $\boldsymbol{u} \in \mathscr{V}$:

$$\boldsymbol{D}\,\boldsymbol{u} = -(\boldsymbol{D}\,\boldsymbol{u})^T, \qquad \boldsymbol{u} \in \mathscr{V}. \tag{12.8}$$

The equations (12.5) and (12.8) enable us to express $\boldsymbol{D}$ in terms of $\boldsymbol{S}$. Indeed, if we take the inner product of (12.5) with $\boldsymbol{w} \in \mathscr{V}$, subtract from the resulting equation the two equations obtained from it by cyclic permutations of $\boldsymbol{u}$, $\boldsymbol{v}$, $\boldsymbol{w}$, and observe (12.8), we find

$$2\,\boldsymbol{u} \cdot (\boldsymbol{D}\,\boldsymbol{w})\,\boldsymbol{v} = \boldsymbol{w} \cdot (\boldsymbol{S}\,\boldsymbol{u})\,\boldsymbol{v} - \boldsymbol{u} \cdot (\boldsymbol{S}\,\boldsymbol{v})\,\boldsymbol{w} - \boldsymbol{v} \cdot (\boldsymbol{S}\,\boldsymbol{w})\,\boldsymbol{u}. \tag{12.9}$$

Since $(\boldsymbol{S}\,\boldsymbol{u})\,\boldsymbol{v} = -(\boldsymbol{S}\,\boldsymbol{v})\,\boldsymbol{u}$, (12.9) is equivalent to

$$(\boldsymbol{D}\,\boldsymbol{u})\,\boldsymbol{v} = \tfrac{1}{2}\{[(\boldsymbol{S}\,\boldsymbol{u}) - (\boldsymbol{S}\,\boldsymbol{u})^T]\,\boldsymbol{v} - (\boldsymbol{S}\,\boldsymbol{v})^T\,\boldsymbol{w}\}. \tag{12.10}$$

Now, since $\boldsymbol{S}$ is determined by the uniform reference $\boldsymbol{K}$, it follows from (12.10), (12.4), and (12.3) that $\overset{*}{\Gamma}$ is uniquely determined by $\boldsymbol{K}$.

To prove the existence of a connection $\overset{*}{\Gamma}$ with the properties (a) and (b), one can define $\overset{*}{\Gamma}$ by (12.3), (12.4), and (12.10) and verify that it has all the required properties. Q.E.D.

Definition 7. *The connection $\overset{*}{\Gamma}$ determined by the conditions* (a) *and* (b) *of Proposition 6 is called the* ***Riemannian connection*** *relative to the uniform reference* $\boldsymbol{K}$. *The field* $\boldsymbol{D}$ *determined by* (12.3) *and* (12.4) *or* (12.5) *and* (12.8) *is called the* ***contortion***[13] *of* $\boldsymbol{K}$.

The term "contortion" will be motivated in Section 13.

[13] It corresponds to what is called "Cosserat structure curvature" or "Nye curvature" in the theory of continuous distributions of dislocations (*cf.* [4]).

In general, the curvature $\overset{*}{\mathfrak{R}}$ of the Riemannian connection is not zero. We can define a field $\overset{*}{\boldsymbol{R}}$ with values $\overset{*}{\boldsymbol{R}}(X)\colon \mathscr{V}\times\mathscr{V}\to\mathscr{L}$ by the condition

$$\overset{*}{\boldsymbol{R}}(\boldsymbol{u},\boldsymbol{v})=\boldsymbol{K}\,\overset{*}{\mathfrak{R}}(\boldsymbol{K}^{-1}\boldsymbol{u},\boldsymbol{K}^{-1}\boldsymbol{v})\,\boldsymbol{K}^{-1} \tag{12.11}$$

for all $\boldsymbol{u},\boldsymbol{v}\in\mathscr{V}$. Let $\boldsymbol{u},\boldsymbol{v},\boldsymbol{w}\in\mathscr{V}$ and put $\mathfrak{h}=\boldsymbol{K}^{-1}\boldsymbol{u}$, $\mathfrak{k}=\boldsymbol{K}^{-1}\boldsymbol{v}$, $\mathfrak{l}=\boldsymbol{K}^{-1}\boldsymbol{w}$. The fields $\mathfrak{h},\mathfrak{k},\mathfrak{l}$ are then materially constant and hence are annihilated by Γ. Recalling that the curvature $\mathfrak{R}$ of Γ vanishes (Theorem 4, Sect. 10), we then infer from (9.22), (10.5), $(11.6)_1$, (12.4), and (12.11) that

$$\begin{aligned}\overset{*}{\boldsymbol{R}}(\boldsymbol{u},\boldsymbol{v})\,\boldsymbol{w}=&-\big(\nabla_{\boldsymbol{K}}(\boldsymbol{D}\,\boldsymbol{v})\,\boldsymbol{u}\big)\,\boldsymbol{w}+\big(\nabla_{\boldsymbol{K}}(\boldsymbol{D}\,\boldsymbol{u})\,\boldsymbol{v}\big)\,\boldsymbol{w}\\&+(\boldsymbol{D}\,\boldsymbol{u})(\boldsymbol{D}\,\boldsymbol{v})\,\boldsymbol{w}-(\boldsymbol{D}\,\boldsymbol{v})(\boldsymbol{D}\,\boldsymbol{u})\,\boldsymbol{w}-\boldsymbol{D}\big((\boldsymbol{S}\,\boldsymbol{u})\,\boldsymbol{v}\big)\,\boldsymbol{w}\end{aligned}$$

and hence

$$\overset{*}{\boldsymbol{R}}(\boldsymbol{u},\boldsymbol{v})=\big((\nabla_{\boldsymbol{K}}\boldsymbol{D})\,\boldsymbol{v}\big)\,\boldsymbol{u}-\big((\nabla_{\boldsymbol{K}}\boldsymbol{D})\,\boldsymbol{u}\big)\,\boldsymbol{v}+(\boldsymbol{D}\,\boldsymbol{u})(\boldsymbol{D}\,\boldsymbol{v})-(\boldsymbol{D}\,\boldsymbol{v})(\boldsymbol{D}\,\boldsymbol{u})-\boldsymbol{D}\big((\boldsymbol{S}\,\boldsymbol{u})\,\boldsymbol{v}\big). \tag{12.12}$$

In view of (12.5) and (12.10), equation (12.12) shows that $\overset{*}{\boldsymbol{R}}$ can be expressed in terms of the contortion $\boldsymbol{D}$ and its gradient relative to $\boldsymbol{K}$ or in terms of the inhomogeneity $\boldsymbol{S}$ and its gradient relative to $\boldsymbol{K}$.

13. Contorted Aeolotropy

Definition 8. *A uniform reference $\boldsymbol{K}$ of class C^{p-1} is called a* ***state of contorted aeolotropy*** *if there exists a configuration $\boldsymbol{\kappa}$ such that the tensor field*

$$\boldsymbol{Q}=(\nabla\boldsymbol{\kappa})\,\boldsymbol{K}^{-1}\in\mathscr{L}_{\mathscr{B}}^{p-1} \tag{13.1}$$

has orthogonal values ($\boldsymbol{Q}(X)\in\mathscr{o}$ for all $X\in\mathscr{B}$).

Assume that $\boldsymbol{K}$ is such a state of contorted aeolotropy. Since the inner product in $\mathscr{V}$ is preserved under orthogonal transformations, the Riemannian structure (12.1) relative to $\boldsymbol{K}$ satisfies.

$$\mathfrak{h}*\mathfrak{k}=\boldsymbol{Q}(\boldsymbol{K}\,\mathfrak{h})\cdot\boldsymbol{Q}(\boldsymbol{K}\,\mathfrak{k})=(\nabla\boldsymbol{\kappa})\,\mathfrak{h}\cdot(\nabla\boldsymbol{\kappa})\,\mathfrak{k} \tag{13.2}$$

for all tangent vector fields $\mathfrak{h}$ and $\mathfrak{k}$. It follows from (13.2) that the Riemannian connection $\overset{*}{\Gamma}$ relative to $\boldsymbol{K}$ is obtained by transporting the gradient operator ∇ from $\boldsymbol{\kappa}(\mathscr{B})$ into $\mathscr{B}$ via $(\nabla\boldsymbol{\kappa})^{-1}$, so that

$$\overset{*}{\Gamma}\mathfrak{k}=(\nabla\boldsymbol{\kappa})^{-1}\,\nabla_{\boldsymbol{\kappa}}\big((\nabla\boldsymbol{\kappa})\,\mathfrak{k}\big)(\nabla\boldsymbol{\kappa}),\qquad \mathfrak{k}\in\mathscr{T}_{\mathscr{B}}^{p-1}. \tag{13.3}$$

Indeed, if $\overset{*}{\Gamma}$ is defined by (13.3), condition (a) of Proposition 6 follows from the symmetry of the second gradient and condition (b) from the rule for the differentiation of inner products. By virtue of (13.1), (13.3) is equivalent to

$$\boldsymbol{K}\overset{*}{\Gamma}_{\mathfrak{h}}\mathfrak{k}=\boldsymbol{Q}^{T}\,\nabla_{\boldsymbol{\kappa}}(\boldsymbol{Q}\,\boldsymbol{K}\,\mathfrak{k})\,\boldsymbol{Q}\,\boldsymbol{K}\,\mathfrak{h},\qquad \mathfrak{h},\mathfrak{k}\in\mathscr{T}_{\mathscr{B}}^{p-1}. \tag{13.4}$$

Now let $\boldsymbol{u},\boldsymbol{v}\in\mathscr{V}$. If we substitute $\mathfrak{h}=\boldsymbol{K}^{-1}\boldsymbol{u}$, $\mathfrak{k}=\boldsymbol{K}^{-1}\boldsymbol{v}$ into (13.4) and observe (12.6), we obtain

$$\boldsymbol{Q}^{T}\,\nabla_{\boldsymbol{\kappa}}(\boldsymbol{Q}\,\boldsymbol{v})\,\boldsymbol{Q}\,\boldsymbol{u}=-(\boldsymbol{D}\,\boldsymbol{u})\,\boldsymbol{v},$$

which, by (13.1) and (8.8), is equivalent to

$$\boldsymbol{D}\boldsymbol{u} = -\boldsymbol{Q}^T(\nabla_{\boldsymbol{K}}\boldsymbol{Q})\boldsymbol{u}, \qquad \boldsymbol{u}\in\mathscr{V}. \tag{13.5}$$

This equation shows that the skew transformation $-\boldsymbol{D}\boldsymbol{u}|_X \in \mathscr{L}$ is the instantaneous rate of change of $\boldsymbol{Q}$ at X in the direction of $\boldsymbol{u}$, if viewed in any configuration belonging to $\boldsymbol{K}(X)$. In other words, $\boldsymbol{D}$ describes the local behavior of the rotation field $\boldsymbol{Q}$, which changes the given state of contorted aeolotropy $\boldsymbol{K}$ into the gradient of a global configuration $\boldsymbol{\kappa}$. It is this property that the term "contortion" for $\boldsymbol{D}$ is meant to express.

Theorem 7. *If* $\boldsymbol{K}$ *is a state of contorted aeolotropy, then the curvature of the Riemannian connection relative to* $\boldsymbol{K}$ *vanishes. Conversely, if the curvature of the Riemannian connection relative to* $\boldsymbol{K}$ *vanishes, then* $\boldsymbol{K}$ *is locally a state of contorted aeolotropy* (*i.e.*, every point in $\mathscr{B}$ has a neighborhood $\mathscr{N}$ such that the restriction of $\boldsymbol{K}$ to $\mathscr{N}$ is a state of contorted aeolotropy for $\mathscr{N}$).

Proof. Assume first that (13.1) holds. It follows from (13.3) that $\overset{*}{\Gamma}\mathfrak{k}=0$ if and only if $(\nabla\boldsymbol{\kappa})\mathfrak{k}$ is constant. Hence we could give a simple direct proof of $\overset{*}{\mathfrak{R}}=0$ by using the same argument as we used in the proof of Theorem 4 (Sect. 10). Another proof can be obtained on the basis of (13.5) as follows:

If γ is an arbitrary configuration and $\boldsymbol{F}=(\nabla\gamma)\boldsymbol{K}^{-1}$, then (13.5) is equivalent to

$$(\nabla_\gamma \boldsymbol{Q})\boldsymbol{h} = -\boldsymbol{Q}\boldsymbol{D}(\boldsymbol{F}^{-1}\boldsymbol{h}), \qquad \boldsymbol{h}\in\mathscr{V}. \tag{13.6}$$

If we take the gradient ∇_γ of (13.6) in the direction of $\boldsymbol{k}\in\mathscr{V}$, we find

$$((\nabla_\gamma^{(2)}\boldsymbol{Q})\boldsymbol{k})\boldsymbol{h} = -\boldsymbol{Q}\{((\nabla_{\boldsymbol{K}}\boldsymbol{D})\boldsymbol{v})\boldsymbol{u} - (\boldsymbol{D}\boldsymbol{v})(\boldsymbol{D}\boldsymbol{u}) + \boldsymbol{D}(((\nabla_\gamma \boldsymbol{F}^{-1})\boldsymbol{k})\boldsymbol{h}), \tag{13.7}$$

where $\boldsymbol{u}=\boldsymbol{F}^{-1}\boldsymbol{h}$, $\boldsymbol{v}=\boldsymbol{F}^{-1}\boldsymbol{k}$. Of course, because of the linearity of the values of the fields $\boldsymbol{D}$, $\nabla_{\boldsymbol{K}}\boldsymbol{D}$, $\nabla_\gamma^{(2)}\boldsymbol{Q}$, etc., (13.6) and (13.7) remain valid if $\boldsymbol{h}$ and $\boldsymbol{k}$ are not fixed vectors but vector fields. In particular, they remain valid when $\boldsymbol{u}$ and $\boldsymbol{v}$ are fixed. If we interchange $\boldsymbol{u}$ and $\boldsymbol{v}$ and hence $\boldsymbol{h}$ and $\boldsymbol{k}$ in (13.7) and subtract the resulting formula from (13.7), we obtain, after observing (11.5) and (12.12),

$$((\nabla_\gamma^{(2)}\boldsymbol{Q})\boldsymbol{k})\boldsymbol{h} - ((\nabla_\gamma^{(2)}\boldsymbol{Q})\boldsymbol{k})\boldsymbol{h} = -\boldsymbol{Q}\overset{*}{\boldsymbol{R}}(\boldsymbol{u},\boldsymbol{v}). \tag{13.8}$$

Thus, $\overset{*}{\boldsymbol{R}}=\boldsymbol{0}$ and hence $\overset{*}{\mathfrak{R}}=0$ follows also from the symmetry of the second gradient $\nabla_\gamma^{(2)}\boldsymbol{Q}$.

Assume now that $\boldsymbol{K}$ is a uniform reference such that $\overset{*}{\boldsymbol{R}}=\boldsymbol{0}$. Let γ be an arbitrary configuration and put $\boldsymbol{F}=(\nabla\gamma)\boldsymbol{K}^{-1}$, as before. We can then regard (13.6) as a differential equation for the determination of $\boldsymbol{Q}$. As we have seen, $\overset{*}{\boldsymbol{R}}=\boldsymbol{O}$ is an integrability condition necessary for the existence of a solution. According to a classical theorem, $\overset{*}{\boldsymbol{R}}=\boldsymbol{0}$ is also sufficient for the existence of a solution that is valid in a simply connected neighborhood $\mathscr{N}'$ of a given point $X_0\in\mathscr{B}$. The solution can be chosen so that for $X_0\in\mathscr{B}$, $\boldsymbol{Q}(X_0)$ has a prescribed value, which we take to be the identity $\boldsymbol{1}$. Since $\boldsymbol{D}\boldsymbol{u}$ is skew for all $\boldsymbol{u}\in\mathscr{V}$, it follows from (13.5) that $\boldsymbol{Q}\boldsymbol{Q}^T$ has gradient zero and hence must be equal to $\boldsymbol{1}$ everywhere in $\mathscr{N}'$. Hence $\boldsymbol{Q}$ has orthogonal values. To summarize: If $\overset{*}{\boldsymbol{R}}=\mathbf{0}$, every point in $\mathscr{B}$ has a neighborhood $\mathscr{N}'$ on which we can find an orthogonal-valued tensor field $\boldsymbol{Q}$ (of class C^{p-1}) such that (13.5) holds.

Assume, then, that (13.5) holds on $\mathscr{N}'$. Combining (12.6) with (13.5) we obtain

$$\boldsymbol{K}\overset{*}{\Gamma}_{\mathfrak{h}}\mathfrak{c}=\boldsymbol{Q}^T((\nabla_{\boldsymbol{K}}\boldsymbol{Q})\boldsymbol{K}\mathfrak{h})\boldsymbol{c}=\boldsymbol{Q}^T\nabla_{\boldsymbol{K}}(\boldsymbol{Q}\boldsymbol{c})\boldsymbol{K}\mathfrak{h}, \tag{13.9}$$

which is valid when $\mathfrak{c}=\boldsymbol{K}^{-1}\boldsymbol{c}$ is materially constant. Consider the affine connection $\bar{\Gamma}$ of class C^{p-1} defined by

$$\bar{\Gamma}\mathfrak{k}=\bar{\boldsymbol{K}}^{-1}\nabla_{\bar{\boldsymbol{K}}}(\bar{\boldsymbol{K}}\mathfrak{k})\bar{\boldsymbol{K}}, \qquad \mathfrak{k}\in\mathscr{T}_{\mathscr{B}}^1, \tag{13.10}$$

where

$$\bar{\boldsymbol{K}}=\boldsymbol{Q}\boldsymbol{K}. \tag{13.11}$$

It is easily seen that (13.9) is equivalent to the statement that

$$\overset{*}{\Gamma}\mathfrak{c}=\bar{\Gamma}\mathfrak{c} \tag{13.12}$$

holds for all materially constant tangent vector fields $\mathfrak{c}$. Using the same argument as in the uniqueness proof of Theorem 3 (Sect. 10) we conclude that $\overset{*}{\Gamma}=\bar{\Gamma}$. Since $\overset{*}{\Gamma}$ has zero torsion, an analogue of Theorem 6 (Sect. 10) shows that every point in $\mathscr{N}'$ must have a neighborhood $\mathscr{N}$ such that $\bar{\boldsymbol{K}}=\nabla\boldsymbol{\kappa}$ for some configuration $\boldsymbol{\kappa}$ of $\mathscr{N}$. Hence, by (13.11), we have $\boldsymbol{Q}=(\nabla\boldsymbol{\kappa})\boldsymbol{K}^{-1}$ on $\mathscr{N}$, *i.e.*, $\boldsymbol{K}$ is a state of contorted aeolotropy on $\mathscr{N}$. Q.E.D.

A special case of contorted aeolotropy is *curvilinear aeolotropy*. It corresponds to the case when there exists an orthogonal coordinate system on $\boldsymbol{\kappa}(\mathscr{B})\subset\mathscr{E}$ with the following property: If $(\boldsymbol{e}_1(X), \boldsymbol{e}_2(X), \boldsymbol{e}_3(X))$ is the orthonormal basis which consists of the unit vectors that point in the direction of the coordinate lines at $\boldsymbol{\kappa}(X)$, then $\boldsymbol{Q}(X)\,\boldsymbol{e}_i(X)$ does not depend on $X\in\mathscr{B}$.

14. Special Types of Materially Uniform Bodies

We consider first the case when the isotropy group $g_{\boldsymbol{K}}$ of $\mathscr{B}$ relative to some — and hence every — uniform reference $\boldsymbol{K}$ is *discrete*. Suppose that $\boldsymbol{K}$ and $\hat{\boldsymbol{K}}$ are two *continuous* uniform references. They must be related by (6.12), where $\boldsymbol{P}(X)\in g_{\boldsymbol{K}}$ must depend continuously on X. Since $g_{\boldsymbol{K}}$ is discrete, this is possible only when $\boldsymbol{P}$ is constant. Thus we can absorb $\boldsymbol{P}$ into $\boldsymbol{L}$, and (6.12) becomes $\hat{\boldsymbol{K}}=\boldsymbol{L}\boldsymbol{K}$, with $\boldsymbol{L}=$const. If we write (6.4) for both $\boldsymbol{K}$ and $\hat{\boldsymbol{K}}$, we see that they correspond to the same continuous material uniformity. Since every continuous material uniformity must be of the form (6.4), where $\boldsymbol{K}$ is a continuous uniform reference, we have the following result:

Theorem 8. *If the isotropy groups of a materially uniform simple body $\mathscr{B}$ are discrete, then $\mathscr{B}$ has at most one continuous material uniformity Φ. Any two continuous uniform references $\boldsymbol{K}$, $\hat{\boldsymbol{K}}$ are related by*

$$\hat{\boldsymbol{K}}=\boldsymbol{L}\boldsymbol{K}, \qquad \boldsymbol{L}=\text{const}\in\ell. \tag{14.1}$$

Since material connections are only associated with differentiable uniformities Φ and not with discontinuous ones, it follows from the uniqueness assertion of

Theorem 8 that in the case when the isotropy groups are discrete, the inhomogeneity $\mathfrak{S}$ is a characteristic of the body. If the isotropy groups are non-discrete Lie groups, however, and if there are any material uniformities of class C^{p-1} at all, there will be many, and hence also many inhomogeneities $\mathfrak{S}$ for one and the same body. This is the case, in particular, for uniform isotropic bodies.

Next, we consider a uniform *isotropic* body $\mathscr{B}$ with an undistorted uniform reference $\boldsymbol{K}$ of class C^{p-1}, so that $o \subset g_{\boldsymbol{K}}$. If $\boldsymbol{K}$ is a state of contorted aeolotropy, so that (13.1) holds, it follows from Theorem 2, (6.12), that $\nabla \kappa$ is again a uniform reference and that $g_{\boldsymbol{K}}$ is also the isotropy group relative to $\nabla \kappa$:

Theorem 9. *If a uniform isotropic body has an undistorted state of contorted aeolotropy, it is homogeneous.*

The conclusion of the theorem becomes false when the qualifier "undistorted" is ommitted; *i.e.*, there are inhomogeneous isotropic bodies with distorted states of contorted aeolotropy.

Finally, suppose there is a natural way to single out, among all uniform references for $\mathscr{B}$, a particular class U with the following property (P): All members of U are of class C^{p-1} and differ from one another by a field of similarity transformations with constant ratio, so that $\boldsymbol{K}, \boldsymbol{K}' \in \mathsf{U}$ implies

$$\boldsymbol{K}' = a\,\boldsymbol{Q}\,\boldsymbol{K}, \tag{14.2}$$

where a is a real constant and $\boldsymbol{Q}$ an orthogonal valued tensor field on $\mathscr{B}$. For example, if $\mathscr{B}$ is a uniform solid body that is either isotropic or has cubic symmetry then the class U of all undistorted references has the property (P). This follows from results proved in reference [9]. Other examples are obtained by letting U be the class of all uniform references $\boldsymbol{K}$ such that the corresponding response functions $\mathfrak{H}_{\boldsymbol{K}}$ satisfy a certain special condition such as $\mathfrak{H}_{\boldsymbol{K}}(\boldsymbol{1}) = 0$. Such references are often called *natural references*. The nature of the response function is often such that the class U of natural references has the property (P).

If (14.2) holds, it follows from (12.1) and the fact that orthogonal transformations preserve inner products that the two Riemannian inner products corresponding to $\boldsymbol{K}'$ and $\boldsymbol{K}$ differ from one another only by the constant factor a^2. Therefore, Proposition 6 shows that the Riemannian connections relative to $\boldsymbol{K}$ and $\boldsymbol{K}'$ are the same, and we have the following result:

Theorem 10. *If $\mathscr{B}$ is a uniform simple body with a distinguished class U of uniform references with the property (P), then the Riemannian connection $\overset{*}{\Gamma}$ and its curvature $\overset{*}{\mathfrak{R}}$ are characteristics of the body.*

The assertion of Theorem 10 applies, in particular, to uniform isotropic solid bodies, for which the curvature, $\overset{*}{\mathfrak{R}}$, defined by the class of undistorted uniform references, is an intrinsic measure of deviation from homogeneity.

15. Cauchy's Equation of Balance

We now derive a new version of CAUCHY's equation of balance, which expresses the fact that the forces acting on every part of a given body $\mathscr{B}$ must add to zero. In order to do so, we first derive a lemma, Proposition 7 below.

Let $\boldsymbol{h}$ be a vector field and $\boldsymbol{T}$ a tensor field of class C^1 on $\mathscr{B}$. We define the *divergence* of these fields *relative to* some *reference* $\boldsymbol{K}$ by

$$\operatorname{div}_{\boldsymbol{K}} \boldsymbol{h} = \operatorname{tr}(\nabla_{\boldsymbol{K}} \boldsymbol{h}) \quad \text{and} \quad (\operatorname{div}_{\boldsymbol{K}} \boldsymbol{T}) \cdot \boldsymbol{u} = \operatorname{div}_{\boldsymbol{K}}(\boldsymbol{T}^T \boldsymbol{u}), \qquad \boldsymbol{u} \in \mathscr{V}, \tag{15.1}$$

respectively. The following product rules are valid for $f \in \mathscr{F}^1_{\mathscr{B}}$, $\boldsymbol{h} \in \mathscr{V}^1_{\mathscr{B}}$, $\boldsymbol{T} \in \mathscr{L}^1_{\mathscr{B}}$:

$$\begin{aligned} \operatorname{div}_{\boldsymbol{K}}(f \boldsymbol{h}) &= (\nabla_{\boldsymbol{K}} f) \cdot \boldsymbol{h} + f \operatorname{div}_{\boldsymbol{K}} \boldsymbol{h}, \\ \operatorname{div}_{\boldsymbol{K}}(\boldsymbol{T} \boldsymbol{h}) &= (\operatorname{div}_{\boldsymbol{K}} \boldsymbol{T}^T) \cdot \boldsymbol{h} + \operatorname{tr}(\boldsymbol{T} \nabla_{\boldsymbol{K}} \boldsymbol{h}). \end{aligned} \tag{15.2}$$

If $\boldsymbol{K} = \nabla \kappa$ is a configuration gradient, we write $\operatorname{div}_\kappa$ instead of $\operatorname{div}_{\boldsymbol{K}}$.

Suppose that a uniform reference $\boldsymbol{K}$ of class C^{p-1} for $\mathscr{B}$ and a tensor field $\boldsymbol{T}$ of class C^1 on $\mathscr{B}$ are given. For any configuration γ of $\mathscr{B}$ we then define another tensor field $\boldsymbol{T}_\gamma$ of class C^1 by

$$\boldsymbol{T}_\gamma = \frac{1}{J} \boldsymbol{T} \boldsymbol{F}^T, \tag{15.3}$$

where

$$\boldsymbol{F} = (\nabla \gamma) \boldsymbol{K}^{-1}, \qquad J = |\det \boldsymbol{F}|. \tag{15.4}$$

Proposition 7. *If*

$$\boldsymbol{K}(X) = \nabla \kappa(X) \tag{15.5}$$

for some $X \in \mathscr{B}$, then the divergences at X relative to $\boldsymbol{K}$ of the tensor field $\boldsymbol{T}$ and of the tensor field $\boldsymbol{T}_\kappa$ defined by (15.3) *and* (15.4) *are related by*

$$\operatorname{div}_{\boldsymbol{K}} \boldsymbol{T}_\kappa |_X = (\operatorname{div}_{\boldsymbol{K}} \boldsymbol{T} + \boldsymbol{T} \boldsymbol{s})|_X, \tag{15.6}$$

where the vector field $\boldsymbol{s}$ is defined, in terms of the inhomogeneity $\boldsymbol{S}$ relative to $\boldsymbol{K}$, by

$$\boldsymbol{s} \cdot \boldsymbol{u} = \operatorname{tr}(\boldsymbol{S} \boldsymbol{u}), \qquad \boldsymbol{u} \in \mathscr{V}. \tag{15.7}$$

Proof. We make use of (15.3) and (15.4) with γ replaced by κ. We then have

$$J(\boldsymbol{T}_\kappa^T \boldsymbol{u}) = \boldsymbol{F}(\boldsymbol{T}^T \boldsymbol{u}), \qquad \boldsymbol{u} \in \mathscr{V}. \tag{15.8}$$

Using the rule

$$\boldsymbol{u} \cdot \nabla_{\boldsymbol{K}}(\det \boldsymbol{F}) = (\det \boldsymbol{F}) \operatorname{tr}[\boldsymbol{F}^{-1}((\nabla_{\boldsymbol{K}} \boldsymbol{F}) \boldsymbol{u})], \qquad \boldsymbol{u} \in \mathscr{V} \tag{15.9}$$

for the differentiation of a determinant, the product rules (15.2), and the definition $(15.1)_2$, we see that taking $\operatorname{div}_{\boldsymbol{K}}$ of (15.8) yields

$$J \operatorname{tr}[\boldsymbol{F}^{-1}((\nabla_{\boldsymbol{K}} \boldsymbol{F})(\boldsymbol{T}_\kappa^T \boldsymbol{u}))] + J(\operatorname{div}_{\boldsymbol{K}} \boldsymbol{T}_\kappa) \cdot \boldsymbol{u} = (\operatorname{div}_{\boldsymbol{K}} \boldsymbol{F}^T) \cdot \boldsymbol{T}^T \boldsymbol{u} + \operatorname{tr}[\boldsymbol{F} \nabla_{\boldsymbol{K}}(\boldsymbol{T}^T \boldsymbol{u})]. \tag{15.10}$$

Since $\boldsymbol{F}(X) = \mathbf{1}$, $J(X) = 1$, and $\boldsymbol{T}_\kappa(X) = \boldsymbol{T}(X)$ by (15.5), (15.4), and (15.3), evaluation of (15.10) at $X \in \mathscr{B}$ gives

$$\{\operatorname{tr}[(\nabla_{\boldsymbol{K}} \boldsymbol{F})(\boldsymbol{T}^T \boldsymbol{u})] + (\operatorname{div}_{\boldsymbol{K}} \boldsymbol{T}_\kappa) \cdot \boldsymbol{u} - (\operatorname{div}_{\boldsymbol{K}} \boldsymbol{F}^T) \cdot \boldsymbol{T}^T \boldsymbol{u} - (\operatorname{div}_{\boldsymbol{K}} \boldsymbol{T}) \cdot \boldsymbol{u}\}_X = 0. \tag{15.11}$$

Using the rule $\operatorname{tr}[(\nabla_{\boldsymbol{K}} \boldsymbol{F}) \boldsymbol{v}] = \nabla_{\boldsymbol{K}}(\operatorname{tr} \boldsymbol{F}) \cdot \boldsymbol{v}$, we see that (15.11) can hold for all $\boldsymbol{u} \in \mathscr{V}$ only if

$$\{\boldsymbol{T}[\nabla_{\boldsymbol{K}}(\operatorname{tr} \boldsymbol{F}) - \operatorname{div}_{\boldsymbol{K}} \boldsymbol{F}^T] + \operatorname{div}_{\boldsymbol{K}} \boldsymbol{T}_\kappa - \operatorname{div}_{\boldsymbol{K}} \boldsymbol{T}\}_X = \mathbf{0}. \tag{15.12}$$

On the other hand, if we evaluate (11.4) at $X \in \mathscr{B}$ and take the trace, we obtain

$$\operatorname{tr}(\boldsymbol{S}\,\boldsymbol{u})|_X = \{(\operatorname{div}_{\boldsymbol{K}} \boldsymbol{F}^T)\cdot \boldsymbol{u} - \nabla_{\boldsymbol{K}}(\operatorname{tr}\boldsymbol{F})\cdot \boldsymbol{u}\}_X,$$

which, in view of (15.7), is equivalent to

$$\boldsymbol{s}|_X = \{\operatorname{div}_{\boldsymbol{K}} \boldsymbol{F}^T - \nabla_{\boldsymbol{K}}(\operatorname{tr}\boldsymbol{F})\}_X. \tag{15.13}$$

The desired result (15.6) follows from (15.12) and (15.13). Q.E.D.

Let us assume now that the body $\mathscr{B}$ is subject to internal contact forces and external body forces[14]. If the forces acting on every part of $\mathscr{B}$ are balanced and if suitable regularity assumptions are satisfied one can prove the following results (*cf.* [*10*]):

(i) With every configuration κ of $\mathscr{B}$ one can associate a stress tensor field $\boldsymbol{T}_\kappa$ of class C^1 and a body force field $\boldsymbol{b}_\kappa$ of class C^0 such that the force $\boldsymbol{f}$ exerted on a part $\mathscr{P}$ of $\mathscr{B}$ by the combined action of a separate part $\mathscr{P}'$ of $\mathscr{B}$ and the external world is given by

$$\boldsymbol{f} = \int_{\kappa(\mathscr{P})} \boldsymbol{b}_\kappa \, dV + \int_{\mathscr{C}} \boldsymbol{T}_\kappa \boldsymbol{n}\, dS, \tag{15.14}$$

where $\mathscr{C}$ is the surface of contact between $\mathscr{P}$ and $\mathscr{P}'$ in the configuration κ and where $\boldsymbol{n}$ is the unit normal to $\mathscr{C}$ directed away from $\kappa(\mathscr{P})$.

(ii) CAUCHY's equation of balance

$$\operatorname{div}_\kappa \boldsymbol{T}_\kappa + \boldsymbol{b}_\kappa = \boldsymbol{0} \tag{15.15}$$

is valid on $\mathscr{B}$ for every configuration κ.

(iii) If κ and γ are two configurations of $\mathscr{B}$, then the stress fields $\boldsymbol{T}_\kappa$ and $\boldsymbol{T}_\gamma$ and the body force fields $\boldsymbol{b}_\kappa$ and $\boldsymbol{b}_\gamma$ are related by

$$\boldsymbol{T}_\gamma = \frac{1}{J}\, \boldsymbol{T}_\kappa \boldsymbol{F}^T, \qquad \boldsymbol{b}_\gamma = \frac{1}{J}\, \boldsymbol{b}_\kappa, \tag{15.16}$$

where

$$\boldsymbol{F} = (\nabla\gamma)(\nabla\kappa)^{-1}, \qquad J = |\det \boldsymbol{F}| \tag{15.17}$$

(*cf.* equation (43A.3) of reference [*2*]).

Let $\boldsymbol{K}$ be a uniform reference of class C^{p-1} for $\mathscr{B}$. Let a particular point $X \in \mathscr{B}$ be given. It is clear from (15.16) and (15.17) that $\boldsymbol{T}_\gamma(X) = \boldsymbol{T}_\kappa(X)$ and $\boldsymbol{b}_\gamma(X) = \boldsymbol{b}_\kappa(X)$ hold whenever both κ and γ belong to the equivalence class by which the local configuration $\boldsymbol{K}(X)$ is defined. Thus, we can define fields $\boldsymbol{T}_{\boldsymbol{K}}$ and $\boldsymbol{b}_{\boldsymbol{K}}$ by the condition that for each $X \in \mathscr{B}$,

$$\boldsymbol{T}_{\boldsymbol{K}}(X) = \boldsymbol{T}_\kappa(X), \qquad \boldsymbol{b}_{\boldsymbol{K}}(X) = \boldsymbol{b}_\kappa(X) \tag{15.18}$$

hold whenever $\nabla\kappa(X) = \boldsymbol{K}(X)$. We call $\boldsymbol{T}_{\boldsymbol{K}}$ and $\boldsymbol{b}_{\boldsymbol{K}}$ the *stress tensor field* and *body force field relative to the reference* $\boldsymbol{K}$. It is clear from (15.16) and (15.17) that

$$\boldsymbol{T}_\gamma = \frac{1}{J}\, \boldsymbol{T}_{\boldsymbol{K}} \boldsymbol{F}^T, \qquad \boldsymbol{b}_\gamma = \frac{1}{J}\, \boldsymbol{b}_{\boldsymbol{K}} \tag{15.19}$$

when $\boldsymbol{F}$ and J are given by (15.4), where γ is an arbitrary configuration.

[14] Inertial forces should be regarded as part of the body forces.

Since $\operatorname{div}_{\kappa} \boldsymbol{T}|_X = \operatorname{div}_{\boldsymbol{K}} \boldsymbol{T}|_X$ whenever $\nabla \kappa(X) = \boldsymbol{K}(X)$, it follows from (15.15) that

$$[\operatorname{div}_{\boldsymbol{K}} \boldsymbol{T}_{\kappa} + \boldsymbol{b}_{\boldsymbol{K}}]_X = \boldsymbol{0} \tag{15.20}$$

holds whenever $\nabla \kappa(X) = \boldsymbol{K}(X)$. On the other hand, Proposition 7 applies when we choose $\boldsymbol{T} = \boldsymbol{T}_{\boldsymbol{K}}$. Thus, by substituting (15.6) with the choice $\boldsymbol{T} = \boldsymbol{T}_{\boldsymbol{K}}$ into (15.20) we obtain the following result:

Theorem 11. *The stress tensor field* $\mathbf{T}_{\boldsymbol{K}}$ *and the body force field* $\mathbf{b}_{\boldsymbol{K}}$ *relative to a uniform reference* $\boldsymbol{K}$ *satisfy the modified equation of balance*

$$\operatorname{div}_{\boldsymbol{K}} \boldsymbol{T}_{\boldsymbol{K}} + \boldsymbol{T}_{\boldsymbol{K}} \boldsymbol{s} + \boldsymbol{b}_{\boldsymbol{K}} = \boldsymbol{0}, \tag{15.21}$$

where $\mathbf{s}$ *is defined in terms of the inhomogeneity* $\mathbf{S}$ *relative to* $\boldsymbol{K}$ *by* (15.7).

The equation (15.21) is much more useful than (15.15) for dealing with inhomogeneous materially uniform bodies. Consider, for example, an elastic body $\mathscr{B}$, for which the set of response descriptors is the set $\mathscr{S} \subset \mathscr{L}$ of symmetric linear transformations. According to Theorem 1 (Sect. 6) we can associate with a given uniform reference $\boldsymbol{K}$ a relative elastic response function $\mathfrak{H}_{\boldsymbol{K}} : \ell \to \mathscr{S}$. In order that a configuration γ be compatible with a given force system, the constitutive equation

$$\mathfrak{H}_{\boldsymbol{K}}(\boldsymbol{F}) = \boldsymbol{T}_{\gamma}, \qquad \boldsymbol{F} = (\nabla \gamma) \boldsymbol{K}^{-1} \tag{15.22}$$

must be satisfied on $\mathscr{B}$, where $\boldsymbol{T}_{\gamma}$ is the stress tensor field for γ. In view of $(15.19)_1$, (15.22) is equivalent to

$$\mathfrak{h}_{\boldsymbol{K}}(\boldsymbol{F}) = \boldsymbol{T}_{\boldsymbol{K}}, \qquad \boldsymbol{F} = (\nabla \gamma) \boldsymbol{K}^{-1}, \tag{15.23}$$

where $\mathfrak{h}_{\boldsymbol{K}} : \ell \to \mathscr{L}$ is defined by

$$\mathfrak{h}_{\boldsymbol{K}}(\boldsymbol{F}) = |\det \boldsymbol{F}| \, \mathfrak{H}_{\boldsymbol{K}}(\boldsymbol{F}) \boldsymbol{F}^{T-1}, \qquad \boldsymbol{F} \in \ell. \tag{15.24}$$

Assume that $\mathfrak{h}_{\boldsymbol{K}}$ is of class C^1, and denote its gradient by $\mathbf{H}_{\boldsymbol{K}}$. For each $\boldsymbol{F} \in \ell$ the value $\mathbf{H}_{\boldsymbol{K}}(\boldsymbol{F})$ is then a linear transformation from $\mathscr{L}$ into $\mathscr{L}$. If we take the gradient of (15.23) relative to $\boldsymbol{K}$, the chain rule yields

$$(\nabla_{\boldsymbol{K}} \boldsymbol{T}_{\boldsymbol{K}}) \boldsymbol{u} = \mathbf{H}_{\boldsymbol{K}}(\boldsymbol{F}) [(\nabla_{\boldsymbol{K}} \boldsymbol{F}) \boldsymbol{u}], \qquad \boldsymbol{u} \in \mathscr{V}. \tag{15.25}$$

It follows that

$$\operatorname{div}_{\boldsymbol{K}} \boldsymbol{T}_{\boldsymbol{K}} = \mathbf{A}_{\boldsymbol{K}}(\boldsymbol{F}) [(\nabla_{\boldsymbol{K}} \boldsymbol{F})], \tag{15.26}$$

where $\mathbf{A}_{\boldsymbol{K}}$ is that function on ℓ whose values $\mathbf{A}_{\boldsymbol{K}}(\boldsymbol{F}) : \mathscr{L}(\mathscr{V}, \mathscr{L}) \to \mathscr{V}$ are determined by the property that $\mathbf{A}_{\boldsymbol{K}}(\boldsymbol{F})[\boldsymbol{Z}] \cdot \boldsymbol{w}$ is the trace of the linear transformation $\boldsymbol{u} \to \{\mathbf{H}_{\boldsymbol{K}}(\boldsymbol{F}) [\boldsymbol{Z} \boldsymbol{u}]\}^T \boldsymbol{w}$ for all $\boldsymbol{w} \in \mathscr{V}$ and all $\boldsymbol{Z} \in \mathscr{L}(\mathscr{V}, \mathscr{L})$. Of course, $\mathbf{A}_{\boldsymbol{K}}$ is determined by the response function $\mathfrak{h}_{\boldsymbol{K}}$. If we substitute (15.26) and (15.23) into (15.21), we obtain[15]

$$\mathbf{A}_{\boldsymbol{K}}(\boldsymbol{F}) [\nabla_{\boldsymbol{K}} \boldsymbol{F}] + \mathfrak{h}_{\boldsymbol{K}}(\boldsymbol{F}) \boldsymbol{s} + \boldsymbol{b}_{\boldsymbol{K}} = \boldsymbol{0}, \quad \boldsymbol{F} = (\nabla \gamma)^{-1} \boldsymbol{K}, \tag{15.27}$$

which is *the differential equation for the determination of configurations* γ *possible in a materially uniform elastic body.* If the body is homogeneous, we can choose

[15] This result, in terms of coordinates, was announced two years ago in reference [2] as equation (44.7).

$K = \nabla \kappa$. Then s vanishes and (15.27) reduces to the classical differential equation of finite elasticity.

Finally, we give another application of Proposition 7. Using the fact that $\mathscr{V}$ is three-dimensional (which was irrelevant up to now), we choose an orientation in $\mathscr{V}$ and consider the associated cross product $\times$. The curl of a vector field h and a tensor field T on $\mathscr{B}$ relative to some configuration γ are defined by

$$\begin{gathered} u \cdot (\nabla_\gamma h)\, v - v \cdot (\nabla_\gamma h)\, u = (\operatorname{curl}_\gamma h) \cdot (u \times v), \\ (\operatorname{curl}_\gamma T)^T u = \operatorname{curl}_\gamma (T u), \end{gathered} \tag{15.28}$$

where $u, v \in \mathscr{V}$. It follows from (15.28) that

$$(\operatorname{curl}_\gamma T^T)(u \times v) = ((\nabla_\gamma T)\, v)\, u - ((\nabla_\gamma T)\, u)\, v. \tag{15.29}$$

Also, we have the rule

$$\operatorname{div}_\gamma \operatorname{curl}_\gamma T = 0. \tag{15.30}$$

The inhomogeneity S reltive to a uniform reference K has the skew symmetry $(S u)\, v = -(S v)\, u$, $u, v \in \mathscr{V}$. Therefore, S determines and is determined by a tensor field A on $\mathscr{B}$ such that[16]

$$(S u)\, v = A(u \times v), \qquad u, v \in \mathscr{V}. \tag{15.31}$$

If we substitute (15.31) into (11.5) and observe the rule (15.29), we see that

$$A(u \times v) = [\operatorname{curl}_\gamma (F^{T^{-1}})]\, (F v \times F u). \tag{15.32}$$

Hence, since $F^T(F v \times F u) = (\det F)(v \times u)$ and since $u \times v$ is arbitrary, (15.32) yields

$$-\operatorname{curl}_\gamma (F^{T^{-1}}) = \frac{1}{J} A F^T, \qquad J = \det F. \tag{15.33}$$

Thus, the tensor field $-\operatorname{curl}_\gamma (F^{T^{-1}})$ is obtained from A by the rules (15.3), (15.4), except that the absolute value signs are omitted in the definition of J, which does not affect the validity of Proposition 7. Since by rule (15.30) we have

$$[\operatorname{div}_K \operatorname{curl}_\kappa (F^{T^{-1}})]_X = [\operatorname{div}_\kappa \operatorname{curl}_\kappa (F^{T^{-1}})]_X = 0,$$

(15.6) yields

$$\operatorname{div}_K A + A s = 0, \tag{15.34}$$

where s is determined by A through

$$(A - A^T)\, u = s \times u, \qquad u \in \mathscr{V}. \tag{15.35}$$

Thus, (15.34) is a differential identity for A and K. One can show that it is equivalent to the Bianchi identity (11.11).

Acknowledgement. The research leading to this paper was supported by the Office of Naval Research under contract NONR-760 (30).

[16] The field A here corresponds to what was denoted by A^T in [2], Sect. 34.

References

[1] NOLL, W., A mathematical theory of the mechanical behavior of continuous media. Arch. Rational Mech. Anal. **2**, 197—226 (1958).
[2] TRUESDELL, C., & W. NOLL, The Non-linear Field Theories of Mechanics. Encyclopedia of Physics, Vol. III/3. Berlin-Heidelberg-New York: Springer 1965.
[3] LANG, S., Introduction to Differentiable Manifolds. New York-London-Sydney: Interscience 1962.
[4] KRÖNER, E.: Allgemeine Kontinuumstheorie der Versetzungen und Eigenspannungen. Arch. Rational Mech. Anal. **4**, 273—334 (1960).
[5] SEEGER, A.: Recent advances in the theory of defects in crystals. Physica status solidi **1**, 669—698 (1961).
[6] NOLL, W., Space-time structures in classical mechanics. Delaware Seminar in the Foundations of Physics, pp. 28—34. New York: Springer 1967.
[7] NOLL, W., Euclidean geometry and Minkowskian chronometry. Amer. Math. Monthly **71**, 129—144 (1964).
[8] WANG, C.-C., On the geometric structures of simple bodies, a mathematical foundation for the theory of continuous distributions of dislocations. Arch. Rational Mech. Anal., following in this issue.
[9] COLEMAN, B. D., & W. NOLL, Material symmetry and thermostatic inequalities in finite elastic deformations. Arch. Rational Mech. Anal. **15**, 87—111 (1964).
[10] NOLL, W., The foundations of classical mechanics in the light of recent advances in continuum mechanics. The Axiomatic Method, with special reference to geometry and physics, pp. 266—281. Amsterdam: North Holland 1959.

Carnegie-Mellon University
Pittsburgh, Pennsylvania

(Received July 22, 1967)

Offprint from "Archive for Rational Mechanics and Analysis",
Volume 27, Number 1, 1967, P. 33–94

Springer-Verlag, Berlin · Heidelberg · New York

On the Geometric Structures of Simple Bodies, a Mathematical Foundation for the Theory of Continuous Distributions of Dislocations

C.-C. WANG

Communicated by W. NOLL & C. TRUESDELL

Contents

1. Introduction

There is a large literature[1] in continuum mechanics on the mathematical representation of the mechanical response of *material particles*. In the physical world, of course, material particles present themselves in various *bodies*. It is the purpose of this research to construct a mathematical theory for such bodies.

1a. Generalized Continuous Distributions of Dislocations

Before laying down in the next section mathematical statements of my basic hypotheses concerning a body, I sketch them here in loose terms:

i) A body is a differentiable manifold, called the *body manifold.* Its elements are called *particles.*

ii) The particles of a body are *simple particles* in the sense of NOLL[2]; moreover, they are *all of the same kind.*

[1] An extensive review of the modern development of continuum mechanics up to 1965 has been written by TRUESDELL & NOLL [*1*].

[2] NOLL [*2*]. A detailed exposition of NOLL's theory is included in the book of TRUESDELL & NOLL [*1*].

iii) The mechanical response of the particles varies *smoothly* over the body manifold.

For definiteness, I call a body satisfying the above three conditions a *smooth materially uniform simple body*. However, for brevity, throughout this paper the terms *simple body* or even *body* will refer only to a *smooth* materially uniform simple body. Physically, the differentiable structure on a body is induced by the Euclidean structure of the physical space, into which the body can be mapped, and it is only in the physical space that we encounter the body. Condition ii) means that a simple body is made up of *one* kind of material particles, such as copper, zinc, water, air, ..., *etc.* so far as their mechanical responses are concerned[3]. In this sense, a simple body is *materially uniform*. Condition iii) represents a physical concept generalizing the notion of *continuous distributions of dislocations* for crystalline bodies. More specifically, for a solid crystal body I require that the crystal axes form *smooth local fields* on the body manifold. Similarly, for a fluid body, I require that the mass density[4] be a smooth field. In general, for solids and for fluid crystals[5] alike, I assume that in a neighborhood of every particle, there exists a smooth field of local reference configurations relative to which the response functional is independent of the particle.

To simplify certain minor technical details, it is convenient to assign a fixed *orientation*[6] to a simple body. Further, we shall consider only those bodies whose isotropy groups are *closed Lie subgroups* of the unimodular group. Physically, this restriction is not severe, since most, if not all simple bodies of interest are of this kind, such as the classical solid crystal classes, the transversely isotropic solids, the isotropic solids and fluids, and many types of anisotropic fluid crystals. Moreover, this restriction is automatically satisfied if the response functional obeys a mild continuity requirement.

In the physical literature, there are several theories of *continuous distributions of dislocations*, proposed by KONDO [*3*], NYE [*4*], BILBY, BULLOUGH & SMITH [*5*], KRÖNER & SEEGER [7], and others. Those theories concern mostly solid crystalline bodies which are endowed with certain *a priori* geometric structures suggested by various forms of so-called atomic lattices (regular arrays of balls). To the extent[7] that such lattices correspond to simple bodies, those theories are all included as special cases of the one presented in this paper. It is not possible to make any precise comparison, however, since the physical literature on dislocation theory

---

[3] In continuum mechanics, particles are distinguished only by their mechanical responses. The chemical composition may give rise to, but does *not* characterize, the mechanical properties of a particle.

[4] Mass density is a basic concept of simple particle, *cf.* NOLL [*2*].

[5] We adopt the usual classification of simple particles: a *solid* particle is one whose isotropy group is conjugate to a subgroup of the orthogonal group, a non-solid particle is a *fluid crystal* particle, and an isotropic fluid crystal particle is a *fluid* particle.

[6] All body manifolds are trivially orientable.

[7] The physical writers sometimes claim that bodies with dislocations are subject to couple stresses or multipolar stresses (*e.g.*, the theory of KRÖNER [*34*, Ch. IV]). To the extent that the theory of simple materials does not allow such stresses, the theory of this paper does not enjoy the generality those writers demand. Since they, on the other hand, do not introduce any definite constitutive equations, they cannot be said to have an exact mechanical continuum theory at all, so that grounds for comparison of their works with the results presented here are wanting.

rarely if ever introduces definite constitutive equations, resting content with heuristic discussions of the body manifolds and seldom taking up the response of bodies to deformation and loading, which is the foundation stone of modern continuum mechanics. Thus the conceptual basis of the present theory is very different from any used by the earlier writers, especially in the following three aspects:

i) *There is no appropriate lattice model for an arbitrary simple body*, since the symmetry of a simple particle, in general, *cannot* be described by a discrete (point) group. In particular, the geometry characterizing the material structure on the body does not always admit a smooth distant parallelism[8]. Many examples of bodies of this general kind will be given below.

ii) *The mechanical properties of a simple body are laid down once and for all in their entirety by the constitutive equations specifying the mechanical responses of its particles* — namely by the contact force that arises at a given particle in consequence of the deformation it has suffered. In particular, the material geometric structure of a simple body is *implied* mathematically by these *mechanical* properties. Therefore, that structure need not, and hence should not, be divined *a priori* by metaphysical or geometric remarks making no use of the concepts of force and deformation.

iii) *We derive exact field equations*[9] *for an elastic body based on the constitutive equation and the geometry characterizing the distribution of the mechanical response of the particles.* No mathematical approximation of any kind is made in this derivation. These exact field equations seem not to have been found by any of the physical approaches to dislocation theory.

Exact solutions for specific elastic bodies are reserved for a future paper.

To determine the material structure from the constitutive equations of the particles is a definite mathematical problem. That is the problem set and solved in this paper, for simple bodies. Therefore, the material structure, some cases of which were motivated by the lattice models in the older theories, is the *outcome* in the present theory. Since the three basic hypotheses defining a simple body are very general, all continuous distributions of dislocations so far considered by others, insofar as they fit upon the framework of NOLL's theory of simple materials, seem to be covered here. For example, the so-called *Moebius crystals*[10] are not excluded.

The research undertaken here grew out of earlier work by NOLL. In a largely unpublished study[11], he constructed a theory of dislocations, derived, like the one presented here, from the constitutive functional defining a simple particle and based

[8] Although distant parallelisms do exist in the material structures of all bodies considered in this paper, in general those parallelisms are *not* smooth.

[9] Equations of this kind were first found by NOLL [*16*] (*cf.* also TRUESDELL & NOLL [*1*, § 44]). Our equations are more general than NOLL's, since the simple bodies considered in this paper are more general than those treated by him.

[10] *Cf.* FRANK [*29*, Fig. 1].

[11] I am indebted to Professor NOLL, who gave me a copy of his unpublished work in 1963. The major results of his theory of that date are summarized in the book of TRUESDELL & NOLL [*1*, § 34]. After my work was complete, Professor NOLL revised his paper, and it is printed just preceding. I did not see this revised paper until mine had been set in type.

on the key concept of material isomorphism, which he had introduced in 1957. However, he adopted in the context of bodies a definition of smoothness which is unreasonably restrictive; namely, he required that the material structure of a body be derivable from some *globally smooth* distant parallelism. As was just remarked in aspect i) above, there are many simple bodies which do not correspond to this assumption, so that NOLL's starting hypothesis excludes them. On the other hand, if the material structure indeed admits such a distant parallelism, then the geometric structure considered by NOLL is included in the present theory. However, that structure may not be the simplest one, and other alternatives should not be ruled out. For example, in the case of an isotropic body, the most convenient geometric structure is a Riemannian one, necessarily torsion-free, but in this case never curvature-free unless the body is locally homogeneous.

1b. Differential Geometry in the Context of Simple Bodies

In this paper, I shall apply to the theory of simple bodies certain concepts and results from the modern theory of differential geometry. These are summarized in detail in the treatises listed as references [*8*–*13*] and in many other texts in this highly developed branch of mathematics. Here I sketch the major ideas they represent in the context of simple bodies.

In the theory of differentiable manifolds, the overall coherence of the tangent spaces of a manifold $\mathscr{M}$ is characterized by two related fibre bundles: the *tangent bundle* $\mathscr{T}(\mathscr{M})$ and the *bundle of linear frames* $\mathscr{E}(\mathscr{M})$. The latter is the *associated principal bundle* of the former. In particular, for a simple body $\mathscr{B}$, the overall geometric structure of the body manifold is characterized by $\mathscr{T}(\mathscr{B})$ and $\mathscr{E}(\mathscr{B})$. But since the body manifold has a very simple structure, namely, it can be identified as an open submanifold of a Euclidean space, the structures of $\mathscr{T}(\mathscr{B})$ and $\mathscr{E}(\mathscr{B})$, likewise, are very simple. Moreover, these two bundles *cannot* characterize the material structure of $\mathscr{B}$, since they are *independent* of the distribution of the response functionals on $\mathscr{B}$.

In order to describe the material structure of a body, we introduce the notion of *material tangent bundles*. Roughly speaking, the material tangent bundles are tangent bundles whose structure groups are the (relative) isotropy groups of the material. If we call the usual tangent bundle of the body manifold the *geometric tangent bundle* of the body, then, in the terminology of fibre bundles, the geometric tangent bundle is *reducible* to the material tangent bundles. We shall construct their associated principal bundles, which we call the *bundles of reference frames*. These last are subbundles of the bundle of linear frames $\mathscr{E}(\mathscr{B})$ of the body manifold.

By hypothesis, the isotropy groups of the simple bodies to be considered in this paper are closed Lie subgroups of the unimodular group $\mathscr{U}$. Hence their Lie algebras can be identified as Lie subalgebras of the Lie algebra $\mathbf{u}$ of $\mathscr{U}$. In Section 6 we give various types[12] of Lie subalgebras of $\mathbf{u}$ and their corresponding connected Lie subgroups of $\mathscr{U}$ in matrix forms. As usual, the Lie algebras of the

[12] *Cf.* NÔNO [*14*], who attributes the classification of the Lie subalgebras of $\mathbf{u}$ to LIE [*15*]. I am indebted to Professor NÔNO for sending me a copy of Section 3 of his forthcoming paper [*14*], which contains the classification of the Lie subalgebras of $\mathbf{u}$.

isotropy groups then give rise to the *fundamental fields* on the bundles of reference frames.

We treat *connections* on the bundles of frames in Sections 8 and 9. Of special interest are those connections on $\mathscr{E}(\mathscr{B})$ which are *reducible* to connections on the bundles of reference frames. Such connections are called *material connections*[13], since the parallel transports relative to them are always material isomorphisms. For a material connection, it is necessary and sufficient that the values of the connection form, restricted to the bundles of reference frames, belong to the fundamental fields induced by the Lie algebras of the isotropy groups.

For a crystalline solid body, whose isotropy groups are discrete, the material connection evidently is unique and coincides with the classical one motivated by the lattice models. Such material connections, of course, are always completely integrable. Hence their curvature tensors vanish. A simple body in general, however, need *not* possess any curvature-free material connection at all, as we shall see in various examples, in Sections 10 and 11.

1c. Summary of the Main Results of this Paper

Besides the explicit mathematical structure of a general simply body, the main results of this paper are three:

First, I give a precise mathematical description of the local inhomogeneity of a body in terms of the curvatures and the torsions of the material connections. I generalize the notions of Burgers vector and the dislocation density.

Second, I show that a simple body is a solid if and only if it can be equipped with an intrinsic Riemannian metric, which is invariant under the parallel transports relative to all material connections. The Riemannian connections associated with such intrinsic metrics, in general, are *not* always material connections. However, if a solid body is isotropic, or if a solid body can be equipped with a *symmetric* material connection, then the Riemannian connection, which is necessarily unique in these cases, is a material connection. Therefore, the material structure of an isotropic solid body can be described by a Riemannian geometry. On the other hand, a fluid crystal body, obviously, cannot be equipped with any intrinsic metric, since the tangent spaces of such a body do not possess any intrinsic inner product. I show also that a solid body is locally homogeneous if and only if it has a flat material connection. Such a condition for local inhomogeneity is false if the body is not a solid.

Third, I derive exact field equations of motion for elastic bodies. As remarked before, the derivation is based on the response function of the particles and the material geometric structure of the body. No approximation or linearization of any nature is used in this analysis. Therefore, any exact solution of these field equations represents precisely a mechanical response of an elastic body which bears a specific continuous distribution of dislocations.

1d. Notations

We use mainly direct vector and tensor notations in the paper. Most symbols are defined when they first occur. If components of vectors and tensors

[13] The notion of material connection was first introduced in the work of NOLL. Our definition is more general than NOLL's, since he requires a material connection to be curvature-free.

occur, then they are referred to the natural bases of local coordinates. For the tangent space, the natural basis of a local coordinate system (x^i) is denoted by $\{\partial/\partial x^i\}$, while for the cotangent space, the natural basis is denoted by $\{dx^i\}$ on the understanding that the standard duality relation holds, namely

$$\left\langle dx^i, \frac{\partial}{\partial x^j} \right\rangle = \delta^i_j. \tag{1.1}$$

We denote the tensor product by the symbol $\otimes$ and the exterior product by the symbol $\wedge$. A tensor of contravariant order r and covariant order s is said to be of *type* (r, s). The standard summation convention is observed.

Part I. The Concept of a Materially Uniform Simple Body

2. The Basic Structure of a Simple Body

We fix once and for all a rectangular Cartesian coordinate system in the physical space $\mathscr{E}$, so that $\mathscr{E}$ is represented in a definite way by the Euclidean space $\mathscr{R}^3$, which consists of ordered triples of real numbers.

Definition 2.1. A *body manifold* $\mathscr{B}$ is an oriented 3-dimensional differentiable manifold which is connected and can be covered by one coordinate neighborhood. That is, there exist diffeomorphisms $\varphi, \psi, \chi, \ldots$, called *configurations*, which map $\mathscr{B}$ into $\mathscr{R}^3$ and are orientation-preserving[14], say

$$\varphi: \mathscr{B} \to \mathscr{R}^3. \tag{2.1}$$

Definition 2.2. Let p be a point in a body manifold $\mathscr{B}$. Then a *local configuration* of p is an orientation-preserving isomorphism of the tangent space $\mathscr{B}_p$ at p onto $\mathscr{R}^3$, say

$$\kappa: \mathscr{B}_p \to \mathscr{R}^3. \tag{2.2}$$

Remark. It is a standard result in the theory of differentiable manifolds that a configuration φ induces a field of local configurations φ_{*p} on $\mathscr{B}$, *viz*,

$$\varphi_{*p}: \mathscr{B}_p \to \mathscr{R}^3. \tag{2.3}$$

We call φ_{*p} the *induced local configuration* of φ at p. Clearly, if p is fixed, then any given local configuration κ of p always can be realized as an induced local configuration, say

$$\kappa = \varphi_{*p}. \tag{2.4}$$

However, a given *field* of local configurations on $\mathscr{B}$, in general, does *not* correspond to an induced field of a configuration unless it satisfies certain compatibility conditions.

Definition 2.3. A *motion* of $\mathscr{B}$ is a 1-parameter family of configurations, say $\varphi(s)$, $s \in (-\infty, \infty)$, where s denotes the *time*. A *local motion* of a point in $\mathscr{B}$ is defined similiarly. The restriction of a (local) motion to the half-open interval $s \in (-\infty, t]$ is called a *(local) history* up to the instant t.

By a previous remark, a motion $\varphi(s)$ induces a field of local motions $\varphi_{*p}(s)$ on $\mathscr{B}$.

---

[14] We assign positive orientation in $\mathscr{R}^3$ to the standard basis $\{(1, 0, 0), (0, 1, 0), (0, 0, 1)\}$.

In continuum mechanics, the contact forces at a point $\varphi(p)$ in a configuration φ are characterized by the *stress tensor* $\boldsymbol{T}$, which is a symmetric second-order tensor[15] on $\mathscr{R}^3$. Thus $\boldsymbol{T}$ corresponds to a linear map,

$$\boldsymbol{T}\colon\ \mathscr{R}^3 \to \mathscr{R}^3. \tag{2.5}$$

Physically, $\boldsymbol{T}$ transforms the unit normal of a smooth surface into the stress vector acting on the surface, at $\varphi(p)$.

Definition 2.4 (Noll). Let $p \in \mathscr{B}$. Suppose that in any motion $\varphi(s)$, and at any instant t, the stress tensor at $\varphi(t, p)$ is uniquely determined by the local history $\varphi_{*\,p}(s)$, $s \in (-\infty, t]$. Then p is called a *simple material particle*, or briefly, a *simple particle*.

A distinguished local configuration $\boldsymbol{r}(p)$ of p is called a *local reference configuration*. Relative to $\boldsymbol{r}(p)$, a local history $\varphi_{*\,p}(s)$ can be represented by the *local deformation history* $\boldsymbol{F}(s)$, which is defined by

$$\boldsymbol{F}(s) \equiv \varphi_{*\,p}(s) \circ \boldsymbol{r}(p)^{-1}, \qquad s \in (-\infty, t]. \tag{2.6}$$

The physical meaning of a local deformation history may be visualized in the following way: Suppose that ψ is a configuration of $\mathscr{B}$ such that

$$\boldsymbol{r}(p) = \psi_{*\,p}. \tag{2.7}$$

Then $\boldsymbol{F}(s)$ is the induced linear map of the diffeomorphism

$$\varphi(s) \circ \psi^{-1}\colon\ \psi(\mathscr{B}) \to [\varphi(s)](\mathscr{B}) \tag{2.8}$$

at the position $\psi(p)$, for each fixed s, *i.e.*,

$$\boldsymbol{F}(s) = [\varphi(s) \circ \psi^{-1}]_{*\,\psi(p)}. \tag{2.9}$$

Since (2.8) represents a *deformation* of the open domain $\psi(\mathscr{B})$ in $\mathscr{R}^3$, $\boldsymbol{F}(s)$ is also called the *deformation gradient*. Notice that, by definition, $\boldsymbol{F}(s)$ is an orientation-preserving isomorphism of $\mathscr{R}^3$,

$$\boldsymbol{F}(s)\colon\ \mathscr{R}^3 \to \mathscr{R}^3, \tag{2.10}$$

so that

$$\det \boldsymbol{F}(s) > 0. \tag{2.11}$$

Now suppose that p is a simple particle. Then the stress tensor is given by a fixed relation

$$\boldsymbol{T}(t, p) = \mathop{\mathfrak{F}}_{s=-\infty}^{t} \left(\varphi_{*\,p}(s), p\right), \tag{2.12}$$

called the *constitutive equation*, where $\mathfrak{F}$ is called the *response functional*. Relative to a fixed reference configuration $\boldsymbol{r}(p)$, (2.12) can be represented by the equation

$$\boldsymbol{T}(t, p) = \mathop{\mathfrak{G}}_{s=-\infty}^{t} \left(\boldsymbol{F}(s), p\right), \tag{2.13}$$

where $\mathfrak{G}$ is called the *response functional relative to* $\boldsymbol{r}(p)$.

[15] We consider $\mathscr{R}^3$ to be equipped with the usual Euclidean inner product.

Note. The dependence of $\mathfrak{G}$ on $\boldsymbol{r}(p)$ has been suppressed in the notation.

Definition 2.5 (NOLL). Let p and q be simple particles. Then they are called *materially isomorphic* if there exist local reference configurations $\boldsymbol{r}(p)$ and $\boldsymbol{r}(q)$ such that

$$\underset{s=-\infty}{\overset{t}{\mathfrak{G}}}(\boldsymbol{F}(s), p) \equiv \underset{s=-\infty}{\overset{t}{\mathfrak{G}}}(\boldsymbol{F}(s), q) \tag{2.14}$$

for all $\boldsymbol{F}(s)$.

Proposition 2.1. *p and q are materially isomorphic if and only if there exists an isomorphism*

$$\boldsymbol{r}(p, q)\colon \mathscr{B}_p \to \mathscr{B}_q \tag{2.15}$$

such that

$$\underset{s=-\infty}{\overset{t}{\mathfrak{F}}}(\varphi_{*q}(s), q) \equiv \underset{s=-\infty}{\overset{t}{\mathfrak{F}}}(\varphi_{*q}(s) \circ \boldsymbol{r}(p, q), p) \tag{2.16}$$

for all motion $\varphi(s)$.

The proof is obvious, and we call such isomorphisms $\boldsymbol{r}(p, q)$ *material isomorphisms* of p and q. Clearly, we have the following

Proposition 2.2. *If $\boldsymbol{r}(p)$ and $\boldsymbol{r}(q)$ are local reference configurations such that the mapping*

$$\boldsymbol{r}(p, q) = \boldsymbol{r}(q)^{-1} \circ \boldsymbol{r}(p) \tag{2.17}$$

is a material isomorphism, then the response functionals of p and q relative to $\boldsymbol{r}(p)$ and $\boldsymbol{r}(q)$ satisfy the identity (2.14). *Conversely,* if (2.14) *holds for some $\boldsymbol{r}(p)$ and $\boldsymbol{r}(q)$, then the mapping $\boldsymbol{r}(p, q)$ defined by* (2.17) *is a material isomorphism of p and q.*

Definition 2.6 (NOLL). Suppose that $\mathscr{B}$ consists entirely of simple particles. Then $\mathscr{B}$ is called *materially uniform* if its particles are pairwise materially isomorphic.

Definition 2.7. Suppose that $\mathscr{B}$ is materially uniform. Then a *reference chart* for $\mathscr{B}$ is a pair $(\mathscr{U}_\alpha, \boldsymbol{r}_\alpha)$ consisting of an open set $\mathscr{U}_\alpha$, called a *reference neighborhood*, and a smooth field $\boldsymbol{r}_\alpha$ of local reference configurations on $\mathscr{U}_\alpha$, called a *reference map*, such that relative to $\boldsymbol{r}_\alpha(p)$, $p \in \mathscr{U}_\alpha$, the response functionals $\mathfrak{G}$ are independent of p. In other words, there exists a functional $\mathfrak{G}_\alpha$, such that

$$\underset{s=-\infty}{\overset{t}{\mathfrak{G}_\alpha}}(\boldsymbol{F}(s)) \equiv \underset{s=-\infty}{\overset{t}{\mathfrak{G}}}(\boldsymbol{F}(s), p), \quad \forall p \in \mathscr{U}_\alpha. \tag{2.18}$$

We call $\mathfrak{G}_\alpha$ the *response functional relative to* $(\mathscr{U}_\alpha, \boldsymbol{r}_\alpha)$.

Remark. A reference map need *not* be an induced field of a configuration of the reference neighborhood. Moreover, from Proposition 2.1, a smooth field $\boldsymbol{r}_\alpha$ is a reference map on $\mathscr{U}_\alpha$ if and only if the mappings

$$\boldsymbol{r}_\alpha(p, q) \equiv \boldsymbol{r}_\alpha(q)^{-1} \circ \boldsymbol{r}_\alpha(p) \tag{2.19}$$

are material isomorphisms for all $p, q \in \mathscr{U}_\alpha$.

Definition 2.8. Two reference charts $(\mathscr{U}_\alpha, \boldsymbol{r}_\alpha)$ and $(\mathscr{U}_\beta, \boldsymbol{r}_\beta)$ are said to be *compatible* if their corresponding response functionals $\mathfrak{G}_\alpha$ and $\mathfrak{G}_\beta$ are identical.

Definition 2.9. A collection $\mathfrak{U}=\{(\mathscr{U}_\alpha, \boldsymbol{r}_\alpha), \alpha\in I\}$ of mutually compatible reference charts is called a *reference atlas* of $\mathscr{B}$ if it is maximal, and if $\{\mathscr{U}_\alpha, \alpha\in I\}$ is an open covering of $\mathscr{B}$.

Clearly, the response functionals $\mathfrak{G}_\alpha$, $\alpha\in I$, corresponding to the reference charts $(\mathscr{U}_\alpha, \boldsymbol{r}_\alpha)$ within a reference atlas $\mathfrak{U}$ are independent of α. We put

$$\mathop{\mathfrak{G}_{\mathfrak{U}}}_{s=-\infty}^{t}(\boldsymbol{F}(s))\equiv \mathop{\mathfrak{G}_{\alpha}}_{s=-\infty}^{t}(\boldsymbol{F}(s)). \tag{2.20}$$

Naturally, $\mathfrak{G}_{\mathfrak{U}}$ is called the *response functional relative to* $\mathfrak{U}$.

Definition 2.10. $\mathscr{B}$ is called a *smoth, materially uniform, simple body* if it can be equipped with a reference atlas.

For brevity, in this paper such bodies are called *simple bodies*, or even *bodies*. The response functionals form a *smooth* field on $\mathscr{B}$ in the sense that the reference maps are smooth. While in the theory of NOLL[16], essentially, a simple body must be covered by one reference neighborhood, the present theory is not subject to such a restriction. Since the reference charts are introduced to characterize the *smoothness* of the field of response functionals, which is obviously a *local* property, the more general definition seems to be a natural one. In Section 4 we shall see examples of simple bodies that cannot be covered by a single reference neighborhood.

Proposition 2.3. *Let $\mathscr{B}$ be a simple body. Suppose that $\mathfrak{U}=\{(\mathscr{U}_\alpha, \boldsymbol{r}_\alpha), \alpha\in I\}$ is a reference atlas of $\mathscr{B}$, and $\boldsymbol{K}$ is an isomorphism of $\mathscr{R}^3$ such that*

$$\det \boldsymbol{K}>0. \tag{2.21}$$

Then the collection

$$\boldsymbol{K}\mathfrak{U}\equiv\{(\mathscr{U}_\alpha, \boldsymbol{K}\circ \boldsymbol{r}_\alpha), \alpha\in I\} \tag{2.22}$$

is also a reference atlas of $\mathscr{B}$. Further, any two reference atlases of $\mathscr{B}$ can be related in this manner, by some $\boldsymbol{K}$.

The proof is obvious. Suppose that $\mathfrak{V}=\{(\mathscr{V}_\beta, s_\beta), \beta\in J\}$ is another reference atlas of $\mathscr{B}$. We pick an arbitrary fixed point $p\in\mathscr{B}$ and let

$$\boldsymbol{K}\equiv s_\beta(p)\circ \boldsymbol{r}_\alpha(p)^{-1}, \tag{2.23}$$

where α and β are chosen in such a way that $p\in\mathscr{U}_\alpha\cap\mathscr{V}_\beta$. Then it is easily verified that

$$\mathfrak{V}=\boldsymbol{K}\mathfrak{U}. \tag{2.24}$$

Obviously, the response functionals $\mathfrak{G}_{\mathfrak{U}}$ and $\mathfrak{G}_{\mathfrak{V}}$ are related by the identity

$$\mathop{\mathfrak{G}_{\mathfrak{U}}}_{s=-\infty}^{t}(\boldsymbol{F}(s)\,\boldsymbol{K})\equiv \mathop{\mathfrak{G}_{\mathfrak{V}}}_{s=-\infty}^{t}(\boldsymbol{F}(s)). \tag{2.25}$$

Remarks. The mapping $\boldsymbol{K}$ in (2.24), in general, is *not* unique, since a reference atlas may be non-trivially related to itself, *i.e.*, there may exist isomorphisms $\boldsymbol{G}$ of $\mathscr{R}^3$, called *reference isomorphisms relative to* $\mathfrak{U}$, such that

$$\mathfrak{U}=\boldsymbol{G}\mathfrak{U}. \tag{2.26}$$

---

[16] *Cf.* TRUESDELL & NOLL [*1*, § 34].

From (2.25), such isomorphisms can be characterized by the condition

$$\underset{s=-\infty}{\overset{t}{\mathfrak{G}_{\mathfrak{U}}}}(F(s)) \equiv \underset{s=-\infty}{\overset{t}{\mathfrak{G}_{\mathfrak{U}}}}(F(s)\,G). \tag{2.27}$$

The collection of all reference isomorphisms relative to $\mathfrak{U}$ forms a group $g(\mathfrak{U})$, called the *isotropy group relative to* $\mathfrak{U}$. We shall consider the structures of the isotropy groups in the next section.

On the Orientations of a Simple Body

In NOLL's original work, while no orientation was introduced for a simple particle or a simple body, the response functional of a simple particle p was assumed to obey the *principle of material frame-indifference, i.e.*,

$$\underset{s=-\infty}{\overset{t}{\mathfrak{G}}}(Q(s)\,F(s), p) \equiv Q(t) \underset{s=-\infty}{\overset{t}{\mathfrak{G}}}(F(s), p)\, Q(t)^T, \tag{2.28}$$

for all orthogonal tensor histories $Q(s)$. In particular,

$$\underset{s=-\infty}{\overset{t}{\mathfrak{G}}}(-F(s), p) \equiv \underset{s=-\infty}{\overset{t}{\mathfrak{G}}}(F(s), p). \tag{2.29}$$

Hence, without loss of generality, all simple particles can be oriented. For the sake of simplicity in some minor technical details to be considered later, in this paper we choose to orient the particles. If the reader prefers to follow NOLL's original formulation, he can simply drop the term *oriented* in our definition 2.1 and then modify the other definitions accordingly. NOLL[17] determined the general solution of the identity (2.28), but it plays only a minor role in this paper.

3. The Isotropy Groups and their Representations

From now on $\mathscr{B}$ will be a simple body.

Definition 3.1 (NOLL). Let $p \in \mathscr{B}$. Then the *isotropy group* $g(p)$ is the group of all material isomorphism of p with itself.

More specifically, an isomorphism

$$h\colon \mathscr{B}_p \to \mathscr{B}_p \tag{3.1}$$

is a member of $g(p)$ if and only if

$$\underset{s=-\infty}{\overset{t}{\mathfrak{F}}}(\varphi_{*\,p}(s), p) \equiv \underset{s=-\infty}{\overset{t}{\mathfrak{F}}}(\varphi_{*\,p}(s) \circ h, p) \tag{3.2}$$

for all motions φ. Clearly, $g(p)$ forms a subgroup of the general linear group $\mathscr{GL}(\mathscr{B}_p)$ of $\mathscr{B}_p$. Following NOLL, we require that[18] $g(p)$ be a subgroup of the special linear group $\mathscr{SL}(\mathscr{B}_p)$, which consists of isomorphisms of $\mathscr{B}_p$ whose determinants are equal to 1.

By a standard procedure in tensor algebra, any element $h \in \mathscr{GL}(\mathscr{B}_p)$ gives rise to various induced transformations of the tensor spaces over $\mathscr{B}_p$.

[17] NOLL [*2*], TRUESDELL & NOLL [*1*, § 29].

[18] See also GURTIN & WILLIAMS [*17*].

Definition 3.2. A tensor at p is called *intrinsic* (or *material*) if it is invariant under all induced transformations of material isomorphisms of p.

Proposition 3.1. *All scalars and volume tensors* (i.e., *elements in* $\Lambda^{3,0}(\mathscr{B}_p)$ *or* $\Lambda^{0,3}(\mathscr{B}_p)$) *are intrinsic.*

The proof is obvious. *Note.* If we do not introduce a fixed orientation for $\mathscr{B}$, then the volume tensors are not intrinsic (but the so-called scalar densities are).

Definition 3.3 (NOLL). Let $\boldsymbol{r}(p)$ be a local reference configuration of p, and let $\mathfrak{G}$ be the response functional relative to $\boldsymbol{r}(p)$. Then an isomorphism

$$\boldsymbol{G}\colon \mathscr{R}^3 \to \mathscr{R}^3 \tag{3.3}$$

is called a *material isomorphism relative to* $\boldsymbol{r}(p)$ if

$$\overset{t}{\underset{s=-\infty}{\mathfrak{G}}}(\boldsymbol{F}(s), p) \equiv \overset{t}{\underset{s=-\infty}{\mathfrak{G}}}(\boldsymbol{F}(s)\,\boldsymbol{G}, p) \tag{3.4}$$

for all $\boldsymbol{F}$. The collection $\mathscr{G}(p)$ of all such isomorphisms forms the *isotropy group relative to* $\boldsymbol{r}(p)$.

Proposition 3.2 (NOLL). *The isotropy groups* $\mathscr{G}(p)$ *and* $g(p)$ *are related by*

$$\mathscr{G}(p) = \boldsymbol{r}(p) \circ g(p) \circ \boldsymbol{r}(p)^{-1}. \tag{3.5}$$

In particular, $\mathscr{G}(p)$ *is a subgroup of the special linear group* $\mathscr{S}\mathscr{L}(3)$ (over $\mathscr{R}^3$).

Proposition 3.3 (NOLL). *Let* $\boldsymbol{r}_\alpha(p)$ *and* $\boldsymbol{r}_\beta(p)$ *be local reference configurations of* p. *Then*

$$\mathscr{G}_\beta(p) = \boldsymbol{A}\,\mathscr{G}_\alpha(p)\,\boldsymbol{A}^{-1}, \tag{3.6}$$

where $\mathscr{G}_\alpha(p)$ *and* $\mathscr{G}_\beta(p)$ *are the isotropy groups relative to* $\boldsymbol{r}_\alpha(p)$ *and* $\boldsymbol{r}_\beta(p)$ *respectively, and where*

$$\boldsymbol{A} = \boldsymbol{r}_\beta(p) \circ \boldsymbol{r}_\alpha(p)^{-1}. \tag{3.7}$$

In particular, if $\boldsymbol{A} \in \mathscr{G}_\alpha(p)$, *then*

$$\mathscr{G}_\alpha(p) = \mathscr{G}_\beta(p). \tag{3.8}$$

Proposition 3.4 (NOLL). *Suppose that* $p, q \in \mathscr{B}$ *and* $\boldsymbol{r}(p, q)$ *is a material isomorphism. Then*

$$g(q) = \boldsymbol{r}(p, q) \circ g(p) \circ \boldsymbol{r}(p, q)^{-1}. \tag{3.9}$$

Consequently, if $\boldsymbol{r}(p)$ *and* $\boldsymbol{r}(q)$ *are local reference configurations which are related by* (2.17), *then*

$$\mathscr{G}(p) = \mathscr{G}(q). \tag{3.10}$$

In particular, the isotropy groups $\mathscr{G}_\alpha(p)$ relative to the reference map $\boldsymbol{r}_\alpha$ are independent of p. We denote this fixed group by $\mathscr{G}_\alpha$, called the *isotropy relative to* $(\mathscr{U}_\alpha, \boldsymbol{r}_\alpha)$. Clearly, $\boldsymbol{G} \in \mathscr{G}_\alpha$ if and only if

$$\overset{t}{\underset{s=-\infty}{\mathfrak{G}_\alpha}}(\boldsymbol{F}(s)) \equiv \overset{t}{\underset{s=-\infty}{\mathfrak{G}_\alpha}}(\boldsymbol{F}(s)\,\boldsymbol{G}). \tag{3.11}$$

Similarly, if $(\mathscr{U}_\alpha, \boldsymbol{r}_\alpha) \in \mathfrak{U}$, then $\mathscr{G}_\alpha$ are independent of α. We put

$$\mathscr{G}(\mathfrak{U}) \equiv \mathscr{G}_\alpha, \tag{3.12}$$

called the *isotropy group relative to* $\mathfrak{U}$. Then $G \in \mathscr{G}(\mathfrak{U})$ if and only if

$$\underset{s=-\infty}{\overset{t}{\mathfrak{G}_{\mathfrak{U}}}}(F(s)) \equiv \underset{s=-\infty}{\overset{t}{\mathfrak{G}_{\mathfrak{U}}}}(F(s)\,G). \tag{3.13}$$

Thus $\mathscr{G}(\mathfrak{U})$ consists of all reference isomorphisms relative to $\mathfrak{U}$.

Proposition 3.5. *The isotropy groups* $\mathscr{G}(\mathfrak{U})$ *satisfy the transformation law*

$$\mathscr{G}(K\,\mathfrak{U}) = K\,\mathscr{G}(\mathfrak{U})\,K^{-1} \tag{3.14}$$

for all (orientation-preserving) isomorphisms K *of* $\mathscr{R}^3$.

For simplicity, we denote the collection of all orientation-preserving isomorphisms of $\mathscr{R}^3$ by $\mathscr{G}\mathscr{L}(3)$. Thus $K \in \mathscr{G}\mathscr{L}(3)$ if and only if it satisfies (2.21).

As before, the material isomorphisms $r(p, q)$ induce various transformations of tensor spaces at p and q.

Definition 3.4. A tensor field μ on $\mathscr{B}$ is called *intrinsic* (or *material*) if it is invariant under all induced transformations by material isomorphisms.

Proposition 3.6. *A tensor field* μ *is intrinsic if and only if its value at any fixed point is an intrinsic tensor, and for each reference chart* $(\mathscr{U}_\alpha, r_\alpha) \in \mathfrak{U}$ *it has the following representation:*

$$\mu(q) = \otimes\, r_\alpha(p, q) \circ \mu(p), \qquad q \in \mathscr{U}_\alpha, \tag{3.15}$$

where p *is a fixed reference point in* $\mathscr{U}_\alpha$, *and* $\otimes\, r_\alpha(p, q)$ *denotes the appropriate induced transformation by* $r_\alpha(p, q)$. *In particular, every intrinsic field is smooth, and any intrinsic tensor at a fixed point can be extended to a unique intrinsic field on* $\mathscr{B}$. *Thus any two intrinsic fields that agree at one point are identical everywhere.*

Combining propositions 3.1 and 3.6, we obtain.

Proposition 3.7. *Every simple body* $\mathscr{B}$ *can be equipped with a (signed) intrinsic volume tensor field* ε^*.

As usual, any two such volume tensor fields differ by a (constant) scalar multiple. Suppose that ε^* is a positive intrinsic volume tensor field on $\mathscr{B}$. Then there exists an *induced (intrinsic) measure* m_{ε^*} on $\mathscr{B}$, which is defined in the following way: Let φ be a configuration of $\mathscr{B}$, say

$$\varphi(x) = (x^1, x^2, x^3). \tag{3.16}$$

Then we can express ε^* in component form relative to (x^i), say

$$\varepsilon^*(x) = 3!\,\bar{\varepsilon}(x)\,dx^1 \wedge dx^2 \wedge dx^3, \tag{3.17}$$

or in the standard tensor form:

$$\varepsilon^*(x) = \varepsilon_{ijk}(x)\,dx^i \otimes dx^j \otimes dx^k, \tag{3.18}$$

where ε_{ijk} is completely skew-symmetric in i, j, k, and

$$\varepsilon_{123} = \bar{\varepsilon}. \tag{3.19}$$

Since ε^* is an intrinsic field, the function $\bar{\varepsilon}$ is smooth. Now let $\mathscr{P}$ be any Borel set in $\mathscr{B}$. Then we define

$$m_{\varepsilon^*}(\mathscr{P}) = \int_{\varphi(\mathscr{P})} \bar{\varepsilon}(x)\, dx^1\, dx^2\, dx^3. \tag{3.20}$$

Clearly, the quantity $m_{\varepsilon^*}(\mathscr{P})$ does not depend on the choice of the coordinate system (x^i), and the additive set function m_{ε^*} is a measure on $\mathscr{B}$.

It is commonly assumed that a simple body $\mathfrak{B}$ has a fixed *mass measure* m. By hypothesis, the Radon-Nikodym derivative[19] of m with respect to m_{ε^*} is a constant on $\mathscr{B}$, *i.e.*,

$$m(\mathscr{P}) = c\, m_{\varepsilon^*}(\mathscr{P}) \tag{3.21}$$

for all Borel sets $\mathscr{P}$ in $\mathscr{B}$. For definiteness, we fix the intrinsic field ε^*, called *the* volume tensor of $\mathscr{B}$, by requiring that the constant c in (3.21) be unity.

By the standard duality of $\wedge^{3.0}$ and $\wedge^{0.3}$, the dual of ε^* is an intrinsic field ε on $\mathscr{B}$, called *the specific volume* of $\mathscr{B}$, such that

$$\langle \varepsilon^*, \varepsilon \rangle = 3!. \tag{3.22}$$

In component form, we then have

$$\varepsilon(x) = \frac{3!}{\bar{\varepsilon}(x)} \frac{\partial}{\partial x^1} \wedge \frac{\partial}{\partial x^2} \wedge \frac{\partial}{\partial x^3}, \tag{3.23}$$

or equivalently,

$$\varepsilon(x) = \varepsilon^{ijk}(x) \frac{\partial}{\partial x^i} \otimes \frac{\partial}{\partial x_j} \otimes \frac{\partial}{\partial x^k}, \tag{3.24}$$

where ε^{ijk} are completely skew in i, j, k, and

$$\varepsilon^{123} = \frac{1}{\bar{\varepsilon}} = \frac{1}{\varepsilon_{123}}, \tag{3.25}$$

so that (3.22) can be written in the form

$$\varepsilon_{ijk}\, \varepsilon^{ijk} = 3!. \tag{3.26}$$

In terms of the isotropy groups, the compatibility condition[20] of reference charts within a fixed reference atlas $\mathfrak{U}$ may be stated in the following way: On the overlaps of any two reference neighborhoods, say $\mathscr{U}_\alpha$ and $\mathscr{U}_\beta$, the corresponding reference maps $\boldsymbol{r}_\alpha$ and $\boldsymbol{r}_\beta$ are related in such a way that the fields

$$\boldsymbol{G}_{\alpha\beta}(p) = \boldsymbol{r}_\alpha(p) \circ \boldsymbol{r}_\beta(p)^{-1} \tag{3.27}$$

are smooth, and their values belong to the isotropy group $\mathscr{G}(\mathfrak{U})$.

Proposition 3.8. *The fields* $\boldsymbol{G}_{\alpha\beta}$ *satisfy the following identities:*

$$\begin{aligned} &\boldsymbol{G}_{\alpha\alpha}(p) = \boldsymbol{1}, \quad p \in \mathscr{U}_\alpha, \\ &\boldsymbol{G}_{\alpha\beta}(p) = \boldsymbol{G}_{\beta\alpha}(p)^{-1}, \quad p \in \mathscr{U}_\alpha \cap \mathscr{U}_\beta, \end{aligned} \tag{3.28}$$

and

$$\boldsymbol{G}_{\alpha\beta}(p)\, \boldsymbol{G}_{\beta\gamma}(p) = \boldsymbol{G}_{\alpha\gamma}(p), \quad p \in \mathscr{U}_\alpha \cap \mathscr{U}_\beta \cap \mathscr{U}_\gamma.$$

---

[19] *Cf.* HALMOS [*18*, § 31].

[20] *Cf.* Definition 2.8.

As we shall see, the fields $G_{\alpha\beta}$ will play the important role of the *coordinate transformations* on the *material tangent bundle* $\mathcal{T}(\mathcal{B}, \mathfrak{U})$ and the *bundle of reference frames* $\mathcal{E}(\mathfrak{B}, \mathfrak{U})$, to be introduced in the next two sections.

Definition 3.5 (NOLL). A simple particle p is called a *solid* particle if there exists a local reference configuration $\boldsymbol{r}(p)$ relative to which the isotropy group $\mathcal{G}(p)$ is contained in the orthogonal group $\mathcal{Q}(3)$ (over $\mathcal{R}^3$).

Special cases of solids are furnished by the classical crystal classes, transversely isotropic solids, and isotropic solids.

Definition 3.6. A non-solid simple particle is called a *fluid crystal* particle[21].

Simple fluids and simple subfluids[22] are among the special cases of fluid crystals.

Evidently, these terms for simple particles can be applied also to materially uniform simple bodies without any ambiguity.

In the various examples of simple particles cited above, the isotropy groups are *closed Lie subgroups* of the special linear group[23]. As was mentioned in the introduction, so as to be able to apply the theory of fibre bundles, we shall restrict attention in this paper to simple bodies that share this property.

We note here a sufficient condition for it. By CARTAN's theorem[24], every closed subgroup of the special linear group is a Lie subgroup. Clearly, the isotropy group $\mathcal{G}(\mathfrak{U})$ is closed if the response functional $\mathfrak{G}_{\mathfrak{U}}$ is continuous in the following sense: Let $\{G_n\}$ be a convergent sequence in $\mathcal{GL}(3)$, say

$$\lim_{n\to\infty} \boldsymbol{G}_n = \boldsymbol{G}. \tag{3.29}$$

Then

$$\lim_{n\to\infty} \mathop{\mathfrak{G}_{\mathfrak{U}}}_{s=-\infty}^{t} (\boldsymbol{F}(s)\,\boldsymbol{G}_n) = \mathop{\mathfrak{G}_{\mathfrak{U}}}_{s=-\infty}^{t} (\boldsymbol{F}(s)\,\boldsymbol{G}) \tag{3.30}$$

for all $\boldsymbol{F}$.

Remark. Certain smoothness conditions on the response functionals of simple particles have been considered in the *theories of fading memory*[25]. The continuity condition (3.30) above is a necessary condition in all such theories proposed so far.

Part II. The Geometric Structure of a Simple Body

4. The Geometric and the Material Tangent Bundles of a Simple Body

In differential geometry, the overall coherence of the tangent spaces of a differentiable manifold $\mathcal{M}$ is characterized by the *(geometric) tangent bundle* $\mathcal{T}(\mathcal{M})$. For a simple body $\mathcal{B}$, the geometric tangent bundle $\mathcal{T}(\mathcal{B})$ alone, however, *cannot* characterize the material structure of $\mathcal{B}$, since the bundle $\mathcal{T}(\mathcal{B})$, being a mathematical structure set up for the body manifold, is independent of the field

---

[21] It is also referred to by the name of *simple liquid crystal particle*, *cf.* COLEMAN [*19*].

[22] *Cf.* WANG [*20*].

[23] We assume that a simple particle is oriented.

[24] CHEVALLEY [*21*, Ch. IV, § XIV], COHN [*22*, Ch. VI, § 6.5].

[25] *Cf.* COLEMAN & NOLL [*23*], WANG [*24*, *25*], COLEMAN & MIZEL [*26*] and MIZEL & WANG [*27*].

of response functionals on $\mathscr{B}$. So as to represent the material structure, we introduce the notion of the *material tangent bundles of* a simple body.

Since some readers may not be familiar with the theory of *fibre bundles*, we first review some basic concepts of the geometric tangent bundle $\mathscr{T}(\mathscr{M})$. By definition, the *bundle space*[26] $\mathscr{T}(\mathscr{M})$ is a differentiable manifold formed by the disjoint union

$$\mathscr{T}(\mathscr{M}) = \bigcup_{p \in \mathscr{M}} \mathscr{M}_p, \tag{4.1}$$

where $\mathscr{M}_p$ denotes the tangent space at p. Thus the generic element of $\mathscr{T}(\mathscr{M})$ is a pair $(p, \boldsymbol{v})$, where $p \in \mathscr{M}$ and $\boldsymbol{v} \in \mathscr{M}_p$. We define the *projection map* $\pi\colon \mathscr{T}(\mathscr{M}) \to \mathscr{M}$ by

$$\pi(p, \boldsymbol{v}) = p, \tag{4.2}$$

whence $\mathscr{M}$ plays the role of the *base space* of $\mathscr{T}(\mathscr{M})$. For any set $\mathscr{U} \subset \mathscr{M}$, we put

$$\mathscr{T}(\mathscr{U}) \equiv \pi^{-1}(\mathscr{U}) = \bigcup_{p \in \mathscr{U}} \mathscr{M}_p. \tag{4.3}$$

In particular,

$$\mathscr{M}_p = \pi^{-1}(p), \tag{4.4}$$

and it is also called the *fibre* of $\mathscr{T}(\mathscr{M})$ at p.

The tangent bundle $\mathscr{T}(\mathscr{M})$ is equipped with a *bundle atlas*, which is a collection

$$\varphi = \{(\mathscr{U}_\alpha, \varphi_\alpha), \alpha \in K\} \tag{4.5}$$

whose elements $(\mathscr{U}_\alpha, \varphi_\alpha)$ are called *bundle charts*. These are pairs consisting of open sets $\mathscr{U}_\alpha \subset \mathscr{M}$, called *coordinate neighborhoods*, and mappings

$$\varphi_\alpha\colon\ \mathscr{U}_\alpha \times \mathscr{R}^n \to \mathscr{T}(\mathscr{U}_\alpha), \tag{4.6}$$

called *bundle maps*. Here n denotes the dimension of $\mathscr{M}$. Such bundle maps are diffeomorphisms satisfying the conditions

$$\varphi_\alpha(\{p\} \times \mathscr{R}^n) = \mathscr{M}_p, \qquad \forall\, p \in \mathscr{U}_\alpha, \tag{4.7}$$

whence there exist mappings

$$\varphi_{\alpha, p}\colon\ \mathscr{R}^n \to \mathscr{M}_p \tag{4.8}$$

for all $p \in \mathscr{U}_\alpha$. Thus $\mathscr{R}^n$ plays the role of the *fibre space*, and all fibres are copies of it via the isomorphisms $\varphi_{\alpha, p}$. On the overlaps of coordinate neighborhoods, say $\mathscr{U}_\alpha \cap \mathscr{U}_\beta$, we define the *coordinate transformations*

$$\boldsymbol{G}_{\alpha\beta}(p) \equiv \varphi_{\alpha, p}^{-1} \circ \varphi_{\beta, p}\colon\ \mathscr{R}^n \to \mathscr{R}^n. \tag{4.9}$$

They are required to be smooth fields with values in $\mathscr{GL}(n)$, which is the *structure group* of $\mathscr{T}(\mathscr{M})$.

We now define the bundle charts $(\mathscr{U}_\alpha, \varphi_\alpha)$ of $\mathscr{T}(\mathscr{M})$. Suppose that the collection

$$\psi = \{(\mathscr{U}_\alpha, \psi_\alpha), \alpha \in J\} \tag{4.10}$$

---

[26] As usual, we denote the fibre bundle and its bundle space by the same symbol.

is the *atlas* of $\mathscr{M}$, *i.e.*, each ψ_α corresponds to a local coordinate system, say $(x^1, \ldots, x^n)$, on the coordinate neighborhood $\mathscr{U}_\alpha$. Then we define the mappings φ_α in (4.6) by

$$\varphi_\alpha(p, v^1, \ldots, v^n) = (p, \boldsymbol{v}), \tag{4.11}$$

where $\boldsymbol{v} \in \mathscr{M}_p$ whose components in $(x^1, \ldots, x^n)$ are $(v^1, \ldots, v^n)$, *viz*

$$\boldsymbol{v} = v^i \left.\frac{\partial}{\partial x^i}\right|_p. \tag{4.12}$$

Such φ_α clearly satisfy the condition (4.7). Moreover, if $(\bar{x}^1, \ldots, \bar{x}^n)$ denotes the local coordinate system corresponding to $(\mathscr{U}_\beta, \psi_\beta)$, then the coordinate transformation $\boldsymbol{G}_{\alpha\beta}$ is given by

$$\boldsymbol{G}_{\alpha\beta}(p) = \left[\left.\frac{\partial x^i}{\partial \bar{x}^j}\right|_p\right] \in \mathscr{GL}(n), \tag{4.13}$$

which forms a smooth field on $\mathscr{U}_\alpha \cap \mathscr{U}_\beta$. Consequently, there exists a unique differentiable structure on $\mathscr{T}(\mathscr{M})$ such that the mappings φ_α become diffeomorphisms. We define the bundle atlas φ by maximizing the collection

$$\varphi' = \{(\mathscr{U}_\alpha, \varphi_\alpha), \alpha \in J\} \tag{4.14}$$

corresponding to the atlas ψ of $\mathscr{M}$. Notice that, the elements $(\mathscr{U}_\alpha, \varphi_\alpha) \in \varphi'$ give rise to local coordinate systems $(x^1, \ldots, x^n, v^1, \ldots, v^n)$ on $\mathscr{T}(\mathscr{U}_\alpha)$, called the *lifted coordinate systems.*

Now let $\mathscr{B}$ be a simple body. Then we can construct its geometric tangent bundle $\mathscr{T}(\mathscr{B})$ in the above manner. The general structure of $\mathscr{T}(\mathscr{B})$ is very simple. In fact, if $(\mathscr{B}, \psi_\alpha)$ induces a global coordinate system on $\mathscr{B}$, then the corresponding bundle map

$$\varphi_\alpha: \mathscr{B} \times \mathscr{R}^3 \to \mathscr{T}(\mathscr{B}) \tag{4.15}$$

is a diffeomorphism of $\mathscr{T}(\mathscr{B})$ with the product manifold $\mathscr{B} \times \mathscr{R}^3$, so that $\mathscr{T}(\mathscr{B})$ is simply a copy of a cylinder in the Euclidean space $\mathscr{R}^6$. Of course, this general structure of $\mathscr{T}(\mathscr{B})$ is not of interest to us, since it *cannot* characterize the material structure of $\mathscr{B}$.

In order to characterize the material structure, we need a special bundle atlas, called a *material atlas*, such that the bundle charts represent the local distributions of the response of particles, and globally, these bundle charts are patched together according to the rule of material isomorphism. More specifically, we state

Definition 4.1. A bundle chart $(\mathscr{U}_\alpha, \varphi_\alpha)$ of $\mathscr{T}(\mathscr{B})$ is called a *material chart* if the mappings

$$\boldsymbol{r}_\alpha(p, q) \equiv \varphi_{\alpha, q} \circ \varphi_{\alpha, p}^{-1}: \mathscr{B}_p \to \mathscr{B}_q \tag{4.16}$$

are material isomorphisms for all $p, q \in \mathscr{U}_\alpha$. Two material charts $(\mathscr{U}_\alpha, \varphi_\alpha)$ and $(\mathscr{U}_\beta, \varphi_\beta)$ are said to be *compatible* if the mappings

$$\boldsymbol{r}_{\alpha\beta}(p, q) \equiv \varphi_{\alpha, p} \circ \varphi_{\beta, q}^{-1}: \mathscr{B}_q \to \mathscr{B}_p \tag{4.17}$$

are material isomorphisms for all $p\in\mathscr{U}_\alpha$ and $q\in\mathscr{U}_\beta$. Finally, a maximal collection of pairwise compatible material charts is called a *material bundle atlas*, or briefly, a *material atlas* of $\mathscr{B}$.

We can visualize the physical meaning of a material atlas in the following way: By definition, any bundle chart $(\mathscr{U}_\alpha, \varphi_\alpha)$ of $\mathscr{T}(\mathscr{B})$ gives rise to a field of (orientation-preserving) isomorphisms

$$\varphi_{\alpha,p}\colon \mathscr{R}^3 \to \mathscr{B}_p, \qquad p\in\mathscr{U}_\alpha. \tag{4.18}$$

Hence their inverses are local configurations, say

$$\varphi_{\alpha,p}^{-1} \equiv \boldsymbol{r}_\alpha(p)\colon \mathscr{B}_p \to \mathscr{R}^3. \tag{4.19}$$

Clearly, such fields $\boldsymbol{r}_\alpha$ are smooth. Conversely, suppose that $\boldsymbol{r}_\alpha$ is a smooth field of local configurations on $\mathscr{U}_\alpha$. Then equation (4.19) *defines* the mapping φ_α, and the pair $(\mathscr{U}_\alpha, \varphi_\alpha)$ is a bundle chart of $\mathscr{T}(\mathscr{B})$. Therefore bundle charts of $\mathscr{T}(\mathscr{B})$ are in one-to-one correspondence with smooth fields of local configurations on coordinate neighborhoods in $\mathscr{B}$. Notice also that a field $\boldsymbol{r}_\alpha$ is the induced field of a configuration if and only if the corresponding bundle chart $(\mathscr{U}_\alpha, \varphi_\alpha)$ is a lifted chart (*i.e.*, $(\mathscr{U}_\alpha, \varphi_\alpha)\in\varphi'$). Hence the lifted charts are in one-to-one correspondence with configurations of coordinate neighborhoods.

In terms of the fields $\boldsymbol{r}_\alpha$, the mappings $\boldsymbol{r}_\alpha(p, q)$ in (4.16) can be represented by

$$\boldsymbol{r}_\alpha(p, q) = \boldsymbol{r}_\alpha(q)^{-1} \circ \boldsymbol{r}_\alpha(p). \tag{4.20}$$

Comparing this equation with (2.19), we see that the pair $(\mathscr{U}_\alpha, \varphi_\alpha)$ is a material chart if and only if the corresponding pair $(\mathscr{U}_\alpha, \boldsymbol{r}_\alpha)$ is a reference chart (*cf.* Section 2). Similarly, from (4.17), two material charts are compatible if and only if their corresponding reference charts are compatible in the earlier sense. Therefore material atlases are in one-to-one correspondence with reference atlases. In particular, *every simple body can be equipped with a material atlas.*

Let $\varphi(\mathfrak{U}) = \{(\mathscr{U}_\alpha, \varphi_\alpha), \alpha\in I\}$ be the material atlas corresponding to the reference atlas $\mathfrak{U} = \{(\mathscr{U}_\alpha, \boldsymbol{r}_\alpha), \alpha\in I\}$ in the above manner. Then the coordinate transformations for $\varphi(\mathfrak{U})$ are given by

$$\boldsymbol{G}_{\alpha\beta}(p) = \varphi_{\alpha,p}^{-1} \circ \varphi_{\beta,p} = \boldsymbol{r}_\alpha(p) \circ \boldsymbol{r}_\beta(p)^{-1}. \tag{4.21}$$

Comparing with equation (3.27), we see that these fields are smooth and have values in the isotropy group $\mathscr{G}(\mathfrak{U})$, which is a closed Lie subgroup of the structure group $\mathscr{GL}(3)$ of $\mathscr{T}(\mathscr{B})$. In the terminology of fibre bundles, we say that $\mathscr{T}(\mathscr{B})$ is *reducible* to a bundle $\mathscr{T}(\mathscr{B}, \mathfrak{U})$ whose structure group is $\mathscr{G}(\mathfrak{U})$. Naturally, we call $\mathscr{T}(\mathscr{B}, \mathfrak{U})$ the *material tangent bundle of* $\mathscr{B}$ *relative to* $\mathfrak{U}$.

We now characterize the relations among material atlases of $\mathscr{B}$. Recall that any two reference atlases are related by a mapping $\boldsymbol{K}$ as shown in (2.24). In view of the one-to-one correspondence between $\varphi(\mathfrak{U})$ and $\mathfrak{U}$, we see that

$$\varphi(\boldsymbol{K}\,\mathfrak{U}) = \{(\mathscr{U}_\alpha, \varphi_\alpha \circ \boldsymbol{K}^{-1}), \alpha\in I\}, \tag{4.22}$$

where $\varphi_\alpha \circ \boldsymbol{K}^{-1}$ are defined by

$$[\varphi_\alpha \circ \boldsymbol{K}^{-1}](p, v^1, v^2, v^3) = \varphi_{\alpha,p} \circ \boldsymbol{K}^{-1}(v^1, v^2, v^3), \tag{4.23}$$

for all $p \in \mathscr{U}_\alpha$. Then the coordinate transformations $\bar{G}_{\alpha\beta}$ for $\varphi(K\mathfrak{U})$ and those for $\varphi(\mathfrak{U})$ (denoted by $G_{\alpha\beta}$) are related by

$$\bar{G}_{\alpha\beta} = K G_{\alpha\beta} K^{-1} \tag{4.24}$$

for all $\alpha, \beta \in I$. This condition reflects the fact that the isotropy groups $G(\mathfrak{U})$ satisfy the transformation law (3.14). Since the relation (2.24) for reference atlases is exhaustive, so is the representation (4.22) for material atlases.

In general, a fibre bundle is said to be *trivial* if the base space can be covered by one coordinate neighborhood of the bundle atlas. For example, the geometric tangent bundle $\mathscr{T}(\mathscr{B})$ is trivial. From (4.23), a material tangent bundle of $\mathscr{B}$ is trivial if and only if all material tangent bundles of $\mathscr{B}$ are trivial. As remarked in Section 2, NOLL considers in his special theory simple bodies which can be covered by one reference neighborhood. Hence via the aforesaid one-to-one correspondence of material charts and reference charts, the material tangent bundles of such simple bodies must be trivial. The simple bodies considered in this paper do not always correspond to this simple situation, as we now see.

Example 4.1

Consider a simple body in the shape of a (thick) spherical shell which is made up of a transversely isotropic material in such a way that the axes of transverse isotropy are in the radial directions at all particles. Then a material atlas of this simple body corresponds to the bundle atlas of the (geometric) tangent bundle of the sphere $\mathscr{S}^2$. It is known[27] that the latter bundle is not trivial.

Example 4.2

Consider a rectangular rod made up of a cubic crystal in such a way that the crystalline axes are parallel to the edges of the rod at every particle. Now suppose that one end of this rod is twisted in the longitudinal direction through a right angle. Then the two ends are joined together smoothly to form a *Moebius crystal*[28]. Clearly, its material tangent bundles are not trivial[29].

Remark. Suppose that the isotropy groups $\mathscr{G}(\mathfrak{U})$ are *discrete*, and suppose that the body manifold $\mathscr{B}$ is *simply connected*. Then the material tangent bundles of $\mathscr{B}$ are necessarily trivial. This situation is assumed in most theories of continuous distributions of dislocations in lattices.

As remarked before, a material chart $(\mathscr{U}_\alpha, \varphi_\alpha)$, in general, need *not* belong to the pre-bundle atlas φ' of $\mathscr{T}(\mathscr{B})$ (cf. eq. (4.14)). We state

Definition 4.2. A simple body $\mathscr{B}$ is said to be *locally homogeneous* if for every point $p \in \mathscr{B}$, there exists a material chart $(\mathscr{U}_\alpha, \varphi_\alpha) \in \varphi'$, such that $\mathscr{U}_\alpha \ni p$.

Physically, the condition for the field of induced local configurations

$$\psi_{\alpha * p} = r_\alpha(p), \qquad p \in \mathscr{U}_\alpha, \tag{4.25}$$

to be a reference map is that relative to the coordinate system (x^1, x^2, x^3) induced by $(\mathscr{U}_\alpha, \psi_\alpha)$ the response functional is *independent* of the coordinates. For a

[27] STEENROD [*28*, § 27].

[28] *Cf.* Fig. 1 in the paper of FRANK [*29*].

[29] For a more detailed proof of this fact, see the remark at the end of the next section.

locally homogeneous simple body, it is necessary and sufficient that every material atlas $\varphi(\mathfrak{U})$ results from maximizing a prebundle atlas $\varphi'(\mathfrak{U})$ contained in φ'.

Similarly, we state the following

Definition 4.3. A simple body $\mathscr{B}$ is called *(globally) homogeneous* if there exists a global material chart $(\mathscr{B}, \varphi_\alpha)$ which belongs to φ'.

Obviously, a homogeneous body is locally homogeneous, and its material tangent bundles are trivial. A locally homogeneous body whose material tangent bundles are trivial, however, need *not* be globally homogeneous.

5. The Bundles of Linear and Reference Frames

By definition, a *principal bundle* is a fibre bundle whose fibre space coincides with the structure group (which acts on itself by the operation of left-multiplication). In general, every fibre bundle $\mathscr{F}$ may be equipped with an *associated principal bundle* $\mathscr{E}$, which is a principal bundle having the following properties[30]:

i) The base spaces and the structure groups of $\mathscr{E}$ and $\mathscr{F}$ are identical.

ii) The bundle charts of $\mathscr{E}$ and $\mathscr{F}$ are in one-to-one correspondence in such a way that a pair of corresponding bundle charts share the same coordinate neighborhood in the common base space.

iii) The coordinate transformations for $\mathscr{E}$ and $\mathscr{F}$ are the same fields on the common overlaps.

In differential geometry, the theory of connections on a fibre bundle is developed for a principal bundle only, since connections on an arbitrary fibre bundle $\mathscr{F}$ are in one-to-one correspondence with those on its associated principal bundle $\mathscr{E}$. In order to treat connections on the tangent bundles of a simple body, we now introduce their associated principal bundles.

It is known[31] that the associated principal bundle of the geometric tangent bundle $\mathscr{T}(\mathscr{M})$ of an arbitrary differentiable manifold $\mathscr{M}$ is the *bundle of linear frames*, denoted by $\mathscr{E}(\mathscr{M})$. We review here its basic structure. Let p be a point in $\mathscr{M}$. Then a *linear frame* at p is an ordered basis, say

$$e_p = \{e_{p,i},\ i=1, \dots, n\}, \tag{5.1}$$

of the tangent space $\mathscr{M}_p$. We denote the collection of all linear frames at p by $\mathscr{E}_p$. Then the *bundle space* $\mathscr{E}(\mathscr{M})$ is a differentiable manifold formed by the disjoint union

$$\mathscr{E}(\mathscr{M}) = \bigcup_{p\in\mathscr{M}} \mathscr{E}_p. \tag{5.2}$$

The *projection*

$$\pi\colon\ \mathscr{E}(\mathscr{M}) \to \mathscr{M} \tag{5.3}$$

is defined in a natural way. Thus $\mathscr{M}$ is again the *base space*. As before, if $\mathscr{U}$ is a subset of $\mathscr{M}$, then we put

$$\mathscr{E}(\mathscr{U}) = \bigcup_{p\in\mathscr{U}} \mathscr{E}_p = \pi^{-1}(\mathscr{U}). \tag{5.4}$$

---

[30] For more details, see the standard treatise by Steenrod [*28*, § 8.1].

[31] *Cf.* Chern [*8*, Ch. IV].

In particular,

$$\mathscr{E}_p = \pi^{-1}(p), \tag{5.5}$$

so that $\mathscr{E}_p$ is the *fibre* of $\mathscr{E}(\mathscr{M})$ at p.

We proceed to describe the *bundle atlas*

$$\xi = \{(\mathscr{U}_\alpha, \xi_\alpha), \alpha \in K\} \tag{5.6}$$

of $\mathscr{E}(\mathscr{M})$. Here, the bundle maps ξ_α, of course, are isomorphisms of the form

$$\xi_\alpha \colon \mathscr{U}_\alpha \times \mathscr{GL}(n) \to \mathscr{E}(\mathscr{U}_\alpha), \tag{5.7}$$

which satisfy the conditions

$$\xi_\alpha(\{p\} \times \mathscr{GL}(n)) = \mathscr{E}_p \tag{5.8}$$

for all $p \in \mathscr{U}_\alpha$. Suppose that $\boldsymbol{G}$ is an arbitrary element on $\mathscr{GL}(n)$, say

$$\boldsymbol{G} = [G^i_j]. \tag{5.9}$$

Then $\boldsymbol{G}$ gives rise to an isomorphism of $\mathscr{E}(\mathscr{M})$,

$$R_{\boldsymbol{G}} \colon \mathscr{E}(\mathscr{M}) \to \mathscr{E}(\mathscr{M}), \tag{5.10}$$

by the condition

$$R_{\boldsymbol{G}}(\boldsymbol{e}_p) = \{\boldsymbol{e}_{p,j}\, G^j_i,\ i = 1, \ldots, n\}, \tag{5.11}$$

for all $\boldsymbol{e}_p \in \mathscr{E}(\mathscr{M})$. We call this operation the *right-multiplication* (or the *right-translation*) of $\boldsymbol{G}$ on $\mathscr{E}(\mathscr{M})$. Clearly, it satisfies the following conditions:

$$R_1(\boldsymbol{e}_p) = \boldsymbol{e}_p, \tag{5.12}$$

and

$$R_{\boldsymbol{G}_2}(R_{\boldsymbol{G}_1}(\boldsymbol{e}_p)) = R_{\boldsymbol{G}_1 \boldsymbol{G}_2}(\boldsymbol{e}_p), \tag{5.13}$$

for all $\boldsymbol{e}_p \in \mathscr{E}(\mathscr{M})$ and $\boldsymbol{G}_1, \boldsymbol{G}_2 \in \mathscr{GL}(n)$. In the terminology of differential geometry, these conditions define $\mathscr{GL}(n)$ to be a *Lie transformation group* on $\mathscr{E}(\mathscr{M})$. Since R_1 is the only right-multiplication which acts as the identity transformation of $\mathscr{E}(\mathscr{M})$, the action of $\mathscr{GL}(n)$ on $\mathscr{E}(\mathscr{M})$ is said to be *effective*.

Let $\boldsymbol{\varphi} = \{(\mathscr{U}_\alpha, \varphi_\alpha), \alpha \in K\}$ be the bundle atlas of $\mathscr{T}(\mathscr{M})$ as before. Then for each $\alpha \in K$ and $p \in \mathscr{M}$, the isomorphism $\varphi_{\alpha,p}$ in (4.8) transforms the *standard frame*

$$\boldsymbol{\imath} = \{(1, 0, \ldots, 0), \ldots, (0, \ldots, 0, 1)\} \tag{5.14}$$

of $\mathscr{R}^n$ onto a linear frame

$$\boldsymbol{e}_p(\alpha) = \varphi_{\alpha,p}(\boldsymbol{\imath}) \tag{5.15}$$

at p. We define the isomorphisms ξ_α in (5.7) by

$$\xi_\alpha(p, \boldsymbol{G}) = R_{\boldsymbol{G}}(\boldsymbol{e}_p(\alpha)), \tag{5.16}$$

for all $p \in \mathscr{U}_\alpha$ and $\boldsymbol{G} \in \mathscr{GL}(n)$.

From (5.15), it is easily verified that the frames $\boldsymbol{e}_p(\alpha)$ satisfy the following transformation law:

$$\boldsymbol{e}_p(\beta) = R_{\boldsymbol{G}_{\alpha\beta}(p)}(\boldsymbol{e}_p(\alpha)), \tag{5.17}$$

for all $p \in \mathscr{U}_\alpha \cap \mathscr{U}_\beta$, where $\boldsymbol{G}_{\alpha\beta}$ denotes the coordinate transformations of the bundle atlas $\boldsymbol{\varphi}$. Combining (5.16) and (5.17), we see that the coordinate transformations of ξ are given by

$$\xi_{\alpha,p}^{-1} \circ \xi_{\beta,p}(\boldsymbol{G}) = \boldsymbol{G}_{\alpha\beta}(p)\,\boldsymbol{G}\,, \tag{5.18}$$

for all $\boldsymbol{G} \in \mathscr{GL}(n)$, or equivalently,

$$\xi_{\alpha,p}^{-1} \circ \xi_{\beta,p} = L_{\boldsymbol{G}_{\alpha\beta}(p)}\,, \tag{5.19}$$

where $L_{\boldsymbol{G}}$ denotes the operation of *left-multiplication* (or *left-translation*) of $\boldsymbol{G}$ on $\mathscr{GL}(n)$. Since by definition $\mathscr{GL}(n)$ is regarded as a Lie transformation group on itself by the operation of left-multiplication, the condition (5.19) implies that the coordinate transformations of ξ are identical to those of φ. In particular, we see that ξ forms a bundle atlas; relative to which $\mathscr{E}(\mathscr{M})$ becomes the associated principal bundle of $\mathscr{T}(\mathscr{M})$.

Evidently, we can apply the above result to the geometric tangent bundle $\mathscr{T}(\mathscr{B})$ of a simple body $\mathscr{B}$. Like the bundle $\mathscr{T}(\mathscr{B})$, the bundle of linear frames $\mathscr{E}(\mathscr{B})$ has a very simple structure. Namely, via a global bundle chart, $\mathscr{E}(\mathscr{B})$ can be identified as the product manifold $\mathscr{B} \times \mathscr{GL}(3)$, which is a cyclinder in the Euclidean space $\mathscr{R}^{12}$. Of course, the general structure of $\mathscr{E}(\mathscr{B})$, again, *cannot* characterize the material structure of $\mathscr{B}$, since it does *not* depend on the distribution of the response functionals on $\mathscr{B}$.

Following exactly the same scheme, we can construct the *bundles of reference frames* $\mathscr{E}(\mathscr{B}, \mathfrak{U})$, which are, by definition, the associated principal bundles of the material tangent bundles $\mathscr{T}(\mathscr{B}, \mathfrak{U})$. Let $\mathfrak{U}$ be a fixed material atlas of $\mathscr{B}$. Then a linear frame $\boldsymbol{e}_p$ is called a *reference frame at p relative* to $\mathfrak{U}$, if there exists a material chart $(\mathscr{U}_\alpha, \varphi_\alpha)$ in $\varphi(\mathfrak{U})$ such that

$$\boldsymbol{e}_p = \boldsymbol{e}_p(\alpha) = \varphi_{\alpha,p}(\dot{\boldsymbol{i}})\,. \tag{5.20}$$

Here $\dot{\boldsymbol{i}}$, of course, is the standard frame of $\mathscr{R}^3$. We denote the collection of all reference frames at p relative to $\mathfrak{U}$ by $\mathscr{E}_p(\mathfrak{U})$. Then we put

$$\mathscr{E}(\mathscr{B}, \mathfrak{U}) \equiv \bigcup_{p \in m} \mathscr{E}_p(\mathfrak{U})\,. \tag{5.21}$$

As before, we have the *projection map*

$$\pi\colon\ \mathscr{E}(\mathscr{B}, \mathfrak{U}) \to \mathscr{B}\,, \tag{5.22}$$

with

$$\mathscr{E}_p(\mathfrak{U}) = \pi^{-1}(p)\,, \tag{5.23}$$

so that $\mathscr{E}_p(\mathfrak{U})$ is the *fibre* of $\mathscr{E}(\mathscr{B}, \mathfrak{U})$ at p.

The bundle atlas

$$\xi(\mathfrak{U}) = \{(\mathscr{U}_\alpha, \xi_\alpha), \alpha \in I\} \tag{5.24}$$

is defined by

$$\xi_\alpha(p, \boldsymbol{G}) = R_{\boldsymbol{G}}(\boldsymbol{e}_p) \tag{5.25}$$

for all $\boldsymbol{e}_p \in \mathscr{E}_p(\mathfrak{U})$ and $\boldsymbol{G} \in \mathscr{G}(\mathfrak{U})$, where $\boldsymbol{e}_p$ denotes the reference frame given by (5.20). It should be noted that the bundle maps ξ_α are now isomorphisms of the form

$$\xi_\alpha\colon\ \mathscr{U}_\alpha \times \mathscr{G}(\mathfrak{U}) \to \pi^{-1}(\mathscr{U}_\alpha) \subset \mathscr{E}(\mathscr{B}, \mathfrak{U})\,. \tag{5.26}$$

By the same argument as before, equations (5.17) and (5.25) imply that the coordinate transformations of $\varphi(\mathfrak{U})$ and $\xi(\mathfrak{U})$ are identical.

Since every reference frame is a linear frame, the bundle spaces $\mathscr{E}(\mathscr{B}, \mathfrak{U})$ are subsets of the bundle space $\mathscr{E}(\mathscr{B})$. Moreover, since the differentiable structures on $\mathscr{E}(\mathscr{B})$ and $\mathscr{E}(\mathscr{B}, \mathfrak{U})$ are characterized by the conditions (5.16) and (5.25) respectively, the inclusion maps of $\mathscr{E}(\mathscr{B}, \mathfrak{U})$ in $\mathscr{E}(\mathscr{B})$ are imbeddings. Thus $\mathscr{E}(\mathscr{B}, \mathfrak{U})$ are submanifolds of $\mathscr{E}(\mathscr{B})$. Also, the structure groups $\mathscr{G}(\mathfrak{U})$ of $\mathscr{E}(\mathscr{B}, \mathfrak{U})$ are Lie subgroups of the structure group $\mathscr{GL}(3)$ of $\mathscr{E}(\mathscr{B})$. In the terminology of fibre bundles, we say that $\mathscr{E}(\mathscr{B}, \mathfrak{U})$ are *subbundles* of $\mathscr{E}(\mathscr{B})$.

We proceed to characterize the relations among the subbundles $\mathscr{E}(\mathscr{B}, \mathfrak{U})$ within the framework of $\mathscr{E}(\mathscr{B})$. Recall that any two material atlases of $\mathscr{B}$ are related by a mapping $\boldsymbol{K}$ as shown in (4.22). Suppose that $\boldsymbol{e}_p \in \mathscr{E}_p(\mathfrak{U})$. Then $R_{\boldsymbol{K}^{-1}}(\boldsymbol{e}_p) \in \mathscr{E}_p(\boldsymbol{K}\mathfrak{U})$, for all $\boldsymbol{K} \in \mathscr{GL}(3)$. We can express this fact by the set-theoretical equation

$$\mathscr{E}(\mathscr{B}, \boldsymbol{K}\mathfrak{U}) = R_{\boldsymbol{K}^{-1}}\big(\mathscr{E}(\mathscr{B}, \mathfrak{U})\big). \tag{5.27}$$

Since $\boldsymbol{K}$ is arbitrary, this equation implies that every linear frame in $\mathscr{E}(\mathscr{B})$ belongs to one and only one subbundle $\mathscr{E}(\mathscr{B}, \mathfrak{U})$. Consequently $\mathscr{E}(\mathscr{B})$ is the disjoint union of the subbundles $\mathscr{E}(\mathscr{B}, \mathfrak{U})$, *viz*,

$$\mathscr{E}(\mathscr{B}) = \bigcup_{\mathfrak{U}} \mathscr{E}(\mathscr{B}, \mathfrak{U}). \tag{5.28}$$

From (4.22) and (5.27), the bundle atlases $\xi(\mathfrak{U})$ satisfy the following transformation law:

$$\xi(\boldsymbol{K}\mathfrak{U}) = \{(\mathscr{U}_\alpha, R_{\boldsymbol{K}^{-1}} \circ \xi_\alpha \circ C_{\boldsymbol{K}^{-1}}), \alpha \in I\}, \tag{5.29}$$

where $C_{\boldsymbol{K}}$ denotes the operation of *conjugation* by $\boldsymbol{K}$ on $\mathscr{GL}(3)$, *i.e.*,

$$C_{\boldsymbol{K}}(\boldsymbol{G}) = \boldsymbol{K}\boldsymbol{G}\boldsymbol{K}^{-1}, \qquad \boldsymbol{G} \in \mathscr{GL}(3). \tag{5.30}$$

Thus the bundle maps $R_{\boldsymbol{K}^{-1}} \circ \xi_\alpha \circ C_{\boldsymbol{K}^{-1}}$ are given by

$$[R_{\boldsymbol{K}^{-1}} \circ \xi_\alpha \circ C_{\boldsymbol{K}^{-1}}](p, \boldsymbol{G}) = R_{\boldsymbol{K}^{-1}}\big(\xi_{\alpha, p}(\boldsymbol{K}^{-1}\boldsymbol{G}\boldsymbol{K})\big), \tag{5.31}$$

for all $p \in \mathscr{U}_\alpha$ and $\boldsymbol{G} \in \mathscr{G}(\mathfrak{U})$ (so that $\boldsymbol{K}^{-1}\boldsymbol{G}\boldsymbol{K} \in \mathscr{G}(\boldsymbol{K}\mathfrak{U})$).

For an arbitrary principal bundle $\mathscr{E}$, a *cross section* in $\mathscr{E}$ over a coordinate neighborhood $\mathscr{U}$ in the base space $\mathscr{M}$ is a smooth mapping

$$\boldsymbol{\sigma}: \mathscr{U} \to \mathscr{E} \tag{5.32}$$

such that the value of $\boldsymbol{\sigma}$ at any point p belongs to the fibre $\mathscr{E}_p$. Evidently, the fields $\boldsymbol{e}_p(\alpha)$ defined by (5.15) and (5.20) are cross sections in $\mathscr{E}(\mathscr{B})$ and $\mathscr{E}(\mathscr{B}, \mathfrak{U})$ respectively, over the coordinate neighborhoods $\mathscr{U}_\alpha$. We put

$$\boldsymbol{\sigma}_\alpha(p) = \boldsymbol{e}_p(\alpha), \qquad p \in \mathscr{U}_\alpha. \tag{5.33}$$

Conversely, if $\boldsymbol{\sigma}_0$ is a cross section in $\mathscr{E}(\mathscr{B})$ or in $\mathscr{E}(\mathscr{B}, \mathfrak{U})$ over $\mathscr{U}_0$, then we define a mapping

$$\xi_0: \mathscr{U}_0 \times \mathscr{GL}(3) \to \pi^{-1}(\mathscr{U}_0) \subset \mathscr{E}(\mathscr{B}), \tag{5.34}$$

or

$$\xi_0: \mathscr{U}_0 \times \mathscr{G}(\mathfrak{U}) \to \pi^{-1}(\mathscr{U}_0) \subset \mathscr{E}(\mathscr{B}, \mathfrak{U}), \tag{5.35}$$

by the condition

$$\xi_0(p, G) = R_G(\sigma_0(p)), \tag{5.36}$$

for all $p \in \mathscr{U}_0$ and $G \in \mathscr{GL}(3)$ or $\mathscr{G}(\mathfrak{U})$. Clearly, the pair $(\mathscr{U}_0, \xi_0)$ then becomes a bundle chart such that σ_0 corresponds to the *section of the identity, i.e.*,

$$\xi_0(p, \mathbf{1}) = \sigma_0(p). \tag{5.37}$$

In this manner, bundle charts are in one-to-one correspondence with cross sections in the bundle space. On the other hand, the bundle charts of $\mathscr{E}(\mathscr{B})$ or $\mathscr{E}(\mathscr{B}, \mathfrak{U})$ are in one-to-one correspondence with those of $\mathscr{T}(\mathscr{B})$ or $\mathscr{T}(\mathscr{B}, \mathfrak{U})$. Therefore, the last are also in one-to-one correspondence with cross sections in $\mathscr{E}(\mathscr{B})$ or $\mathscr{E}(\mathscr{B}, \mathfrak{U})$, whence the $\mathscr{T}(\mathscr{B}, \mathfrak{U})$ are trivial if and only if the $\mathscr{E}(\mathscr{B}, \mathfrak{U})$ admit some global cross sections. By virtue of this fact, it is easily seen that the material tangent bundles of the simple body in Example 4.2 are not trivial.

6. The Lie Algebras of the Isotropy Groups

By hypothesis, the isotropy groups $\mathscr{G}(\mathfrak{U})$ of a simple body $\mathscr{B}$ are closed Lie subgroups of the special linear group $\mathscr{SL}(3)$. In particular, $\mathscr{G}(\mathfrak{U})$ are Lie groups. In the general theory of Lie groups, it is known[32] that every Lie group $\mathscr{G}$ is endowed with a *Lie algebra* g, which is a vector space equipped with a *bracket operation*. The elements of g are *left-invariant* vector fields on $\mathscr{G}$, and the bracket of two elements $\boldsymbol{u}, \boldsymbol{v} \in \mathsf{g}$ is defined by

$$[\boldsymbol{u}, \boldsymbol{v}] = \underset{\boldsymbol{u}}{\pounds}\, \boldsymbol{v}, \tag{6.1}$$

where the symbol £ stands for the usual operation of the *Lie derivative*.

At any point $p \in \mathscr{G}$, the values of the left-invariant fields coincide with the whole tangent space $\mathscr{G}_p$, and the restriction map

$$|_p \colon \mathsf{g} \to \mathscr{G}_p \tag{6.2}$$

is an isomorphism. Thus

$$\dim \mathsf{g} = \dim \mathscr{G}. \tag{6.3}$$

We call the restriction map at the identity element $e \in \mathscr{G}$ the *standard representation* of g.

Let $\mathscr{H}$ be a Lie subgroup of a Lie group $\mathscr{G}$. Then $\mathscr{H}$ is endowed with a Lie algebra h consisting of left-invariant vector fields on $\mathscr{H}$. By the classical Frobenius theorem, we can identify h as a Lie subalgabra of g. Conversely, every Lie subalgebra of g corresponds to the Lie algebra of a least one Lie subgroup of $\mathscr{G}$ in this manner (that Lie subgroup is unique if it is required to be connected). From this fact, the Lie algebras $\mathsf{g}(\mathfrak{U})$ of $\mathscr{G}(\mathfrak{U})$ are Lie subalgebras of $\mathsf{sl}(3)$ (the Lie algebra of $\mathscr{SL}(3)$). Likewise, $\mathsf{sl}(3)$ is itself a Lie subalgebra of $\mathsf{gl}(3)$ (the Lie algebra of $\mathscr{GL}(3)$).

The general linear group $\mathscr{GL}(3)$ forms an open set in the Euclidean space $\mathscr{R}^9$ consisting of all 3×3 matrices. Hence the tangent space at any point of $\mathscr{GL}(3)$ is a copy of $\mathscr{R}^9$. Consequently, via the standard representation, $\mathsf{gl}(3)$ can be

[32] *Cf.* Chevalley [*21*, Ch. IV, § II], Cohn [*22*, Ch. III, § 3.2].

regarded as a copy of $\mathscr{R}^9$. In this sense, the bracket operation on $\mathfrak{gl}(3)$ is given by

$$[U, V] = UV - VU, \tag{6.4}$$

where the expressions on the left-hand side are the ordinary matrix products.

To consider the structure of $\mathfrak{gl}(3)$ and its various Lie subalgebras, we introduce the notion of the *exponential map*. In general, if $\mathscr{G}$ is an arbitrary Lie group, and $\boldsymbol{v}$ is an element in $\mathfrak{g}$, then $\boldsymbol{v}$ induces a 1-parameter group $\lambda_{\boldsymbol{v}}(t)$, $t \in (-\infty, \infty)$, in $\mathscr{G}$ such that

$$\dot{\lambda}_{\boldsymbol{v}}(0) = \boldsymbol{v}(e). \tag{6.5}$$

Namely, $\lambda_{\boldsymbol{v}}(t)$ is the integral curve of the vector field $\boldsymbol{v}$ satisfying the above initial condition. We define the *exponential map*

$$\exp\colon \mathfrak{g} \to \mathscr{G} \tag{6.6}$$

by the condition

$$\exp(\boldsymbol{v}) = \lambda_{\boldsymbol{v}}(1), \tag{6.7}$$

for all $\boldsymbol{v} \in \mathfrak{g}$. For the general linear group $\mathscr{GL}(3)$, it is easily verified that

$$\lambda_V(t) = \mathbf{1} + Vt + \frac{1}{2!} V^2 t^2 + \cdots + \frac{1}{n!} V^n t^n + \cdots \tag{6.8}$$

for all $V \in \mathfrak{gl}(3)$, whence[33]

$$\exp(V) = \mathbf{1} + V + \frac{1}{2!} V^2 + \cdots + \frac{1}{n!} V^n + \cdots. \tag{6.9}$$

Notice that we have observed the aforesaid standard representation of $\mathfrak{gl}(3)$ by $\mathscr{R}^9$ in the above equations.

From (6.9), it is easily seen that

$$\exp(G V G^{-1}) = G \exp(V) G^{-1} \tag{6.10}$$

for all $V \in \mathfrak{gl}(3)$ and $G \in \mathscr{GL}(3)$. Thus[34]

$$\det[\exp(V)] = \exp \operatorname{tr}(V). \tag{6.11}$$

Consequently, the Lie algebra $\mathfrak{sl}(3)$ consists of all 3×3 matrices whose traces vanish.

For a fixed simple body $\mathscr{B}$, the isotropy groups $\mathscr{G}(\mathfrak{U})$ satisfy the following transformation law:

$$\mathscr{G}(K\mathfrak{U}) = K \mathscr{G}(\mathfrak{U}) K^{-1}. \tag{3.14}$$

Then from (6.10), the corresponding Lie algebras $\mathfrak{g}(\mathfrak{U})$, regarded as Lie subalgebras of $\mathfrak{gl}(3)$, satisfy the transformation law:

$$\mathfrak{g}(K\mathfrak{U}) = K \mathfrak{g}(\mathfrak{U}) K^{-1}. \tag{6.12}$$

Clearly, the last operation of conjugation defines an equivalence relation among all Lie subalgebras of $\mathfrak{gl}(3)$. In the terminology of classical groups, an equivalence class relative to this equivalence relation is called a *type* of Lie subalgebras. In this sense, the Lie algebras $\mathfrak{g}(\mathfrak{U})$ of a simple body $\mathscr{B}$ form a type of Lie subalgebras.

---

[33] *Cf.* CHEVALLEY [*21*, Ch. IV, § III].

[34] CHEVALLEY [*21*, Ch. I, § II].

Remark. The term *type* here refers to an equivalence class of Lie subalgebras. Elsewhere in the literature of continuum mechanics that term usually means an equivalence class of subgroups of $\mathcal{SL}(3)$ (or the unimodular group $\mathcal{U}(3)$ if orientation is not fixed) with respect to the conjugation operation in (3.14). Clearly, a type of Lie subgroup in the latter sense determines a unique type of Lie algebras in the former sense. Conversely, however, a type of Lie subalgebras, in general, corresponds to many types of Lie subgroups, since a Lie subgroup may fail to be connected.

We now give the classification of the Lie subalgebras[35] of **sl**(3) and their corresponding connected Lie subgroups of $\mathcal{SL}(3)$ in matrix forms, in the following table.

Table 6.1. *Connected Lie subgroups of $\mathcal{SL}(3)$ and their Lie algebras*

No.	Lie subgroups (det G= 1)	Lie subalgebras (tr H= 0)	dim.
1	$\begin{bmatrix} e^{\alpha a} & 0 & 0 \\ b & e^{\beta a} & 0 \\ c & d & e^{\gamma a} \end{bmatrix}$	$\begin{bmatrix} \alpha a & 0 & 0 \\ b & \beta a & 0 \\ c & d & \gamma a \end{bmatrix}$	5, 4, 3,
2	$\begin{bmatrix} e^{\alpha a} & 0 & 0 \\ b & e^{\beta a} & 0 \\ c & 0 & e^{\gamma a} \end{bmatrix}$	$\begin{bmatrix} \alpha a & 0 & 0 \\ b & \beta a & 0 \\ c & 0 & \gamma a \end{bmatrix}$	4, 3, 2,
3	$\begin{bmatrix} e^{\alpha a} & b & c \\ 0 & e^{\beta a} & 0 \\ 0 & 0 & e^{\gamma a} \end{bmatrix}$	$\begin{bmatrix} \alpha a & b & c \\ 0 & \beta a & 0 \\ 0 & 0 & \gamma a \end{bmatrix}$	4, 3, 2,
4	$\begin{bmatrix} e^{\alpha a} & 0 & 0 \\ b & e^{\beta a} & 0 \\ 0 & 0 & e^{\gamma a} \end{bmatrix}$	$\begin{bmatrix} \alpha a & 0 & 0 \\ b & \beta a & 0 \\ 0 & 0 & \gamma a \end{bmatrix}$	3, 2, 1,
5	$\begin{bmatrix} e^{\alpha a} & 0 & 0 \\ 0 & e^{\beta a} & 0 \\ 0 & 0 & e^{\gamma a} \end{bmatrix}$	$\begin{bmatrix} \alpha a & 0 & 0 \\ 0 & \beta a & 0 \\ 0 & 0 & \gamma a \end{bmatrix}$	2, 1, 0,
6	$\begin{bmatrix} e^{\alpha a} \operatorname{Cos} a & e^{\alpha a} \operatorname{Sin} a & 0 \\ -e^{\alpha a} \operatorname{Sin} a & e^{\alpha a} \operatorname{Cos} a & 0 \\ b & c & e^{\beta a} \end{bmatrix}$	$\begin{bmatrix} \alpha a & a & 0 \\ -a & \alpha a & 0 \\ b & c & \beta a \end{bmatrix}$	4, 3,
7	$\begin{bmatrix} e^{\alpha a} \operatorname{Cos} a & e^{\alpha a} \operatorname{Sin} a & b \\ -e^{\alpha a} \operatorname{Sin} a & e^{\alpha a} \operatorname{Cos} a & c \\ 0 & 0 & e^{\beta a} \end{bmatrix}$	$\begin{bmatrix} \alpha a & a & b \\ -a & \alpha a & c \\ 0 & 0 & \beta a \end{bmatrix}$	4, 3,
8	$\begin{bmatrix} e^{\alpha a} \operatorname{Cos} a & e^{\alpha a} \operatorname{Sin} a & 0 \\ -e^{\alpha a} \operatorname{Sin} a & e^{\alpha a} \operatorname{Cos} a & 0 \\ 0 & 0 & e^{\beta a} \end{bmatrix}$	$\begin{bmatrix} \alpha a & a & 0 \\ -a & \alpha a & 0 \\ 0 & 0 & \beta a \end{bmatrix}$	2, 1,
9	$\begin{bmatrix} a & b & c \\ d & e & f \\ g & h & i \end{bmatrix}$	$\begin{bmatrix} a & b & c \\ d & e & f \\ g & h & i \end{bmatrix}$	8
10	$\begin{bmatrix} a & b & c \\ d & e & f \\ 0 & 0 & g \end{bmatrix}$	$\begin{bmatrix} a & b & c \\ d & e & f \\ 0 & 0 & g \end{bmatrix}$	6
11	$\begin{bmatrix} a & b & 0 \\ c & d & 0 \\ e & f & g \end{bmatrix}$	$\begin{bmatrix} a & b & 0 \\ c & d & 0 \\ e & f & g \end{bmatrix}$	6

[35] *Cf.* Nôno [*14*], who attributes this result to Lie [*15*].

No.	Lie subgroups (det $G=1$)	Lie subalgebras (tr $H=0$)	dim.
12	$\begin{bmatrix} a & b & c \\ d & e & f \\ 0 & 0 & 1 \end{bmatrix}$	$\begin{bmatrix} a & b & c \\ d & e & f \\ 0 & 0 & 0 \end{bmatrix}$	5
13	$\begin{bmatrix} a & b & 0 \\ c & d & 0 \\ e & f & 1 \end{bmatrix}$	$\begin{bmatrix} a & b & 0 \\ c & d & 0 \\ e & f & 0 \end{bmatrix}$	5
14	$\begin{bmatrix} a & b & 0 \\ c & d & 0 \\ 0 & 0 & e \end{bmatrix}$	$\begin{bmatrix} a & b & 0 \\ c & d & 0 \\ 0 & 0 & e \end{bmatrix}$	4
15	$\begin{bmatrix} a & b & 0 \\ c & d & 0 \\ 0 & 0 & 1 \end{bmatrix}$	$\begin{bmatrix} a & b & 0 \\ c & d & 0 \\ 0 & 0 & 0 \end{bmatrix}$	3
16	The proper orthogonal group.	$\begin{bmatrix} 0 & -a & -b \\ a & 0 & -c \\ b & c & 0 \end{bmatrix}$	3
17	$\begin{bmatrix} e^{a} & 0 & 0 \\ b & e^{-2a} & c \\ a e^{a} & 0 & e^{a} \end{bmatrix}$	$\begin{bmatrix} a & 0 & 0 \\ b & -2a & c \\ a & 0 & a \end{bmatrix}$	3
18	$\begin{bmatrix} e^{a} & b & 0 \\ 0 & e^{-2a} & 0 \\ a e^{a} & c & e^{a} \end{bmatrix}$	$\begin{bmatrix} a & b & 0 \\ 0 & -2a & 0 \\ a & c & a \end{bmatrix}$	3
19[a]	$\exp \begin{bmatrix} 0 & a & b \\ b & c & 0 \\ a & 0 & -c \end{bmatrix}$	$\begin{bmatrix} 0 & a & b \\ b & c & 0 \\ a & 0 & -c \end{bmatrix}$	3

[a] The explicit form of this subgroup is quite complex, namely,

$$G_{11}=1+\frac{2ab}{(c^2+2ab)^2}\left[\operatorname{Cosh}(c^2+2ab)-1\right],$$

$$G_{12}=\frac{a}{(c^2+2ab)}\operatorname{Sinh}(c^2+2ab)+\frac{ac}{(c^2+2ab)^2}\left[\operatorname{Cosh}(c^2+2ab)-1\right],$$

$$G_{13}=\frac{b}{(c^2+2ab)}\operatorname{Sinh}(c^2+2ab)-\frac{bc}{(c^2+2ab)^2}\left[\operatorname{Cosh}(c^2+2ab)-1\right],$$

$$G_{21}=\frac{b}{(c^2+2ab)}\operatorname{Sinh}(c^2+2ab)+\frac{bc}{(c^2+2ab)^2}\left[\operatorname{Cosh}(c^2+2ab)-1\right],$$

$$G_{22}=1+\frac{c}{(c^2+2ab)}\operatorname{Sinh}(c^2+2ab)+\frac{c^2+ab}{(c^2+2ab)^2}\left[\operatorname{Cosh}(c^2+2ab)-1\right],$$

$$G_{23}=\frac{b^2}{(c^2+2ab)^2}\left[\operatorname{Cosh}(c^2+2ab)-1\right],$$

$$G_{31}=\frac{a}{(c^2+2ab)}\operatorname{Sinh}(c^2+2ab)-\frac{ac}{(c^2+2ab)^2}\left[\operatorname{Cosh}(c^2+2ab)-1\right],$$

$$G_{32}=\frac{a^2}{(c^2+2ab)^2}\left[\operatorname{Cosh}(c^2+2ab)-1\right],$$

$$G_{33}=1-\frac{c}{(c^2+2ab)}\operatorname{Sinh}(c^2+2ab)+\frac{c^2+ab}{(c^2+2ab)^2}\left[\operatorname{Cosh}(c^2+2ab)-1\right].$$

No.	Lie subgroups ($\det \boldsymbol{G}=1$)	Lie subalgebras ($\operatorname{tr} \boldsymbol{H}=0$)	dim.
20	$\begin{bmatrix} 1 & 0 & a \\ ae^b & e^b & c \\ 0 & 0 & e^{-b} \end{bmatrix}$	$\begin{bmatrix} 0 & 0 & a \\ a & b & c \\ 0 & 0 & -b \end{bmatrix}$	3
21	$\begin{bmatrix} e^a & 0 & 0 \\ b & e^{-2a} & 0 \\ ae^a & 0 & e^a \end{bmatrix}$	$\begin{bmatrix} a & 0 & 0 \\ b & -2a & 0 \\ a & 0 & a \end{bmatrix}$	2
22	$\begin{bmatrix} e^a & 0 & 0 \\ 0 & e^{-2a} & 0 \\ ae^a & b & e^a \end{bmatrix}$	$\begin{bmatrix} a & 0 & 0 \\ 0 & -2a & 0 \\ a & b & a \end{bmatrix}$	2
23	$\begin{bmatrix} 1 & 0 & a \\ ae^b & e^b & \frac{1}{2}a^2e^b \\ 0 & 0 & e^{-b} \end{bmatrix}$	$\begin{bmatrix} 0 & 0 & a \\ a & b & 0 \\ 0 & 0 & -b \end{bmatrix}$	2
24	$\begin{bmatrix} 1 & a & 0 \\ 0 & 1 & 0 \\ a & b & 1 \end{bmatrix}$	$\begin{bmatrix} 0 & a & 0 \\ 0 & 0 & 0 \\ a & b & 0 \end{bmatrix}$	2
25	$\begin{bmatrix} e^a & 0 & 0 \\ 0 & e^{-2a} & 0 \\ ae^a & 0 & e^a \end{bmatrix}$	$\begin{bmatrix} a & 0 & 0 \\ 0 & -2a & 0 \\ a & 0 & a \end{bmatrix}$	1
26	$\begin{bmatrix} 1 & a & 0 \\ 0 & 1 & 0 \\ a & \frac{1}{2}a^2 & 1 \end{bmatrix}$	$\begin{bmatrix} 0 & a & 0 \\ 0 & 0 & 0 \\ a & 0 & 0 \end{bmatrix}$	1

In the above table, $a, b, c, \ldots,$ are arbitrary real numbers, and α, β, γ, are arbitrary real parameters which can take on either fixed values or arbitrary values, subject only to the natural restrictions $\det \boldsymbol{G}=1$ and $\operatorname{tr}\boldsymbol{H}=0$. Thus let $\alpha=\beta=\gamma=0$ in family No. 1. Then we obtain a type of 3-dimensional subgroup and subalgebra whose components are

$$\begin{bmatrix} 1 & 0 & 0 \\ b & 1 & 0 \\ c & d & 1 \end{bmatrix} \quad \begin{bmatrix} 0 & 0 & 0 \\ b & 0 & 0 \\ c & d & 0 \end{bmatrix}$$

where $b, c, d,$ are arbitrary. Similarly, if we choose $\alpha=1$, $\beta=2$, and $\gamma=-3$ (say), then a type of 4-dimensional subgroup and subalgebra having components

$$\begin{bmatrix} e^a & 0 & 0 \\ b & e^{2a} & 0 \\ c & d & e^{-3a} \end{bmatrix} \quad \begin{bmatrix} a & 0 & 0 \\ b & 2a & 0 \\ c & d & -3a \end{bmatrix}$$

results. Finally, if we let α, β, and γ be arbitrary to within the condition $\alpha+\beta+\gamma=0$, then we obtain a 5-dimensional type:

$$\begin{bmatrix} e^a & 0 & 0 \\ c & e^b & 0 \\ d & f & e^{-(a+b)} \end{bmatrix} \quad \begin{bmatrix} a & 0 & 0 \\ c & b & 0 \\ d & f & -(a+b) \end{bmatrix}$$

where $a, b, c, d, f,$ are arbitrary. It is easily verified that each of the familes No. 1 – 8 contains ∞^1-types; the rest are single types.

There are three types of solids – isotropic, transversely isotropic, and crystalline, in the above table. They belong respectively to type No. 16, family No. 8 (with the parameter $\alpha=0$), and family No. 5 (with $\alpha=\beta=\gamma=0$). All other types are fluid crystals. For example, the various types of subfluids considered by WANG[36] are included, some of them in families No. 1–5, the rest in types No. 9, 10, 11, and 14. The fluid crystals treated by COLEMAN[36] belong to family No. 2 (with $\alpha=\beta=\gamma=0$) or to type No. 13.

7. Fundamental Fields on the Bundles of Frames

Recall that the fibres of a principal bundle are copies of the structure group, which is, by definition, a Lie group. Thus it is equipped with a Lie algebra consisting of left-invariant vector fields on the group. It is known that there is a natural way to map these fields onto vector fields on the fibres. The totalities of the last are then called the *fundamental fields* of the principal bundle[37]. Naturally, we are interested mainly in the applications of this result to the bundles of frames associated with simple bodies.

We review briefly the basic concepts of the fundamental fields on an arbitrary principal bundle $\mathscr{E}$. Let $\mathscr{G}$ be the structure group, and let

$$\xi=\{(\mathscr{U}_\alpha, \xi_\alpha), \alpha\in K\} \tag{7.1}$$

be the bundle atlas. Then each bundle chart $(\mathscr{U}_\alpha, \xi_\alpha)$ induces a field of isomorphisms

$$\xi_{\alpha,p}\colon \mathscr{G}\to\mathscr{E}_p \tag{7.2}$$

on $\mathscr{U}_\alpha$. Suppose that $\boldsymbol{v}\in\mathfrak{g}$ is a left-invariant vector field on $\mathscr{G}$. Then the induced linear maps $\xi_{\alpha,p*}$ carry $\boldsymbol{v}$ onto vector fields on the fibres $\mathscr{E}_p$, say

$$\xi_{\alpha,p*}(\boldsymbol{v})=\bar{\boldsymbol{v}}_\alpha(p,\cdot). \tag{7.3}$$

We claim that these fields are independent of the choice of bundle chart, so that there exists a well-defined vector field $\bar{\boldsymbol{v}}$ on $\mathscr{E}$ such that

$$\bar{\boldsymbol{v}}(p,\cdot)=\bar{\boldsymbol{v}}_\alpha(p,\cdot), \tag{7.4}$$

for all α and p. To see this fact, suppose that $(\mathscr{U}_\beta, \xi_\beta)$ is another bundle chart such that $p\in\mathscr{U}_\beta$. Then by definition

$$\xi_{\beta,p}=\xi_{\alpha,p}\circ L_{G_{\alpha\beta}(p)}. \tag{7.5}$$

Using the chain rule, we get

$$\xi_{\beta,p*}(\boldsymbol{v})=\xi_{\alpha,p*}\circ L_{G_{\alpha\beta}(p)*}(\boldsymbol{v})=\xi_{\alpha,p*}(\boldsymbol{v}), \tag{7.6}$$

where the last equation is a consequence of the fact that $\boldsymbol{v}$ is left-invariant. Thus the field $\bar{\boldsymbol{v}}$ exists.

For an arbitrary $\boldsymbol{v}\in\mathfrak{g}$, the corresponding field $\bar{\boldsymbol{v}}$ on $\mathscr{E}$ defined in this manner is called a *fundamental field* on $\mathscr{E}$. We denote their union by $\bar{\mathfrak{g}}$. From (7.5), the fundamental fields are smooth and possess the following two important properties:

[36] WANG [*20*], COLEMAN [*19*].
[37] *Cf.* KOBAYASHI & NOMIZU [*11*, Ch. II].

i) The fundamental fields lie in the fibre directions, *i.e.*,

$$\pi_*(\bar{v})=\mathbf{0} \tag{7.7}$$

for all $\bar{v}\in\bar{\mathfrak{g}}$. Here π denotes the projection map defined by the condition

$$\mathscr{E}_p=\pi^{-1}(p) \tag{7.8}$$

for all p. From (6.2), the restriction map

$$|_x\colon\ \bar{\mathfrak{g}}\to\ker\pi_{*\,x} \tag{7.9}$$

is an isomorphism at each fixed $x\in\mathscr{E}$.

ii) We define the bracket operation of two elements $\bar{u}$ and $\bar{v}$ in $\bar{\mathfrak{g}}$ again by (6.1). Then $\bar{\mathfrak{g}}$ is closed with respect to this operation. Moreover, the bracket is preserved by the "bar" isomorphism of $\mathfrak{g}$ and $\bar{\mathfrak{g}}$, *i.e.*,

$$[\bar{u},\bar{v}]=\overline{[u,v]}\,. \tag{7.10}$$

Thus $\bar{\mathfrak{g}}$ has the structure of a Lie algebra, called the *Lie algebra of* $\mathscr{E}$.

We now apply the above results to the bundles of frames $\mathscr{E}(\mathscr{B})$ and $\mathscr{E}(\mathscr{B},\mathfrak{U})$ of a simple body $\mathscr{B}$. Then $\mathscr{E}(\mathscr{B})$ is endowed with the Lie algebra $\overline{\mathfrak{gl}}(3)$, and $\mathscr{E}(\mathscr{B},\mathfrak{U})$ are endowed with the Lie algebras $\bar{\mathfrak{g}}(\mathfrak{U})$. Since $\mathscr{E}(\mathscr{B},\mathfrak{U})$ are subbundles of $\mathscr{E}(\mathscr{B})$, the fundamental fields on $\mathscr{E}(\mathscr{B},\mathfrak{U})$ are restrictions of the fundamental fields of $\mathscr{E}(\mathscr{B})$ in the same way that $\bar{\mathfrak{g}}(\mathfrak{U})$ are regarded as Lie subalgebras of $\overline{\mathfrak{gl}}(3)$.

We proceed to characterize the relations among the fundamental fields on the subbundles $\mathscr{E}(\mathscr{B},\mathfrak{U})$ within the framework of the bundle $\mathscr{E}(\mathscr{B})$. Recall that $\mathscr{E}(\mathscr{B},\mathfrak{U})$ satisfy the transformation law:

$$\mathscr{E}(\mathscr{B},\boldsymbol{K}\mathfrak{U})=R_{\boldsymbol{K}^{-1}}\big(\mathscr{E}(\mathscr{B},\mathfrak{U})\big)\,. \tag{5.27}$$

Hence if $\bar{v}$ is a fundamental field on $\mathscr{E}(\mathscr{B},\mathfrak{U})$, then $R_{\boldsymbol{K}^{-1}*}(\bar{v})$ is a fundamental field on $\mathscr{E}(\mathscr{B},\boldsymbol{K}\mathfrak{U})$, for all $\boldsymbol{K}\in\mathscr{GL}(3)$. Of course, $\bar{v}$ and $R_{\boldsymbol{K}^{-1}*}(\bar{v})$ are the restrictions of some fundamental fields, say $\bar{v}_0$ and $R_{\boldsymbol{K}^{-1}*}(\bar{v}_0)$ on $\mathscr{E}(\mathscr{B})$. Then via the bar isomorphism of $\mathfrak{gl}(3)$ and $\overline{\mathfrak{gl}}(3)$, $\bar{v}_0$ and $R_{\boldsymbol{K}^{-1}*}(\bar{v}_0)$ correspond to left-invariant fields v_0 and $R_{\boldsymbol{K}^{-1}*}(v_0)$ on $\mathscr{GL}(3)$[38]. Finally, via the standard representation of $\mathfrak{gl}(3)$, v_0 and $R_{\boldsymbol{K}^{-1}*}(v_0)$ correspond to two 3×3 matrices, say $\boldsymbol{V}$ and $\boldsymbol{U}$, respectively. We claim that

$$\boldsymbol{U}=\boldsymbol{K}\boldsymbol{V}\boldsymbol{K}^{-1}. \tag{7.11}$$

The proof of this relation is obvious. By the chain rule and the fact that v_0 is left-invariant, we have

$$C_{\boldsymbol{K}*}(v_0)=R_{\boldsymbol{K}^{-1}*}\circ L_{\boldsymbol{K}*}(v_0)=R_{\boldsymbol{K}^{-1}*}(v_0)\,, \tag{7.12}$$

where $C_{\boldsymbol{K}}$ denotes the operation of conjugation by $\boldsymbol{K}$ on $\mathscr{GL}(3)$, *i.e.*,

$$C_{\boldsymbol{K}}=R_{\boldsymbol{K}^{-1}}\circ L_{\boldsymbol{K}}\,. \tag{7.13}$$

The relation (7.11) then follows from (7.12) and (6.10).

[38] Notice that the bar isomorphism commutes with the operations $R_{\boldsymbol{K}^{-1}*}$, for all $\boldsymbol{K}\in\mathscr{GL}(3)$.

For an arbitrary Lie group $\mathscr{G}$, the restrictions of the induced linear maps C_{K*} to the elements of the Lie algebra $\mathfrak{g}$ are called the operations of the *adjoint representation* by K, denoted by the notation $\operatorname{ad}(K)$. Clearly, these operations are linear isomorphisms of $\mathfrak{g}$, so that they belong to $\mathscr{GL}(\mathfrak{g})$. The operator ad thus defined is called the *adjoint representation* of $\mathscr{G}$, *viz*,

$$\operatorname{ad}: \mathscr{G} \to \mathscr{GL}(\mathfrak{g}). \tag{7.14}$$

Evidently, it is smooth and satisfies the condition

$$\operatorname{ad}(K_1 K_2) = \operatorname{ad}(K_1) \circ \operatorname{ad}(K_2). \tag{7.15}$$

Thus ad defines a smooth homomorphism of the Lie groups $\mathscr{G}$ into $\mathscr{GL}(\mathfrak{g})$.

In terms of the operations of adjoint representation, equation (7.11) then takes on the form

$$U = [\operatorname{ad}(K)](V). \tag{7.15}$$

Similarly, the transformation law of the Lie algebras $\mathfrak{g}(\mathfrak{U})$ shown in (6.12) becomes

$$\mathfrak{g}(K\mathfrak{U}) = [\operatorname{ad}(K)]\,\mathfrak{g}(\mathfrak{U}). \tag{7.16}$$

Finally, via the bar isomorphism, $\operatorname{ad}(K)$ can be identified as operators on $\overline{\mathfrak{gl}}(3)$. Thus the Lie subalgebras $\bar{\mathfrak{g}}(\mathfrak{U})$ satisfy the transformation law

$$\bar{\mathfrak{g}}(K\mathfrak{U}) = [\operatorname{ad}(K)]\,\bar{\mathfrak{g}}(\mathfrak{U}) \tag{7.17}$$

within the framework of $\overline{\mathfrak{gl}}(3)$. It should be noted that there exists *no* operation of conjugation by elements in $\mathscr{GL}(3)$ for the bundle space $\mathscr{E}(\mathscr{B})$. However, since the fundamental fields are images of left-invariant fields, we do have the adjoint representations

$$\operatorname{ad}(K) = R_{K^{-1}*} \tag{7.18}$$

as operators on $\overline{\mathfrak{gl}}(3)$.

8. Material Connections. I. Abstract Formulation

In this section, we introduce the notion of *material connections* on the bundle of linear frames of a simple body. In contrast with the fundamental fields, which characterize the local structure of *individual fibres* of a principal bundle, *connections* represent *relations* of fibres with their neighboring fibres. Thus in the terminology of differential geometry, the values of the fundamental fields form the *vertical subspaces* of a principal bundle, while a connection is, by definition, a smooth field of *horizontal subspaces.*

We review briefly the basic concepts of connections[39] on an arbitrary principal bundle. For definiteness, let $\mathscr{E}$ be the bundle space, $\mathscr{M}$ be the base space, $\mathscr{G}$ be the structure group (and also the fibre space), and π be the projection. We put

$$m = \dim \mathscr{M}, \tag{8.1}$$

and

$$n = \dim \mathscr{G}. \tag{8.2}$$

[39] For more details see the book by KOBAYASHI & NOMIZU [*11*, Ch. V].

Then we define the *vertical subspace* $\mathscr{V}_x$ at any point $x \in \mathscr{E}$ by

$$\mathscr{V}_x = \ker \pi_{*x}. \tag{8.3}$$

From (7.9),

$$\mathscr{V}_x = \bar{\mathfrak{g}}|_x, \tag{8.4}$$

whence

$$\dim \mathscr{V}_x = n. \tag{8.5}$$

Suppose that $\mathscr{H}_x$ is an m-dimensional subspace of $\mathscr{E}_x$ such that $\mathscr{E}_x$ can be decomposed into the direct sum

$$\mathscr{E}_x = \mathscr{V}_x \oplus \mathscr{H}_x. \tag{8.6}$$

Then $\mathscr{H}_x$ is called a *horizontal subspace* at x. It should be noted that, unlike the vertical subspace, horizontal subspaces are *not* unique. As was already mentioned, we define a *connection* on $\mathscr{E}$ to be a smooth field whose values are horizontal subspaces.

A connection $\mathscr{H}$ can be represented by a tensor field in the following way: Relative to the decomposition (8.6), $\mathscr{H}$ corresponds to a field of projection maps, say

$$\mathbf{v}_x\colon\ \mathscr{E}_x \to \mathscr{V}_x. \tag{8.7}$$

From the property i) of the fundamental fields, the restriction map $|_x$ at x is a canonical isomorphism of $\bar{\mathfrak{g}}$ and $\mathscr{V}_x$ (cf. (7.9) and (8.3)). Hence there exists a field

$$\omega(x) = |_x^{-1} \circ \mathbf{v}_x\colon\ \mathscr{E}_x \to \bar{\mathfrak{g}}. \tag{8.8}$$

By a standard canonical isomorphism in tensor algebra, we can regard ω as a field with values in the tensor products $\mathscr{E}_x^* \otimes \bar{\mathfrak{g}}$, where $\mathscr{E}_x^*$ denotes the cotangent space at x. In this sense, ω is a $\bar{\mathfrak{g}}$-valued 1-form on $\mathscr{E}$, called the *connection form* of $\mathscr{H}$. We can recover the connection $\mathscr{H}$ from its connection form ω by the relation

$$\mathscr{H}_x = \ker \omega(x). \tag{8.9}$$

Now suppose that a connection $\mathscr{H}$ is defined on $\mathscr{E}$. Then, from (8.3) and (8.6), the restriction maps

$$\pi_{*x}|_{\mathscr{H}_x} \equiv \boldsymbol{\eta}_x\colon\ \mathscr{H}_x \to \mathscr{M}_{\pi(x)} \tag{8.10}$$

are isomorphisms for all $x \in \mathscr{E}$. Thus for each fixed vector $\boldsymbol{v} \in \mathscr{M}_p$, there exists a smooth field $\tilde{\boldsymbol{v}}$ on the fibre $\mathscr{E}_p$ by

$$\tilde{\boldsymbol{v}}(x) \equiv \boldsymbol{\eta}_x^{-1}(\boldsymbol{v}), \tag{8.11}$$

called the *horizontal lift* of $\boldsymbol{v}$ relative to the connection $\mathscr{H}$. Similarly, suppose that λ is a smooth curve in $\mathscr{M}$. Then we define the *horizontal lift* of λ to be the collection of all horizontal curves $\tilde{\boldsymbol{\lambda}}$ in $\mathscr{E}$ such that

$$\pi(\tilde{\boldsymbol{\lambda}}(t)) = \lambda(t). \tag{8.12}$$

In other words, the tangent vectors $\dot{\tilde{\boldsymbol{\lambda}}}(t)$ are the values of the horizontal lift of $\dot{\boldsymbol{\lambda}}(t)$, for all t. Sometimes we say that $\tilde{\boldsymbol{\lambda}}$ are *horizontal curves lying above* λ.

Clearly, the curves $\tilde{\lambda}$ are characterized by the conditions (8.12) and

$$\langle \omega(\tilde{\lambda}(t)), \dot{\tilde{\lambda}}(t) \rangle = \mathbf{0}. \tag{8.13}$$

Since by hypothesis ω is a smooth field in $\mathscr{E}$, we can solve the above differential equation for each given initial condition, say a point $\tilde{\lambda}(0) \in \mathscr{E}_{\lambda(0)}$. The general solution of this equation corresponds to a 1-parameter family of mappings among the fibres above λ, say

$$\rho_t\colon\ \mathscr{E}_{\lambda(0)} \to \mathscr{E}_{\lambda(t)}, \tag{8.14}$$

such that

$$\rho_t(\tilde{\lambda}(0)) = \tilde{\lambda}(t) \tag{8.15}$$

for all solutions $\tilde{\lambda}$. Since the solution depends smoothly on the initial data, these mappings are all diffeomorphisms, called the *parallel transports along* λ relative to the connection $\mathscr{H}$.

An important criterion for a class of connections $\mathscr{H}$ on $\mathscr{E}$ now arises naturally. Via any bundle chart $(\mathscr{U}_\alpha, \xi_\alpha)$ such that $\lambda \in \mathscr{U}_\alpha$, the parallel transports ρ_t determine a class of transformations of the structure group $\mathscr{G}$, namely

$$\xi_{\alpha,\lambda(t)}^{-1} \circ \rho_t \circ \xi_{\alpha,\lambda(0)}\colon\ \mathscr{G} \to \mathscr{G}. \tag{8.16}$$

If $\mathscr{H}$ is arbitrary, then such transformations, in general, *cannot* be identified as operations of left-multiplication by elements in $\mathscr{G}$. Suppose a connection $\mathscr{H}$ obeys the criterion that[40] along any smooth curve in $\mathscr{M}$ all such transformations are operations of left-multiplication on $\mathscr{G}$. Then $\mathscr{H}$ is called a *$\mathscr{G}$-connection.* As we shall see, on the bundles of frames of simple bodies, only $\mathscr{G}$-connections are of physical interest.

We now explore the structures of $\mathscr{G}$-connections on the bundle of linear frames $\mathscr{E}(\mathscr{B})$ of a simple body $\mathscr{B}$. First, we give a necessary and sufficient condition for its $\mathscr{G}$-connections. Let $\mathscr{H}$ be an arbitrary connection on $\mathscr{E}(\mathscr{B})$. Then along any smooth curve λ in $\mathscr{B}$, the parallel transports ρ_t transform the linear frames $\tilde{\lambda}(0)$ at $\lambda(0)$ onto the linear frames $\tilde{\lambda}(t)$ at $\lambda(t)$. Since a linear frame is an ordered basis, we can extend the transformations ρ_t to a family of *linear isomorphisms, viz*

$$\rho_t(\tilde{\lambda})\colon\ \mathscr{B}_{\lambda(0)} \to \mathscr{B}_{\lambda(t)} \tag{8.17}$$

in such a way that if

$$\tilde{\lambda}(\tau) = \{e_i(\tau, \tilde{\lambda}),\ i = 1, 2, 3\}, \tag{8.18}$$

then

$$[\rho_t(\tilde{\lambda})](e_i(0, \tilde{\lambda})) \equiv e_i(t, \tilde{\lambda}), \qquad i = 1, 2, 3. \tag{8.19}$$

We claim that $\mathscr{H}$ is a $\mathscr{G}$-connection if and only if the mappings $\rho_t(\tilde{\lambda})$ are independent of $\tilde{\lambda}$, so that there exist well-defined induced *parallel transports of the tangent spaces* along λ.

To see this result, suppose that $\tilde{\lambda}_1$, and $\tilde{\lambda}_2$ are any two horizontal curves lying above λ. Then we define mappings $\boldsymbol{K}(t) \in \mathscr{GL}(3)$ by the condition

$$\tilde{\lambda}_2(t) = R_{\boldsymbol{K}(t)}(\tilde{\lambda}_1(t)), \tag{8.20}$$

[40] This criterion is self-consistent, since the condition that the transformations (8.16) belong to $\mathscr{G}$ is a property of $\mathscr{H}$ *independent* of the choice of bundle chart.

for all t. From (8.19), it is easily verified that

$$\rho_t(\tilde{\lambda}_1)=\rho_t(\tilde{\lambda}_2) \tag{8.21}$$

if and only if

$$\boldsymbol{K}(t)=\boldsymbol{K}(0)\equiv\boldsymbol{K} \tag{8.22}$$

for all t. Let $(\mathscr{U}_\alpha, \xi_\alpha)$ be a bundle chart such that $\lambda\in\mathscr{U}_\alpha$. Then

$$\mathscr{E}_{\lambda(t)}=\xi_{\alpha,\lambda(t)}(\mathscr{GL}(3)), \tag{8.23}$$

say

$$\tilde{\lambda}_1(t)=\xi_{\alpha,\lambda(t)}(\boldsymbol{G}_1(t)), \tag{8.24}$$

and

$$\tilde{\lambda}_2(t)=\xi_{\alpha,\lambda(t)}(\boldsymbol{G}_2(t)). \tag{8.25}$$

From (8.20) and the above representations of $\tilde{\lambda}_1$ and $\tilde{\lambda}_2$, we have

$$\boldsymbol{G}_2(t)=\boldsymbol{G}_1(t)\,\boldsymbol{K}(t). \tag{8.26}$$

But by definition, if $\mathscr{H}$ is a $\mathscr{G}$-connection, then the mappings $\xi^{-1}_{\alpha,\lambda(t)}\circ\rho_t\circ\xi_{\alpha,\lambda(0)}$ correspond to certain operations of left-multiplication on $\mathscr{GL}(3)$, say

$$\xi^{-1}_{\alpha,\lambda(t)}\circ\rho_t\circ\xi_{\alpha,\lambda(0)}=\boldsymbol{F}(t). \tag{8.27}$$

Applying these operations to $\boldsymbol{G}_1(0)$ and $\boldsymbol{G}_2(0)$ respectively, we obtain

$$\boldsymbol{G}_1(t)=\boldsymbol{F}(t)\,\boldsymbol{G}_1(0), \tag{8.28}$$

and

$$\boldsymbol{G}_2(t)=\boldsymbol{F}(t)\,\boldsymbol{G}_2(0). \tag{8.29}$$

Consequently,

$$\boldsymbol{K}(t)=\boldsymbol{G}_1(t)^{-1}\,\boldsymbol{G}_2(t)=\boldsymbol{G}_1(0)^{-1}\,\boldsymbol{G}_2(0)=\boldsymbol{K}(0), \tag{8.30}$$

for all t.

Conversely, if (8.22) holds for arbitrary curves $\boldsymbol{\lambda}_1$, and $\tilde{\lambda}_2$, then from (8.20), (8.24), and (8.25),

$$\boldsymbol{G}_1(t)\,\boldsymbol{G}_1(0)^{-1}=\boldsymbol{G}_2(t)\,\boldsymbol{G}_2(0)^{-1}. \tag{8.31}$$

Thus the mapping $\boldsymbol{F}(t)$ exist and belong to $\mathscr{GL}(3)$; the proof is complete.

From (8.22) and (8.20), the tangent vectors $\dot{\tilde{\lambda}}_1$ and $\dot{\tilde{\lambda}}_2$ are related by

$$\dot{\tilde{\lambda}}_2=R_{\boldsymbol{K}*}(\dot{\tilde{\lambda}}_1). \tag{8.32}$$

Since such tangent vectors form the horizontal lift of $\dot{\lambda}$, varying λ arbitrarily, we obtain

$$R_{\boldsymbol{K}*}(\mathscr{H})=\mathscr{H}. \tag{8.33}$$

Thus another necessary and sufficient condition for a $\mathscr{G}$-connection $\mathscr{H}$ is that $\mathscr{H}$ be stable under the operations $R_{\boldsymbol{K}*}$ for all $\boldsymbol{K}$. In terms of the field of projection maps $\boldsymbol{v}$ introduced in (8.7), the above condition reduces to

$$R_{\boldsymbol{K}*}\circ\boldsymbol{v}=\boldsymbol{v}\circ R_{\boldsymbol{K}*}. \tag{8.34}$$

Then from (8.8), the conncetion form $\boldsymbol{\omega}$ satisfies the condition

$$R_{\boldsymbol{K}}^*(\omega)=[\mathrm{ad}(\boldsymbol{K}^{-1})](\omega), \tag{8.35}$$

which means that

$$\langle\omega(R_{\boldsymbol{K}}(\boldsymbol{x})), R_{\boldsymbol{K}*\boldsymbol{x}}(\boldsymbol{v})\rangle=\mathrm{ad}(\boldsymbol{K}^{-1})\langle\omega(\boldsymbol{x}), \boldsymbol{v}\rangle \tag{8.36}$$

for all $\boldsymbol{x}\in\mathscr{E}(\mathscr{B})$ and $\boldsymbol{x}\in\mathscr{E}(\mathscr{B})_{\boldsymbol{x}}$.

So far we have considered $\mathscr{G}$-connections on $\mathscr{E}(\mathscr{B})$. Such connections correspond to the classical *affine connections* on the body manifold $\mathscr{B}$, since their induced parallel transports of tangent spaces are linear isomorphisms. For a simple body $\mathscr{B}$, the $\mathscr{G}$-connections on $\mathscr{E}(\mathscr{B})$ of special interest are those whose induced parallel transports are material isomorphisms. We state

Definition 8.1. A *material connection* of a simple body $\mathscr{B}$ is a $\mathscr{G}$-connection on $\mathscr{E}(\mathscr{B})$ whose induced parallel transports are always material isomorphisms.

We proceed to characterize a material connection. From conditions (4.16), (4.17), and (5.20), the parallel transports ρ_t defined by (8.19) are material isomorphisms if and only if each horizontal curve in $\mathscr{E}(\mathscr{B})$ is contained in one and only one of the bundles of reference frames $\mathscr{E}(\mathscr{B}, \mathfrak{U})$, here regarded as subbundles of $\mathscr{E}(\mathscr{B})$. This condition is self-consistent, since the subbundles $\mathscr{E}(\mathscr{B}, \mathfrak{U})$, like the horizontal curves $\tilde{\lambda}$, are related to each other by right-multiplications of elements in $\mathscr{GL}(3)$. In the terminology of connections, a $\mathscr{G}$-connection satisfying a condition of this kind is said to be *reducible* to $\mathscr{G}$-connections on the subbundles. Thus *material connections are $\mathscr{G}$-connections on $\mathscr{E}(\mathscr{B})$ which are reducible to $\mathscr{G}$-connections on the bundles of reference frames $\mathscr{E}(\mathscr{B}, \mathfrak{U})$.*

Now suppose that $\mathscr{H}(\mathfrak{U})$ is a $\mathscr{G}$-connection on a bundle of reference frames $\mathscr{E}(\mathscr{B}, \mathfrak{U})$. Then we can extend $\mathscr{H}(\mathfrak{U})$ to a unique material connection $\mathscr{H}$ in the following way: Let $\boldsymbol{x}$ be an arbitrary point in $\mathscr{E}(\mathscr{B}, \mathfrak{U})$. Then the tangent space $\mathscr{E}(\mathscr{B}, \mathfrak{U})_{\boldsymbol{x}}$ can be identified as a subspace of $\mathscr{E}(\mathscr{B})_{\boldsymbol{x}}$. In particular, the value of $\mathscr{H}(\mathfrak{U})$ at $\boldsymbol{x}$ corresponds to a subspace of $\mathscr{E}(\mathscr{B})_{\boldsymbol{x}}$. Clearly, it is a horizontal subspace at $\boldsymbol{x}$. We put

$$\mathscr{H}_{\boldsymbol{x}}\equiv\mathscr{H}(\mathfrak{U})_{\boldsymbol{x}}. \tag{8.37}$$

Then this field $\mathscr{H}$ can be extended to the whole bundle space $\mathscr{E}(\mathscr{B})$ by the condition (8.33). Since $\mathscr{H}(\mathfrak{U})$ is a $\mathscr{G}$-connection on the subbundle $\mathscr{E}(\mathscr{B}, \mathfrak{U})$, the extension $\mathscr{H}$ is well-defined. We express the relation between $\mathscr{H}$ and $\mathscr{H}(\mathfrak{U})$ by

$$\mathscr{H}(\mathfrak{U})=\mathscr{H}|_{\mathscr{E}(\mathscr{B},\mathfrak{U})}, \tag{8.38}$$

and this *restriction operation* is an isomorphism of material connections and $\mathscr{G}$-connections on $\mathscr{E}(\mathscr{B}, \mathfrak{U})$, for each fixed $\mathfrak{U}$.

By virtue of this isomorphism, a $\mathscr{G}$-connection $\mathscr{H}$ on $\mathscr{E}(\mathscr{B})$ is a material connection if and only if the restrictions of the connection form ω on $\mathscr{E}(\mathscr{B}, \mathfrak{U})$ are $\bar{\mathfrak{g}}(\mathfrak{U})$-valued[41], where $\bar{\mathfrak{g}}(\mathfrak{U})$ are regarded as Lie subalgebras of $\overline{\mathfrak{gl}}(3)$.

We mention here that since the body manifold $\mathscr{B}$ is paracompact, $\mathscr{G}$-connections on the bundles $\mathscr{E}(\mathscr{B}, \mathfrak{U})$ exist. Consequently, via the isomorphism (8.38), material connections exist. A standard way to construct a $\mathscr{G}$-connection on $\mathscr{E}(\mathscr{B}, \mathfrak{U})$ is as follows: Let $\xi(\mathfrak{U})=\{(\mathscr{U}_\alpha, \xi_\alpha), \alpha\in I\}$ be the bundle atlas of $\mathscr{E}(\mathscr{B}, \mathfrak{U})$. Then for

[41] *Cf.* KOBAYASHI & NOMIZU [*11*, Ch. II, § 6].

each coordinate neighborhood $\mathscr{U}_\alpha$, there exists a $\mathscr{G}$-connection $\mathscr{H}(\mathfrak{U}, \alpha)$ on the set

$$\pi^{-1}(\mathscr{U}_\alpha) \subset \mathscr{E}(\mathscr{B}, \mathfrak{U}). \tag{8.38}$$

For instance, we can choose $\mathscr{H}(\mathfrak{U}, \alpha)$ to be the completely integrable $\mathscr{G}$-connection such that the cross section $\boldsymbol{\sigma}_\alpha$ is its integral manifold. Let

$$f = \{f_\alpha, \alpha \in I\} \tag{8.39}$$

be a smooth locally finite *partition of unity*[42] subordinate to the open covering

$$\{\mathscr{U}_\alpha, \alpha \in I\}, \tag{8.40}$$

so that

$$\operatorname{supp} f_\alpha \subset \mathscr{U}_\alpha, \tag{8.41}$$

and each point $p \in \mathscr{B}$ has a neighborhood which meets only finitely many of $\operatorname{supp} f_\alpha$. Here $\operatorname{supp} f_\alpha$ denotes the *support* of the function f_α, *i.e.*, the closure (in $\mathscr{B}$) of the set $\{q: f_\alpha(q) \neq 0\}$. The functions f_α satisfy also the conditions

$$0 \leqq f_\alpha(p) \leqq 1, \tag{8.42}$$

and

$$\sum_\alpha f_\alpha(p) = 1, \tag{8.43}$$

for all $p \in \mathscr{B}$. Notice that the left-hand side of (8.43) is a finite sum. Now we put

$$[\omega(\mathfrak{U})](\boldsymbol{x}) \equiv \sum_\alpha f_\alpha(\pi(\boldsymbol{x}))\,[\omega(\mathfrak{U}, \alpha)](\boldsymbol{x}) \tag{8.44}$$

for all $\boldsymbol{x} \in \mathscr{E}(\mathscr{B}, \mathfrak{U})$, where $\omega(\mathfrak{U}, \alpha)$ denotes the connection forms of the connections $\mathscr{H}(\mathfrak{U}, \alpha)$ on the sets $\pi^{-1}(\mathscr{U}_\alpha)$ in $\mathscr{E}(\mathscr{B}, \mathfrak{U})$. Clearly, $\omega(\mathfrak{U})$ is a smooth field; moreover, it satisfies the condition (8.35), since so do the fields $\omega(\mathfrak{U}, \alpha)$. By the condition (8.43), at each $\boldsymbol{x} \in \mathscr{E}(\mathscr{B}, \mathfrak{U})$ the $\bar{\mathfrak{g}}(\mathfrak{U})$-valued covector $[\omega(\mathfrak{U})](\boldsymbol{x})$ acts as the inverse of the restriction map at $\boldsymbol{x}$, *viz*,

$$[\omega(\mathfrak{U})](\boldsymbol{x}) = |_{\boldsymbol{x}}^{-1}: \mathscr{E}(\mathscr{B}, \mathfrak{U})_{\boldsymbol{x}} \to \bar{\mathfrak{g}}(\mathfrak{U}). \tag{8.45}$$

Thus $\omega(\mathfrak{U})$ is the connection form of a $\mathscr{G}$-connection $\mathscr{H}(\mathfrak{U})$ on $\mathscr{E}(\mathscr{B}, \mathfrak{U})$, namely,

$$\mathscr{H}(\mathfrak{U}) \equiv \ker \omega(\mathfrak{U}). \tag{8.46}$$

It should be noted that the $\mathscr{G}$-connection $\mathscr{H}(\mathfrak{U})$ constructed in this manner, however, need *not* be completely integrable any more, since the functions f_α are not constants.

Remark. It is interesting to note that the condition characterizing a material connection depends on the isotropy groups only through the Lie algebras $\bar{\mathfrak{g}}(\mathfrak{U})$. This fact is not surprising, since $\bar{\mathfrak{g}}(\mathfrak{U})$ uniquely determine a connected component of $\mathscr{E}(\mathscr{B}, \mathfrak{U})$ in $\mathscr{E}(\mathscr{B})$, and vice versa. In general, if a smooth curve lies in a subbundle, then locally it must stay in one and only one component of that subbundle. Consequently, if two subbundles share a fixed component (hence their Lie algebras are isomorphic), then a $\mathscr{G}$-connection is reducible on one of them if and only if it is reducible on both. In view of this fact, so far as material connections are concerned it suffices to consider simple bodies whose isotropy groups are connected.

[42] *Cf.* Kobayashi & Nomizu [*11*, Apendix 3].

9. Material Connections. II. Component Forms

Let ω be the connection form of a $\mathscr{G}$-connection $\mathscr{H}$ on $\mathscr{E}(\mathscr{B})$. Then by definition ω must satisfy the following two conditions:

i)
$$\langle\omega(x), \bar{v}(x)\rangle = \bar{v} \tag{9.1}$$
for all $\bar{v}\in\overline{\mathbf{gl}}(3)$, and

ii)
$$\langle\omega(R_K(x)), R_{K*x}(z)\rangle = \operatorname{ad}(K^{-1})\langle\omega(x), z\rangle \tag{9.2}$$

for all $K\in\mathscr{G}\mathscr{L}(3)$, $z\in\mathscr{E}(\mathscr{B})_x$, and $x\in\mathscr{E}(\mathscr{B})$.

From these conditions, if ω is known at any point in $\mathscr{E}(\mathscr{B})$, say x, then it is completely determined on the whole fibre $\mathscr{E}(\mathscr{B})_{\pi(x)}$ containing x. This fact is the basic principle to be used in representing ω in component form as we now proceed.

Let $\{e^i_j,\ i,j=1,2,3\}$ be a basis of $\mathbf{gl}(3)$ and $\{\bar{e}^i_j,\ i,j=1,2,3\}$ be the corresponding basis of $\overline{\mathbf{gl}}(3)$. Then we can represent ω in component form relative to $\{\bar{e}^i_j\}$, say
$$\omega = \bar{e}^i_j \otimes \omega^j_i, \tag{9.3}$$

where ω^i_j are some ordinary 1-forms on $\mathscr{E}(\mathscr{B})$. From (9.1),
$$\langle\omega^i_j(x), \bar{e}^k_l(x)\rangle = \delta^i_l\,\delta^k_j. \tag{9.4}$$

Thus ω^i_j are linearly independent. By definition,
$$\langle\omega^i_j(x), z\rangle = 0 \tag{9.5}$$

for all $z\in\mathscr{H}_x$. Combining (9.4) and (9.5), we see that $\{\omega^i_j(x),\ i,j=1,2,3\}$ spans the orthogonal complement of $\mathscr{H}_x$.

Let $(\mathscr{U}_\alpha, \xi_\alpha)$ be a bundle chart, and let σ_α be the corresponding cross-section (*cf.* (5.33)). Then the induced linear map σ^*_α transforms the 1-forms ω^i_j onto 1-forms $\underset{\alpha}{\omega}{}^i_j$ on $\mathscr{U}_\alpha$ in the usual way. We define
$$\underset{\alpha}{\omega} = e^i_j \otimes \underset{\alpha}{\omega}{}^j_i, \tag{9.6}$$

which is a $\mathbf{gl}(3)$-valued 1-form on $\mathscr{U}_\alpha$. Evidently, $\underset{\alpha}{\omega}$ is independent of the choice of the basis $\{e^i_j\}$.

Since we can recover the field ω on $\mathscr{E}(\mathscr{U}_\alpha)$ from the field $\underset{\alpha}{\omega}$, the latter is called the *representation* of the former relative to the bundle chart $(\mathscr{U}_\alpha, \xi_\alpha)$. By construction, the form $\underset{\alpha}{\omega}$ has the property that
$$\overline{\langle\underset{\alpha}{\omega}(p), z\rangle} = \langle\omega(\sigma_\alpha(p)), \sigma_{\alpha*p}(z)\rangle \tag{9.7}$$

for all $z\in\mathscr{B}_p$, $p\in\mathscr{U}_\alpha$. Hence relative to the basis $\{\bar{e}^i_j\}$,
$$v(\sigma_{\alpha*p}(z)) = \langle\underset{\alpha}{\omega}{}^i_j(p), z\rangle\,\bar{e}^j_i(\sigma_\alpha(p)). \tag{9.8}$$

Since σ_α is a cross section, the image $\sigma_{\alpha*p}(\mathscr{B}_p)$ forms a horizontal subspace at $\sigma_\alpha(p)$ (although not necessarily coincide with the horizontal subspace $\mathscr{H}_{\sigma_\alpha(p)}$). By linearity and conditions (9.1) and (9.8), ω is completely determined on the set

$\sigma_\alpha(\mathscr{U}_\alpha)$. Then by the basic principle of $\mathscr{G}$-connections remarked before, that $\boldsymbol{\omega}(x)$ determines $\boldsymbol{\omega}$ completely on $\mathscr{E}(\mathscr{B})_{\pi(x)}$, we see that $\underset{\alpha}{\boldsymbol{\omega}}$ indeed determines $\boldsymbol{\omega}$ over the set $\mathscr{U}_\alpha$.

Let λ be a smooth curve in $\mathscr{U}_\alpha$. Then any smooth curve $\boldsymbol{\lambda}^*$ in $\mathscr{E}(\mathscr{U}_\alpha)$ lying directly above λ can be expressed by

$$\boldsymbol{\lambda}^* = R_{\boldsymbol{K}(t)}(\boldsymbol{\sigma}_\alpha(\lambda(t))), \tag{9.9}$$

where $\boldsymbol{K}(t)$ is a smooth curve in $\mathscr{GL}(3)$. From (9.2), (9.8), and the chain rule, the vertical component of the tangent vector $\dot{\boldsymbol{\lambda}}^*(t)$ is given by

$$\mathbf{v}(\dot{\boldsymbol{\lambda}}^*(t)) = \mathrm{ad}(\boldsymbol{K}(t)^{-1})\langle\underset{\alpha}{\boldsymbol{\omega}}(\lambda(t)), \dot{\lambda}(t)\rangle|_{\boldsymbol{\lambda}^*(t)} + \boldsymbol{\theta}\circ\dot{\boldsymbol{K}}(t)|_{\boldsymbol{\lambda}^*(t)}, \tag{9.10}$$

where $\boldsymbol{\theta}$ denotes the canonical form on $\mathscr{GL}(3)$ characterized by

$$\boldsymbol{\theta}(\boldsymbol{G}) \equiv |_{\boldsymbol{G}}^{-1}: \; \mathscr{GL}(3)_{\boldsymbol{G}} \to \mathbf{gl}(3) \tag{9.11}$$

for all $\boldsymbol{G}\in\mathscr{GL}(3)$. Applying (9.10) to the curve

$$\boldsymbol{\sigma}_\beta(\lambda(t)) = R_{\boldsymbol{G}_{\alpha\beta}(\lambda(t))}(\boldsymbol{\sigma}_\alpha(\lambda(t))), \tag{9.12}$$

where $\boldsymbol{\sigma}_\beta$ denotes the cross section of another bundle chart $(\mathscr{U}_\beta, \xi_\beta)$, we see that the representations $\underset{\alpha}{\boldsymbol{\omega}}$ and $\underset{\beta}{\boldsymbol{\omega}}$ satisfy the following transformation law[43]:

$$\underset{\beta}{\boldsymbol{\omega}} = \mathrm{ad}(\boldsymbol{G}_{\alpha\beta}^{-1})\,\underset{\alpha}{\boldsymbol{\omega}} + \boldsymbol{\theta}\circ\boldsymbol{G}_{\alpha\beta\,*} \tag{9.13}$$

on the overlap $\mathscr{U}_\alpha \cap \mathscr{U}_\beta$.

Now suppose that the curve $\boldsymbol{\lambda}^*$ is horizontal. Then $\boldsymbol{K}$ satisfies the differential equation

$$\mathrm{ad}(\boldsymbol{K}(t)^{-1})\langle\underset{\alpha}{\boldsymbol{\omega}}(\lambda(t)), \dot{\lambda}(t)\rangle + \boldsymbol{\theta}\circ\dot{\boldsymbol{K}}(t) = \mathbf{0}, \tag{9.14}$$

which is a representation of the equation of parallel transport in $(\mathscr{U}_\alpha, \xi_\alpha)$.

In practice, it is convenient to use the lifted bundle charts $(\mathscr{U}_\alpha, \xi_\alpha)$ associated with local coordinate systems on $\mathscr{B}$. For such charts, the cross-sections $\boldsymbol{\sigma}_\alpha$ are the fields of natural frames, *viz*

$$\boldsymbol{\sigma}_\alpha(p) = \left\{ \left.\frac{\partial}{\partial x^i}\right|_p, \; i = 1, 2, 3 \right\}. \tag{9.15}$$

Also, we choose $\{\boldsymbol{e}_j^i\}$ to be the standard basis of $\mathbf{gl}(3)$. Then $\underset{\alpha}{\omega}$ can be expressed in the matrix form

$$\underset{\alpha}{\omega} = \boldsymbol{e}_j^i \otimes \underset{\alpha}{\omega}{}_i^j = [\underset{\alpha}{\omega}{}_i^j], \tag{9.16}$$

where $\underset{\alpha}{\omega}{}_j^i$ are ordinary 1-forms on $\mathscr{U}_\alpha$. Of course, they too can be expressed in component form, say

$$\underset{\alpha}{\omega}{}_j^i = \Gamma_{jk}^i\, dx^k, \tag{9.17}$$

---

[43] Kobayashi & Nomizu [*11*, Ch. II, §1].

where the components Γ^i_{jk} are smooth functions on $\mathscr{U}_\alpha$, called the *connection symbols* relative to the local coordinate systems (x^i).

For the lifted charts, the coordinate transformations are given by the Jacobian matrices (*cf.* (4.13)). Then the transformation law (9.13) reduces to the form

$$\bar{\Gamma}^i_{jk}=\Gamma^l_{mn}\frac{\partial \bar{x}^i}{\partial x^l}\frac{\partial x^m}{\partial \bar{x}^j}\frac{\partial x^n}{\partial \bar{x}^k}+\frac{\partial \bar{x}^i}{\partial x^l}\frac{\partial^2 x^l}{\partial \bar{x}^j \partial \bar{x}^k}. \tag{9.18}$$

Similarly, the equation of parallel transport (9.14) becomes

$$\dot{e}^i_j(t)+\Gamma^i_{kl}(\lambda(t))\, e^k_j(t)\, \dot{\lambda}^l(t)=0, \tag{9.19}$$

where

$$\boldsymbol{K}(t)=[e^i_j(t)], \tag{9.20}$$

i.e., the horizontal curves $\tilde{\lambda}$ are determined by the frames

$$\tilde{\lambda}(t)=\left\{e^i_j(t)\left.\frac{\partial}{\partial x^i}\right|_{\lambda(t)},\quad j=1,2,3\right\}. \tag{9.21}$$

By the canonical one-to-one correspondence between $\mathscr{G}$-connections on $\mathscr{E}(\mathscr{B})$ and on $\mathscr{T}(\mathscr{B})$, the equations of parallel transport of the tangent spaces have the following component forms:

$$\dot{v}^i(t)+\Gamma^i_{kl}(\lambda(t))\, v^k(t)\, \dot{\lambda}^l(t)=0, \tag{9.22}$$

which are, of course, the classical equations of parallel transport relative to an affine connection characterized by the connection symbols Γ^i_{jk}.

We proceed to express the conditions characterizing material connections in component forms. Suppose that $(\mathscr{U}_\alpha, \xi_\alpha)$ is a lifted chart corresponding to the local coordinate system (x^1, x^2, x^3). Then a point $(p, \boldsymbol{e}_p)\in\mathscr{E}(\mathscr{U}_\alpha)$ has local coordinates $(x^i(p), e^j_k(p), i, j, k=1,2,3)$ relative to $(\mathscr{U}_\alpha, \xi_\alpha)$, where $[e^j_k(p)]$ is the matrix in $\mathscr{GL}(3)$ such that

$$\boldsymbol{e}_p=R_{[e^j_k(p)]}(\boldsymbol{\sigma}_\alpha(p))=\left\{e^j_k(p)\left.\frac{\partial}{\partial x^j}\right|_p,\quad k=1,2,3\right\}. \tag{9.23}$$

Put $\boldsymbol{x}=(p, \boldsymbol{e}_p)$; then the tangent space $\mathscr{E}(\mathscr{B})_{\boldsymbol{x}}$ is spanned by the natural basis

$$\left\{\left.\frac{\partial}{\partial x^i}\right|_{\boldsymbol{x}}, \left.\frac{\partial}{\partial e^j_k}\right|_{\boldsymbol{x}},\quad i,j,k=1,2,3\right\}. \tag{9.24}$$

Evidently, the vertical subspace $\mathscr{V}_{\boldsymbol{x}}$ is spanned by the set

$$\left\{\left.\frac{\partial}{\partial e^j_k}\right|_{\boldsymbol{x}},\quad j,k=1,2,3\right\}. \tag{9.25}$$

From the equations of parallel transport (9.19), it is easily verified that the horizontal subspace $\mathscr{H}_{\boldsymbol{x}}$ is spanned by the set

$$\left\{\left.\frac{\partial}{\partial x^i}\right|_{\boldsymbol{x}}-\Gamma^j_{mi}(p)\, e^m_k(p)\left.\frac{\partial}{\partial e^j_k}\right|_{\boldsymbol{x}},\quad i=1,2,3\right\}. \tag{9.26}$$

For definiteness, we fix the reference atlas $\mathfrak{U}$ in such a way that the Lie algebra $\mathbf{g}(\mathfrak{U})$ has the standard form given in Table 6-1. To determine the values of $\boldsymbol{\omega}$ on the subbundle $\mathscr{E}(\mathscr{B}, \mathfrak{U})$, we must express the tangent spaces of $\mathscr{E}(\mathscr{B}, \mathfrak{U})$ as subspaces of the tangent spaces of $\mathscr{E}(\mathscr{B})$. Suppose that $\boldsymbol{\mu}$ is a cross-section in $\mathscr{E}(\mathscr{B}, \mathfrak{U})$ above $\mathfrak{U}_\alpha$, say

$$\boldsymbol{\mu}(p)=\left\{f_j^i(p)\left.\frac{\partial}{\partial x^i}\right|_p,\quad j=1,2,3\right\}. \tag{9.27}$$

Then the induced linear map $\boldsymbol{\mu}_{*\,p}$ transforms the natural basis of (x^1, x^2, x^3) at p onto the set

$$\left\{\left.\frac{\partial}{\partial x^i}\right|_{\boldsymbol{\mu}(p)}+\left.\frac{\partial f_k^j}{\partial x^i}\right|_p\left.\frac{\partial}{\partial e_k^j}\right|_{\boldsymbol{\mu}(p)},\quad i=1,2,3\right\} \tag{9.28}$$

in $\mathscr{E}(\mathscr{B})_{\boldsymbol{\mu}(p)}$. From (9.26), we see that the vertical components of the vectors in this set are

$$\left\{\left(\left.\frac{\partial f_k^j}{\partial x^i}\right|_p+\Gamma^j_{m\,i}(p)f_k^m(p)\right)\left.\frac{\partial}{\partial e_k^j}\right|_{\boldsymbol{\mu}(p)},\quad i=1,2,3\right\}. \tag{9.29}$$

Since the point $\boldsymbol{\mu}(p)$ is related to the natural frame $\boldsymbol{\sigma}_\alpha(p)$ by

$$\boldsymbol{\mu}(p)=R_{\boldsymbol{F}}(\boldsymbol{\sigma}_\alpha(p)), \tag{9.30}$$

where $\boldsymbol{F}$ denotes the matrix $[f_j^i(p)]$ in (9.27), we conclude that the values of $\boldsymbol{\omega}$ at $\boldsymbol{\mu}(p)$ restricted to the subbundle $\mathscr{E}(\mathscr{B}, \mathfrak{U})$ form a subspace of $\overline{\mathbf{gl}}(3)$, which under the bar isomorphism corresponds to the subspace in $\mathbf{gl}(3)$ spanned by the set

$$\left\{\left[\overset{-1}{f_j^i}(p)\left(\left.\frac{\partial f_k^j}{\partial x^m}\right|_p+\Gamma^j_{s\,m}(p)f_k^s(p)\right)\right],\quad m=1,2,3\right\}, \tag{9.31}$$

expressed in the standard components of $\mathbf{gl}(3)$. Therefore, $\mathscr{H}$ is a material connection on $\mathscr{E}(\mathscr{B})$ if and only if

$$\left\{\left[\overset{-1}{f_j^i}(p)\left(\left.\frac{\partial f_k^j}{\partial x^m}\right|_p+\Gamma^j_{s\,m}(p)f_k^s(p)\right)\right],\quad m=1,2,3\right\}\subset\mathbf{g}(\mathfrak{U}) \tag{9.32}$$

for all $p\in\mathscr{B}$.

Notice that the above condition does not depend on the choice of the cross-section $\boldsymbol{\mu}$. Suppose that $\bar{\boldsymbol{\mu}}$ is another cross-section of $\mathscr{E}(\mathscr{B}, \mathfrak{U})$ over $\mathscr{U}_\alpha$, say

$$\bar{\boldsymbol{\mu}}(p)=\left\{\bar{f}_j^i(p)\left.\frac{\partial}{\partial x^i}\right|_p,\quad j=1,2,3\right\}. \tag{9.33}$$

Then the matrices $[f_j^i]$ and $[\bar{f}_j^i]$ are related by

$$[\bar{f}_j^i(p)]=[f_j^i(p)][G_j^i(p)], \tag{9.34}$$

where

$$\boldsymbol{G}(p)\equiv[G_j^i(p)]\in\mathscr{G}(\mathfrak{U}). \tag{9.35}$$

By the chain rule, we can express the set (9.31) for $\bar{\boldsymbol{\mu}}$ in the following form:

$$\left\{\left[\overset{-1}{G_s^i}\,\overset{-1}{f_j^s}\left(\frac{\partial f_i^j}{\partial x^m}+\Gamma^j_{t\,m}f_r^t\right)G_k^r+\overset{-1}{G_r^i}\frac{\partial G_k^r}{\partial x^m}\right],\quad m=1,2,3\right\}. \tag{9.36}$$

The matrices

$$\left[\overset{-1}{G}{}^i_r \frac{\partial G^r_k}{\partial x^m}\right], \qquad m=1,2,3, \tag{9.37}$$

clearly belong to $\mathfrak{g}(\mathfrak{U})$, since they are given by

$$\left[\overset{-1}{G}{}^i_r \frac{\partial G^r}{\partial x^m}\right]=[\boldsymbol{\theta}\circ \boldsymbol{G}_*]\left(\frac{\partial}{\partial x^m}\right), \qquad m=1,2,3, \tag{9.38}$$

where $\boldsymbol{G}$ denotes the field

$$\boldsymbol{G}\colon \mathscr{U}_\alpha \to \mathscr{G}(\mathfrak{U}), \tag{9.39}$$

shown in (9.35). On the other hand, the first terms in (9.36), *i.e.*, the matrices

$$\left[\overset{-1}{G}{}^i_s \overset{-1}{f}{}^s_j\left(\frac{\partial f^j_i}{\partial x^m}+\Gamma^j_{t\,m} f^t_r\right) G^r_k\right], \qquad m=1,2,3, \tag{9.40}$$

regarded as elements in $\mathfrak{gl}(3)$, are related to those in (9.31) by

$$\left[\overset{-1}{G}{}^i_s \overset{-1}{f}{}^s_j\left(\frac{\partial f^j_i}{\partial x^m}+\Gamma^j_{t\,m} f^t_r\right) G^r_k\right]=\operatorname{ad}(\boldsymbol{G}^{-1})\left[\overset{-1}{f}{}^i_j\left(\frac{\partial f^j_k}{\partial x^m}+\Gamma^j_{s\,m} f^s_k\right)\right]. \tag{9.41}$$

Therefore the set (9.31) is contained in $\mathfrak{g}(\mathfrak{U})$ if and only if the set (9.36) is so contained. Thus the condition (9.32) does not depend on the coice of $\boldsymbol{\mu}$.

For solid crystal bodies, the Lie algebra $\mathfrak{g}(\mathfrak{U})$ is the trivial type, *i.e.*

$$\mathfrak{g}(\mathfrak{U})=\{\mathbf{0}\}, \tag{9.42}$$

so that material connections for such bodies are characterized by the conditions

$$\frac{\partial f^j_k}{\partial x^m}+\Gamma^j_{i\,m} f^i_k=0. \tag{9.43}$$

In this case the connection symbols[44] are unique and are determined by

$$\Gamma^j_{i\,m}=-\overset{-1}{f}{}^k_i \frac{\partial f^j_k}{\partial x^m}. \tag{9.44}$$

In general, if a connection admits integral manifolds in the neighborhood of every point, then it is said to be *completely integrable*. In the case of solid crystals, comparing (9.43) with (9.19), we see that every smooth curve in $\boldsymbol{\mu}$ is horizontal. Hence $\boldsymbol{\mu}$ is an integral manifold of the material connection. Moreover, from (8.33), the cross sections $R_{\boldsymbol{K}}(\boldsymbol{\mu})$ are all integral manifolds, so that the connection is completely integrable. Conversely, suppose that $\mathscr{H}$ is completely integrable. Then we can choose the cross-section $\boldsymbol{\mu}$ to be an integral manifold of $\mathscr{H}$. The condition (9.43) again holds, so that the connection symbols are given by (9.44). It should be noted that a simple body, in general, need not have any completely integrable material connection at all. Examples of such bodies will be given in the next two sections.

Remark. In NOLL's theory of simple bodies, all material connections are defined to be completely integrable[45].

---

[44] *Cf.* BILBY, BULLOUGH & SMITH [*5*], KONDO [*3*], BILBY [*6*].

[45] *Cf.* TRUESDELL & NOLL [*1*, § 34].

Part III. Applications of the Geometric Theory

10. Covariant Derivatives, Curvatures, and Torsions

In Section 8 we have shown that the parallel transports relative to a $\mathscr{G}$-connection $\mathscr{H}$ on $\mathscr{E}(\mathscr{B})$ give rise to unique parallel transports in $\mathscr{T}(\mathscr{B})$. In component forms, these parallelisms are characterized by the equations (9.19) and (9.22). Since the parallel transports of tangent spaces are linear isomorphisms, by a standard operation in tensor algebra they induce linear isomorphisms of the tensor spaces, say

$$\rho_t^{r,s}\colon \mathscr{T}^{r,s}_{\lambda(0)} \to \mathscr{T}^{r,s}_{\lambda(t)}, \tag{10.1}$$

called the *parallel transports of tensor spaces of types* (r, s) *along* λ. In component forms such parallelisms are characterized by the equations

$$\begin{aligned} \dot{y}^{i_1 \dots i_r}_{j_1 \dots j_s} + (\Gamma^{i_1}_{jk}\, y^{j\, i_2 \dots i_r}_{j_1 \dots j_s} + \cdots + \Gamma^{i_r}_{jk}\, y^{i_1 \dots i_{r-1} j}_{j \dots j_s} \\ - \Gamma^{i}_{j_1 k}\, y^{i_1 \dots i_r}_{i\, j_2 \dots j_s} - \cdots - \Gamma^{i}_{j_s k}\, y^{i_1 \dots i_r}_{j_1 \dots j_{s-1} i})\, \dot{\lambda}^k = 0, \end{aligned} \tag{10.2}$$

where $y^{i_1 \dots i_r}_{j_1 \dots j_s}(t)$ are the components of a tensor field $\boldsymbol{y}(t)$ satisfying the condition

$$\boldsymbol{y}(t) = \rho_t^{r,s}(\boldsymbol{y}(0)) \tag{10.3}$$

on λ. We say that $\boldsymbol{y}$ is *displaced by parallel transports along* λ.

The parallel transports induce the operations of *covariant differentiation* in the standard way. Let $\boldsymbol{z}$ be an arbitrary smooth tensor field of type (r, s) on λ. Then the *covariant derivative of* $\boldsymbol{z}$ is defined to be the field

$$\left.\frac{D\boldsymbol{z}}{dt}\right|_t = \lim_{\Delta t \to 0} \frac{1}{\Delta t}\left[\boldsymbol{z}(t) - \rho_t^{r,s}(\rho_{t-\Delta t}^{r,s})^{-1}\, \boldsymbol{z}(t - \Delta t)\right], \tag{10.4}$$

so that $\boldsymbol{z}$ is displaced by parallel transport if and only if the covariant derivative of $\boldsymbol{z}$ vanishes. In component form

$$\begin{aligned} \left(\frac{D\boldsymbol{z}}{dt}\right)^{i_1 \dots i_r}_{j_1 \dots j_s} = \dot{z}^{i_1 \dots i_r}_{j_1 \dots j_s} + (\Gamma^{i_1}_{jk}\, z^{j\, i_2 \dots i_r}_{j_1 \dots j_s} + \cdots + \Gamma^{i_r}_{jk}\, z^{i_1 \dots i_{r-1} j}_{j_1 \dots j_s} \\ - \Gamma^{i}_{j_1 k}\, z^{i_1 \dots i_r}_{i\, j_2 \dots j_s} - \cdots - \Gamma^{i}_{j_s k}\, z^{i_1 \dots i_r}_{j_1 \dots j_{s-1} i})\, \dot{\lambda}^k. \end{aligned} \tag{10.5}$$

Now suppose that $\boldsymbol{z}$ is a smooth tensor field defined on some open set in $\mathscr{B}$. Then along any smooth curve λ in the domain of $\boldsymbol{z}$, the restriction

$$\hat{\boldsymbol{z}}(t) = \boldsymbol{z}(\lambda(t)) \tag{10.6}$$

is a smooth field on λ. Hence the covariant derivative $(D\hat{\boldsymbol{z}})/(dt)$ is defined. By the chain rule,

$$\begin{aligned} \left(\frac{D\hat{\boldsymbol{z}}}{dt}\right)^{i_1 \dots i_r}_{j_1 \dots j_s} = \Bigg(\frac{\partial z^{i_1 \dots i_r}_{j_1 \dots j_s}}{\partial x^k} + \Gamma^{i_1}_{jk}\, z^{j\, i_2 \dots i_r}_{j_1 \dots j_s} + \cdots + \Gamma^{i_r}_{jk}\, z^{i_1 \dots i_{r-1} j}_{j_1 \dots j_s} \\ - \Gamma^{i}_{j_1 k}\, z^{i_1 \dots i_r}_{i\, j_2 \dots j_s} - \cdots - \Gamma^{i}_{j_s k}\, z^{i_1 \dots i_r}_{j_1 \dots j_{s-1} i}\Bigg)\, \dot{\lambda}^k. \end{aligned} \tag{10.7}$$

Thus there exists a tensor field Dz, called the *covariant derivative* of z, such that for any smooth curve λ

$$\frac{D\hat{z}}{dt} = \langle Dz, \dot{\lambda}\rangle = D_{\dot{\lambda}} z. \tag{10.8}$$

From (10.7), the components of Dz are

$$\begin{aligned}(Dz)^{i_1\dots i_r}_{j_1\dots j_s k} = z^{i_1\dots i_r}_{j_1\dots j_s|k} = \frac{\partial z^{i_1\dots i_r}_{j_1\dots j_s}}{\partial x^k} + \Gamma^{i_1}_{jk} z^{j\, i_2\dots i_r}_{j_1\dots j_s} + \cdots \\ + \Gamma^{i_r}_{jk} z^{i_1\dots i_{r-1} j}_{j_1\dots j_s} - \Gamma^{i}_{j_1 k} z^{i_1\dots i_r}_{i\, j_2\dots j_s} - \cdots - \Gamma^{i}_{j_s k} z^{i_1\dots i_r}_{j_1\dots j_{s-1} i}.\end{aligned} \tag{10.9}$$

Proposition 10.1. *If z is an intrinsic field on $\mathscr{B}$, then the covariant derivative of z relative to any material connection of $\mathscr{B}$ vanishes. Conversely, if a field z is an intrinsic tensor at a point $p\in\mathscr{B}$, and if the covariant derivative of z relative to some material connection vanishes, then z is an intrinsic field.*

In general, an affine connection is said to be *(locally) flat* if every point has a local coordinate system in which the connection symbols vanish.

Proposition 10.2. *If $\mathscr{B}$ has a flat material connection, then it is locally homogeneous.*

This fact is obvious. Since the connection is spanned by the set (9.26), if the connection symbols vanish, then the natural frames of the corresponding local coordinate system are reference frames.

The converse of this proposition, however, is false in general, although it is true for solid bodies. The first part of this assertion can be seen in the following example; the second part will be proved in the next section.

Example 10.1

Consider a simple material whose isotropy group is of type No. 14 in Table 6.1. This material is a subfluid having one preferred axis and one preferred plane not containing the preferred axis. We can construct a locally homogeneous body made up of this subfluid in the shape of a spherical shell such that the preferred axis is in the radial direction, and the preferred plane is perpendicular to the preferred axis, at every particle. Let $\mathscr{A}$ be a homogeneous circular disc with radius a and thickness b consisting of the subfluid such that the axis of the disc coincides with the preferred axis and is normal to the preferred plane. In a cylindrical coordinate system, $\mathscr{A}$ occupies the domain $\{(r, \theta, z), 0\leqq r\leqq a, 0\leqq\theta\leqq 2\pi, 0\leqq z\leqq b\}$. We now map $\mathscr{A}$ onto a hemispherical shell $\mathscr{B}_1$ with inner radius c and thickness b. Let (R, Θ, Φ) be the spherical coordinates of a point in $\mathscr{B}_1$. Then the mapping from $\mathscr{A}$ to $\mathscr{B}_1$ is given by

$$R = c + z, \qquad \Theta = \theta, \qquad \Phi = \frac{\pi r}{2a}. \tag{10.10}$$

We construct the bottom hemispherical shell $\mathscr{B}_2$ in exactly the same way. Then the shell $\mathscr{B}$ is defined to be the interior of the union of $\mathscr{B}_1$ and $\mathscr{B}_2$. Evidently, $\mathscr{B}$ has the desired properties. Since $\mathscr{B}$ is simply connected, a completely integrable material connection, if it existed, would induce a distant parallelism, contradicting the fact that the sphere $\mathscr{S}^2$ admits no such parallelism. Thus this body $\mathscr{B}$ cannot be equipped with any such connection.

We do have a partial converse of Proposition 10.2 for simple bodies in general, namely,

Proposition 10.3. *If a simple body $\mathscr{B}$ is locally homogeneous, then for every particle $p \in \mathscr{B}$, there exists a material connection $\mathscr{H}$ (depending on p) such that the connection symbols of $\mathscr{H}$ vanish in some local coordinate system near p.*

Since the connection $\mathscr{H}$ in this proposition depends on the particle p, it is useful only for local problems. The proof of this proposition is a trivial consequence of the existence theorem for $\mathscr{G}$-connections, proved in Section 8.

A flat connection, clearly, is completely integrable. In differential geometry, the condition of integrability for $\mathscr{G}$-connections is characterized by the *curvature forms*. We review briefly this condition for $\mathscr{G}$-connections on $\mathscr{E}(\mathscr{B})$. From (9.26), the orthogonal complement of $\mathscr{H}$ is spanned by the set of covectors

$$\{d e_k^j + \Gamma_{m\,i}^j e_k^m \, d x^i, \; j, k=1,2,3\}. \tag{10.11}$$

For simplicity, set

$$\zeta_k^j = d e_k^j + \Gamma_{m\,i}^j e_k^m \, d x^i \tag{10.12}$$

which are ordinary 1-forms on the set $\mathscr{E}(\mathscr{U}_\alpha)$, where $\mathscr{U}_\alpha$ is the coordinate neighborhood of (x^1, x^2, x^3). Taking the exterior derivative of (10.12), we get

$$d\zeta_k^j = \tfrac{1}{2} e_k^m R_{m\,r\,i}^j \, d x^r \wedge d x^i + \Gamma_{m\,i}^j \, d e_k^m \wedge d x^i, \tag{10.13}$$

where

$$R_{m\,r\,i}^j = \frac{\partial \Gamma_{m\,i}^j}{\partial x^r} - \frac{\partial \Gamma_{m\,r}^j}{\partial x^i} + \Gamma_{s\,r}^j \Gamma_{m\,i}^s - \Gamma_{s\,i}^j \Gamma_{m\,r}^s, \tag{10.14}$$

called the *curvature symbols of $\mathscr{H}$* relative to the local coordinate system (x^i). By virtue of (9.14), $R_{m\,r\,i}^j$ are the components of a tensor field $\boldsymbol{R}$, called the *curvature tensor* of $\mathscr{H}$.

By the classical Frobenius theorem[46], $\mathscr{H}$ is completely integrable if and only if

$$d\zeta_k^j \wedge \zeta_1^1 \wedge \zeta_2^1 \wedge \cdots \wedge \zeta_3^3 = \mathbf{0}, \qquad j, k=1,2,3. \tag{10.15}$$

Thus from (10.13), we have the following condition of integrability:

$$\boldsymbol{R} \equiv \mathbf{0}. \tag{10.16}$$

Now suppose that $\mathscr{H}$ is completely integrable. Let $\boldsymbol{\mu}$ be a horizontal cross-section over $\mathscr{U}_\alpha$, say

$$\boldsymbol{\mu}(p) = \{\boldsymbol{f}_i(p), \; i=1,2,3\}. \tag{10.17}$$

Then the components f_j^i of $\boldsymbol{f}_j$ relative to (x^1, x^2, x^3) satisfy the condition (9.43). Hence the Poisson brackets $[\boldsymbol{f}_i, \boldsymbol{f}_j]$ are given by

$$[\boldsymbol{f}_i, \boldsymbol{f}_j]^k = \frac{\partial f_j^k}{\partial x^m} f_i^m - \frac{\partial f_i^k}{\partial x^m} f_j^m = T_{m\,r}^k f_i^m f_j^r, \tag{10.18}$$

---

[46] *Cf.* Lang [*12*, Ch. VI, §1], Sternberg [*9*, Ch. III, § 5], and Chern [*8*, Ch. III, § 2].

where T^k_{mr}, called the *torsion symbols* of $\mathscr{H}$ relative to (x^i), are given by

$$T^k_{mr} = \Gamma^k_{mr} - \Gamma^k_{rm}. \tag{10.19}$$

From (9.14), it is evident that T^k_{mr} are the components of a tensor field $\boldsymbol{T}$, called the *torsion tensor* of $\mathscr{H}$.

Suppose that $\boldsymbol{T}$ vanishes also. Then from (10.18),

$$[\boldsymbol{f}_i, \boldsymbol{f}_j] = \boldsymbol{0}. \tag{10.20}$$

In this case, again by the Frobenius theorem, there exists a local coordinate system, say $(\bar{x}^1, \bar{x}^2, \bar{x}^3)$, such that

$$\boldsymbol{f}_i = \frac{\partial}{\partial \bar{x}^i} = \delta^j_i \frac{\partial}{\partial \bar{x}^j}. \tag{10.21}$$

Obviously, the connection symbols relative to $(\bar{x}^i)$ vanish, since from (9.44)

$$\bar{\Gamma}^j_{im} = -\delta^k_i \frac{\partial \delta^j_k}{\partial \bar{x}^m} = 0. \tag{10.22}$$

Thus we have shown that a $\mathscr{G}$-connection $\mathscr{H}$ on $\mathscr{E}(\mathscr{B})$ is flat if and only if its curvature tensor and torsion tensor both vanish.

Since the existence of a flat material conncetion is related to the local homogeneity of $\mathscr{B}$, the above result implies that, in some sense, the curvature tensors and the torsion tensors of material connections represent the local inhomogeneity of $\mathscr{B}$. Indeed, in the theory[47] of continuous distributions of dislocations in lattices, the torsion tensor of the unique completely integrable material connection characterizes the *local dislocation density*. For arbitrary simple bodies $\mathscr{B}$, the curvature tensors and the torsion tensors of material connections, in general, are both non-vanishing. Examples of simple bodies corresponding to this general situation will be given in the next section.

Proposition 10.4. *A simple body $\mathscr{B}$ can be equipped with a torsion-free material connection if and only if for each particle $p \in \mathscr{B}$ there exists a torsion-free material connection $\mathscr{H}_{\mathscr{U}}$ on $\mathscr{B}(\mathscr{U})$ for some neighborhood $\mathscr{U} \ni p$.*

Proof. Necessity is trivial. Sufficiency is a simple consequence of the existence theorem of $\mathscr{G}$-connections, given in Section 8. From (8.44), if the material connections corresponding to the connection forms $\omega(\mathfrak{U}, \alpha)$ are all torsion-free, then so is the material connection corresponding to $\omega(\mathfrak{U})$. Q.E.D.

By virtue of this proposition, every locally homogeneous simple body has a torsion-free material connection.

Since the vanishing of the curvature tensor is the integrability condition for the connection, we have the following

Proposition 10.5. *A simple body $\mathscr{B}$ can be equipped with a curvature-free material connection if the bundles of reference frames $\mathscr{E}(\mathscr{B}, \mathfrak{U})$ are trivial.*

Further, if $\mathscr{B}$ is simply connected, then this last condition is also necessary. In particular, we have

[47] *Cf.* BILBY, BULLOUGH & SMITH [5], BILBY [6].

Proposition 10.6. *Every star-shaped simple body $\mathscr{B}$ can be equipped with a curvature-free material connection.*

The geometric meanings of the curvature tensor and the torsion tensor can be visualized most easily from the following *Ricci identities:*

$$f_{|i|j} - f_{|j|i} = T^k_{ji} f_{|k}, \tag{10.23}$$

and

$$z^i_{\ |j|k} - z^i_{\ |k|j} = R^i_{lkj} z^l + T^l_{kj} z^i_{\ |l}, \tag{10.24}$$

for all smooth scalar fields f and vector fields z. More generally, for a smooth tensor field z of type (r, s), we have

$$\begin{aligned} z^{i_1 \dots i_r}_{j_1 \dots j_s |k| l} - z^{i_1 \dots i_r}_{j_1 \dots j_s |l| k} = R^{i_1}_{jlk} z^{j i_2 \dots i_r}_{j_1 \dots j_s} + \cdots \\ + R^{i_r}_{jlk} z^{i_1 \dots i_{r-1} j}_{j_1 \dots j_s} + R^i_{j_1 kl} z^{i_1 \dots i_r}_{i j_2 \dots j_s} + \cdots + R^i_{j_s kl} z^{i_1 \dots i_r}_{j_1 \dots j_{s-1} i} + T^i_{lk} z^{i_1 \dots i_r}_{j_1 \dots j_s | i}. \end{aligned} \tag{10.25}$$

The geometric meanings of the covariant derivatives, of course, are given by (10.4) and (10.8).

Using the Ricci identities, SCHOUTEN[48] has given the following geometric interpretation of the curvature tensor. Let p be a fixed point in $\mathscr{B}$ and $\mathscr{S}$ be a two-dimensional submanifold in $\mathscr{B}$ containing p. Suppose that (α, β) is a local coordinate system on $\mathscr{S}$ such that p has the coordinates $(0, 0)$. We put

$$\boldsymbol{u} = \left.\frac{\partial}{\partial \alpha}\right|_p, \tag{10.26}$$

and

$$\boldsymbol{v} = \left.\frac{\partial}{\partial \beta}\right|_p. \tag{10.27}$$

Then $\boldsymbol{u}$, $\boldsymbol{v}$ can be identified as vectors in $\mathscr{B}_p$. Now we consider a closed coordinate circuit λ:

$$(0,0) - (\alpha, 0) - (\alpha, \beta) - (0, \beta) - (0,0) \tag{10.28}$$

on $\mathscr{S}$. Suppose that z is a fixed vector in $\mathscr{B}_p$. Then the parallel transport of z around the circuit λ transforms z into another vector $\bar{z}$ at p. It turns out that for sufficiently small α and β, we have

$$\bar{z} - z = -\alpha \beta [\boldsymbol{R}(p)] (z, \boldsymbol{u}, \boldsymbol{v}) + o(\alpha^2 + \beta^2), \tag{10.29}$$

where $\boldsymbol{R}(p)$ is the curvature tensor at p, and

$$[\boldsymbol{R}(z, \boldsymbol{u}, \boldsymbol{v})]^i = R^i_{jkl} z^j u^k v^l. \tag{10.30}$$

Thus the curvature tensor characterizes the *local holonomy group* of the connection. The exact relation between the holonomy group and the curvature tensor is given by a theorem of AMBROSE & SINGER[49], which we do not consider here.

The torsion tensor can be interpreted in a similar way. Let p be a point and $\mathscr{S}$ be a two-dimensional surface containing p as before. Suppose that λ is a closed

[48] *Cf.* SCHOUTEN [*30*, Ch. III, § 4]. He attributes the proof of (10.29) to MORINAGA.

[49] AMBROSE & SINGER [*31*], see also NOMIZU [*10*, Ch. II, § 7] and KOBAYASHI & NOMIZU [*11*, Ch. II, § 8].

circuit on $\mathscr{S}$ passing through the point p, say

$$\lambda(0)=\lambda(1)=p. \tag{10.31}$$

We consider a field of linear frames

$$\boldsymbol{e}(t)=\{\boldsymbol{e}_i(t),\ i=1,2,3\} \tag{10.32}$$

which is displaced by parallel transport along λ. At any point $\lambda(t)$, we can express the tangent vector $\dot{\boldsymbol{\lambda}}(t)$ in component form relative to $\boldsymbol{e}(t)$, say

$$\dot{\boldsymbol{\lambda}}(t)=c^i(t)\,\boldsymbol{e}_i(t). \tag{10.33}$$

We put

$$c^i=-\int_0^1 c^i(t)\,dt. \tag{10.34}$$

Then the vector

$$\boldsymbol{c}\equiv c^i\,\boldsymbol{e}_i(0)\in\mathscr{B}_p \tag{10.35}$$

is called the *Cartan displacement* of the circuit at p. Notice that the quantities c^i depend only on the sense of the parametrization of λ. Moreover, the vector $\boldsymbol{c}$ does *not* depend on the choice of the parallel-displaced linear frames $\boldsymbol{e}$. Thus the Cartan displacement is an intrinsic property of the oriented circuit λ and the connection. For the coordinate circuit (10.28), we have

$$\boldsymbol{c}=-\alpha\beta[\boldsymbol{T}(p)](\boldsymbol{u},\boldsymbol{v})+o(\alpha^2+\beta^2), \tag{10.36}$$

where $\boldsymbol{T}(p)$ is the torsion tensor at p, and

$$[\boldsymbol{T}(\boldsymbol{u},\boldsymbol{v})]^i=T^i_{jk}\,u^j\,v^k. \tag{10.37}$$

Combining (10.29) and (10.36), we see that the Cartan displacement of the vector field $\boldsymbol{z}$ around the circuit λ is

$$\boldsymbol{d}=\boldsymbol{c}-(\bar{\boldsymbol{z}}-\boldsymbol{z})=\alpha\beta\{[\boldsymbol{R}(p)](\boldsymbol{z},\boldsymbol{u},\boldsymbol{v})-[\boldsymbol{T}(p)](\boldsymbol{u},\boldsymbol{v})\}+o(\alpha^2+\beta^2). \tag{10.38}$$

For more details, we refer to the book by SCHOUTEN[50].

In the theory of continuous distributions of dislocations for solid crystals, the Cartan displacement of a circuit relative to the unique material connection is called the *Burgers vector*[51]. We can visualize its physical meaning in the following way: Let λ be a closed circuit drawn in a crystal body with dislocations. If we repeat the corresponding crystallographic steps of λ in a perfect crystal, then the resulting circuit, in general, is not closed, and the closing link, oriented from the end point to the starting point, is the *Burgers vector* of the circuit. For bodies with continuous distributions of dislocations, we can define *local Burgers vectors* for infinitesimal circuits, such as the coordinate circuit λ in (10.28) with infinitesimal α and β. In this sense, the torsion tensor characterizes the *local Burgers vector densities* on arbitrary smooth surfaces in B.

Remark. If the connection is not curvature-free, then the Cartan displacement of the circuit λ, parametrized in the sense opposite to that considered before, is

[50] SCHOUTEN [*20*, Ch. III, §§ 2—4].

[51] *Cf*. BILBY, BULLOUGH & SMITH [*5*], and BILBY [*6*].

the vector

$$\bar{c} = -c^i \, e_i(1) \in \mathscr{B}_p, \tag{10.39}$$

which is *not* the opposite vector of c in (10.35), because the frames $e(0)$ and $e(1)$ need not be identical. However, for infinitesimal circuits, such as the coordinate circuit in (10.28), the equation (10.29) implies that $e(0)$ and $e(1)$ are related by a matrix of the form $[\delta^i_j + o(\alpha^2+\beta^2)]$. Consequently, we still have the same approximation:

$$\bar{c} = -\alpha\beta[T(p)](v, u) + o(\alpha^2+\beta^2). \tag{10.40}$$

In the above analysis, the point p is regarded as a fixed one. Then the curvature tensor $R(p)$ and the torsion tensor $T(p)$ can take on arbitrary values for arbitrary simple bodies. The curvature *field* R and the torsion *field* T, however, are no longer arbitrary. Rather, they must satisfy certain compatibility conditions, called the *Bianchi identities*. In component form, the identities are

$$\left(\frac{\partial T^k_{ml}}{\partial x^s} + R^k_{sml} + T^i_{ml}\,\Gamma^k_{is}\right) dx^s \wedge dx^m \wedge dx^l = 0, \tag{10.41}$$

and

$$\left(\frac{\partial R^m_{kqp}}{\partial x^s} + R^l_{kpq}\,\Gamma^m_{ls} + R^m_{lqp}\,\Gamma^l_{ks}\right) dx^s \wedge dx^p \wedge dx^q = 0. \tag{10.42}$$

Now let ε be the specific volume of $\mathscr{B}$ defined by (3.22). Then we introduce the tensor field A by

$$A^{ij} = \varepsilon^{ikl}\, T^j_{kl}. \tag{10.43}$$

This relation is invertible; by (3.26),

$$T^i_{jk} = \tfrac{1}{2}\,\varepsilon_{ljk}\, A^{li}. \tag{10.44}$$

Similarly, for the curvature tensor R, we introduce the field B by

$$B^{ij}_k = \varepsilon^{ipq}\, R^j_{kpq}. \tag{10.45}$$

Then the inverse relation is

$$R^j_{kpq} = \tfrac{1}{2}\,\varepsilon_{lpq}\, B^{lj}_k. \tag{10.46}$$

Using the fact that ε and ε^* are intrinsic fields on $\mathscr{B}$, we can express the fields

$$\frac{\partial \varepsilon^{ijk}}{\partial x^s} \quad \text{and} \quad \frac{\partial \varepsilon_{ijk}}{\partial x^s}$$

as functions of ε_{ijk} and the connection symbols. Then the identities (10.41) and (10.42) can be expressed in terms of the tensors A and B. We omit the detail of this result. In the theory of dislocations for lattices, the tensor field A is called the *local dislocation density*[52]. The Bianchi identities (10.41) then yield a compatibility condition for A.

11. Intrinsic Riemannian Metrics on Solid Bodies

By definition, a simple body $\mathscr{B}$ is a solid if for some reference atlas $\mathfrak{U}$ the isotropy group is contained in the orthogonal group $\mathscr{Q}(3)$. Such reference atlases $\mathfrak{U}$

[52] *Cf.* Bilby [6], Bilby, Bullough & Smith [5].

are then said to be *undistorted*. Since the group $\mathcal{Q}(3)$ is the group of all isomorphisms of $\mathcal{R}^3$ which preserve the standard Euclidean inner product, evidently we have the following

Proposition 11.1. *Every undistorted reference atlas* $\mathfrak{U}$ *induces an intrinsic Riemannian metric* $g_{\mathfrak{U}}$ on $\mathcal{B}$.

Proof. Let p be a fixed point. We choose a reference chart $(\mathcal{U}_\alpha, \boldsymbol{r}_\alpha)$ in $\mathfrak{U}$ such that $\mathcal{U}_\alpha \ni p$. Then we define the inner product $g_{\mathfrak{U}}(p)$ by

$$[g_{\mathfrak{U}}(p)](\boldsymbol{u}, \boldsymbol{v}) = \boldsymbol{r}_{\alpha, p}(\boldsymbol{u}) \cdot \boldsymbol{r}_{\alpha, p}(\boldsymbol{v}), \tag{11.1}$$

for all $\boldsymbol{u}, \boldsymbol{v} \in \mathcal{B}_p$, where the dot on the right-hand side denotes the Euclidean inner product for $\mathcal{R}^3$. Clearly, $g_{\mathfrak{U}}$ can be identified as an intrinsic tensor in $\mathcal{T}_p^{0,2}$. By Proposition 3.6, there exists a unique extension $g_{\mathfrak{U}}$, which forms an intrinsic Riemannian metric on $\mathcal{B}$. Q.E.D.

Definition 11.1. The intrinsic Riemannian metric $g_{\mathfrak{U}}$ is called the *induced metric* of the undistorted reference $\mathfrak{U}$.

The converse of the preceding proposition is also valid; we have

Proposition 11.2. *Suppose that a simple body* $\mathcal{B}$ *possesses an intrinsic Riemannian metric* g. *Then* $\mathcal{B}$ *is a solid body; moreover, there exists an undistorted reference atlas* $\mathfrak{U}$ *such that*

$$g = g_{\mathfrak{U}}. \tag{11.2}$$

Proof. Since g is an intrinsic field, at any point p g_p is an intrinsic tensor in $\mathcal{T}_p^{0,2}$. Thus if $\boldsymbol{h} \in g(p)$, then

$$g_p(\boldsymbol{u}, \boldsymbol{v}) = g_p(\boldsymbol{h}(\boldsymbol{u}), \boldsymbol{h}(\boldsymbol{v})) \tag{11.3}$$

for all $\boldsymbol{u}, \boldsymbol{v} \in \mathcal{B}_p$. Let $\boldsymbol{r}(p)$ be a local reference configuration of p such that

$$g_p(\boldsymbol{u}, \boldsymbol{v}) = \boldsymbol{r}_p(\boldsymbol{u}) \cdot \boldsymbol{r}_p(\boldsymbol{v}). \tag{11.4}$$

Clearly such local configuration exists. Then from (11.3) and (11.4),

$$\boldsymbol{r}_p \circ \boldsymbol{h} \circ \boldsymbol{r}_p^{-1}(\boldsymbol{a}) \cdot \boldsymbol{r}_p \circ \boldsymbol{h} \circ \boldsymbol{r}_p^{-1}(\boldsymbol{b}) = \boldsymbol{a} \cdot \boldsymbol{b} \tag{11.5}$$

for all $\boldsymbol{a}, \boldsymbol{b} \in \mathcal{R}^3$. Thus

$$\boldsymbol{r}_p \circ g(p) \circ \boldsymbol{r}_p^{-1} \subset \mathcal{Q}(3). \tag{11.6}$$

From (3.5), we see that $\mathcal{B}$ is a solid body.

Now let $\mathfrak{U} = \{(\mathcal{U}_\alpha, \boldsymbol{r}_\alpha), \alpha \in I\}$ be a reference atlas such that for some α, with $p \in \mathcal{U}_\alpha$,

$$\boldsymbol{r}_p = \boldsymbol{r}_\alpha(p). \tag{11.7}$$

Then $\mathfrak{U}$ is undistorted; moreover, at the point p

$$g_p = g_{\mathfrak{U}}(p). \tag{11.8}$$

By proposition 3.6, g and $g_{\mathfrak{U}}$ are identical. Q.E.D.

We now characterize the relations among the intrinsic Riemannian metrics on $\mathcal{B}$.

Proposition 11.3. *Let* $\mathfrak{U}$ *be an undistorted reference atlas and* $\boldsymbol{K}$ *be an element in* $\mathcal{GL}(3)$ *whose polar decomposition is*

$$\boldsymbol{K}=\boldsymbol{R}\,\boldsymbol{U}, \tag{11.9}$$

where $\boldsymbol{R}$ *is orthogonal and* $\boldsymbol{U}$ *is positive-definite and symmetric over* $\mathcal{R}^3$. Then the *reference atlas* $\boldsymbol{K}\mathfrak{U}$ *is undistorted if and only if* $\boldsymbol{U}$ *commutes with all elements in* $\mathcal{G}(\mathfrak{U})$.

Since the transformation law of the isotropy groups of a simple body is identical with that of a simple particle, the proof of this proposition is exactly the same as that of a theorem of COLEMAN & NOLL[53].

Proposition 11.4. *Let* $\mathfrak{U}$ *be an undistorted reference atlas of* $\mathcal{B}$. *Then*

$$g_{\mathfrak{U}}=g_{\boldsymbol{R}\mathfrak{U}} \tag{11.10}$$

for all $\boldsymbol{R}\in\mathcal{Q}(3)$.

The proof is obvious. Combining the above two propositions, we see that the intrinsic Riemannian metrics on $\mathcal{B}$ are all of the forms $g_{\boldsymbol{U}\mathfrak{U}}$ where $\mathfrak{U}$ is any fixed undistorted reference atlas of $\mathcal{B}$, and $\boldsymbol{U}$ are arbitrary positive-definite symmetric transformations in $\mathcal{GL}(3)$ which commute with all elements of $\mathcal{G}(\mathfrak{U})$. This fact implies the following

Proposition 11.5. *Suppose that* $\boldsymbol{g}$ *is an intrinsic Riemannian metric on* $\mathcal{B}$. *Then a field* $\bar{\boldsymbol{g}}$ *is an intrinsic Riemannian metric on* $\mathcal{B}$ *if and only if there exists an intrinsic field* $\boldsymbol{\mu}$ *of type* (1.1), *which is positive-definite and symmetric with respect to* $\boldsymbol{g}$, *such that*

$$\bar{g}_p(\boldsymbol{u},\boldsymbol{v})=g_p(\boldsymbol{\mu}_p(\boldsymbol{u}),\boldsymbol{\mu}_p(\boldsymbol{v})) \tag{11.11}$$

for all $p\in\mathcal{B}$ *and* $\boldsymbol{u},\boldsymbol{v}\in\mathcal{B}_p$.

Proof. From Proposition 11.2, we can assume that

$$g=g_{\mathfrak{U}} \tag{11.12}$$

for some undistorted reference atlas $\mathfrak{U}=\{(\mathcal{U}_\alpha,\boldsymbol{r}_\alpha)\}$. Suppose that the intrinsic field $\boldsymbol{\mu}$ exists. Then we claim that

$$\boldsymbol{r}_\alpha(p)\circ\boldsymbol{\mu}(p)\circ\boldsymbol{r}_\alpha(p)^{-1}=\boldsymbol{U}\in\mathcal{GL}(3), \tag{11.13}$$

independent of α and p. This fact is obvious. Since $\boldsymbol{\mu}$ is intrinsic, it commutes with all material isomorphisms. Thus for arbitrary $p\in\mathcal{U}_\alpha$ and $q\in\mathcal{U}_\beta$,

$$\begin{aligned}\boldsymbol{r}_\beta(q)\circ\boldsymbol{\mu}(q)\circ\boldsymbol{r}_\beta(q)^{-1}&=\boldsymbol{r}_\alpha(p)\circ\boldsymbol{r}_{\alpha\beta}(p,q)\circ\boldsymbol{\mu}(q)\circ\boldsymbol{r}_{\alpha\beta}(p,q)^{-1}\circ\boldsymbol{r}_\alpha(p)^{-1}\\&=\boldsymbol{r}_\alpha(p)\circ\boldsymbol{\mu}(p)\circ\boldsymbol{r}_\alpha(p)^{-1},\end{aligned} \tag{11.14}$$

where $\boldsymbol{r}_{\alpha\beta}(p,q)$ are material isomorphisms defined by

$$\boldsymbol{r}_{\alpha\beta}(p,q)=\boldsymbol{r}_\alpha(p)^{-1}\circ\boldsymbol{r}_\beta(q). \tag{11.15}$$

[53] COLEMAN & NOLL [32].

From (11.13), U is positive-definite symmetric and commutes with all elements in $\mathscr{G}(\mathfrak{U})$, so that $U\mathfrak{U}$ is undistorted. Now it is clear that $\bar{g}$ is an intrinsic Riemannian metric on $\mathscr{B}$, since from (11.11) and (11.13),

$$\bar{g} = g_{U\mathfrak{U}}; \tag{11.16}$$

sufficiently has been proved.

Necessity can be proved by a similar argument. Suppose that $\bar{g}$ is an intrinsic Riemannian metric. From Propositions 11.2 and 11.3, we can assume that $\bar{g}$ is given by (11.16) for some positive-definite symmetric U which commutes with all elements of $\mathscr{G}(\mathfrak{U})$. The field μ is then determined by the condition (11.13). Q.E.D.

Remark. COLEMAN & NOLL[54] have found the most general forms of the tensors U which commute with all elements of $\mathscr{G}(\mathfrak{U})$ for various types of solids. In particular, they showed that for cubic crystals and isotropic solids

$$U = c\,\mathbf{1}, \qquad c > 0. \tag{11.17}$$

This condition implies that for such bodies the intrinsic Riemannian metrics are unique up to a constant factor.

Since an intrinsic Riemannian metric is an intrinsic field, its covariant derivatives relative to all material connections vanish, *viz*,

$$g_{ij|k} = \frac{\partial g_{ij}}{\partial x^k} - \Gamma^l_{ik}\, g_{lj} - \Gamma^l_{jk}\, g_{il} = 0. \tag{11.18}$$

By definition, the *Riemannian connection* relative to a Riemannian metric g is the (unique) torsion-free $\mathscr{G}$-connection, which satisfies the above condition (11.18). Thus we have the following result:

Proposition 11.6. *A solid body $\mathscr{B}$ can be equipped with at most one torsion-free material connection. If such material connection exists, then it coincides with the Riemannian connections relative to all intrinsic Riemannian metrics on $\mathscr{B}$.*

Hence if the Riemannian connection of an intrinsic Riemannian metric is not a material connection, then none of the material connections is torsion-free. This fact enables us to construct an example of a solid body whose material connections all have nonvanishing curvatures and torsions.

Example 11.1

Consider the simple body given in Example 4.1. Suppose that in a certain configuration of the body, an intrinsic Riemannian metric coincides with the Euclidean metric. Then this body cannot be equipped with any torsion-free or curvature-free material connection.

From Proposition 10.4, we know that every locally homogeneous simple body can be equipped with a torsion-free material connection. Consequently, we have the following

---

[54] COLEMAN & NOLL [*32*].

Proposition 11.7. *A solid body is locally homogeneous if and only if it has a flat material connection.*

Proof. The sufficiency has been shown in proposition 10.2. We proceed to show the necessity. Suppose that $\mathscr{B}$ is locally homogeneous. Let $\mathfrak{U}=\{(\mathscr{U}_\alpha, \boldsymbol{r}_\alpha), \alpha\in I\}$ be an undistorted reference atlas of $\mathscr{B}$. Then for each point $p\in\mathscr{B}$, there exists a reference chart $(\mathscr{U}_\alpha, \boldsymbol{r}_\alpha)\in\mathfrak{U}$ such that $\mathscr{U}_\alpha\ni p$, and $\boldsymbol{r}_\alpha$ corresponds to the field of induced local configuration of a coordinate map ψ_α on $\mathscr{U}_\alpha$, as shown in (4.25). From (11.1), the components of the intrinsic Riemannian metric $\boldsymbol{g}_{\mathfrak{U}}$ relative ψ_α are δ_{ij}. Hence the Riemannian connection associated with $\boldsymbol{g}_{\mathfrak{U}}$ is flat. But from Propositions 10.4 and 11.6, this Riemannian connection must be a material connection. Thus the proof is complete.

Remark. From Proposition 11.6, the flat material connection in the above proposition is unique.

As was remarked before, for an isotropic solid body, the intrinsic Riemannian metrics are unique up to a scalar factor. Thus the Riemannian connection is unique. Since the isotropy group $\mathscr{g}(p)$ is the orthogonal group relative to any intrinsic metric, we have the following

Proposition 11.8. *The unique Riemannian connection associated with the intrinsic Riemannian metrics on an isotropic solid body is a material connection.*

The connection symbols of the Riemannian connection, of course, are the classical Christoffel symbols, *viz,*

$$\Gamma^i_{jk}=\begin{Bmatrix} i \\ j\,k \end{Bmatrix}=\frac{1}{2}\,g^{ia}\left(\frac{\partial g_{ka}}{\partial x^j}+\frac{\partial g_{ja}}{\partial x^k}-\frac{\partial g_{jk}}{\partial x^a}\right), \tag{11.19}$$

where g_{ij} and g^{ij} are the components of the Riemannian metric. It should be noted that the Christoffel symbols, and hence the Riemannian connections, are invariant under multiplication of the Riemannian metric by a constant.

Isotropic solid bodies are the only type of solid bodies for which the (unique) Riemannian connections of the intrinsic Riemannian metrics are always material connections. For other types of solid bodies, the intrinsic Riemannian metrics, in general, do not give rise to unique Riemannian connections. In this case, by Proposition 11.6, the bodies do not have any symmetric material connection, and hence none of the Riemannian connections is a material connection. From (11.18), the connection symbols of an arbitrary material connection and the Christoffel symbols of a Riemannian metric of a solid body $\mathscr{B}$ are related by

$$\Gamma^i_{jk}=\begin{Bmatrix} i \\ j\,k \end{Bmatrix}+\tfrac{1}{2}\,g^{ia}(T_{jka}-T_{ajk}+T_{kaj}) \tag{11.20}$$

where $\boldsymbol{T}$ is the torsion tensor of the material connection, and

$$T_{jka}=T^i_{jk}\,g_{ia}. \tag{11.21}$$

For transversely isotropic solid bodies, we have the following partial converse of Proposition 11.6:

Proposition 11.9. *Suppose that $\mathscr{B}$ is a transversely isotropic solid body such that all of its intrinsic Riemannian metrics give rise to a unique Riemannian connection $\mathscr{H}$. Then $\mathscr{H}$ is a material connection.*

Proof. We must show that all parallel transports relative to $\mathscr{H}$ are material isomorphisms. Since the parallel transports preserve all intrinsic Riemannian metrics on $\mathscr{B}$, they are material isomorphisms if and only if they preserve the axes of transverse isotropy. Let p and q be any two points in $\mathscr{B}$, and let $\boldsymbol{a}_p$ and $\boldsymbol{a}_q$ be the axes of transverse isotropy at p and q, respectively. Suppose that along some smooth curve λ joining p and q, the parallel transport relative to $\mathscr{H}$ displaces $\boldsymbol{a}_p$ to some vector $\bar{\boldsymbol{a}}_q$ at q. Then relative to *all* intrinsic metrics on $\mathscr{B}$

$$\|\boldsymbol{a}_p\| = \|\bar{\boldsymbol{a}}_q\| . \tag{11.22}$$

But by Proposition 11.5, this condition is impossible unless $\bar{\boldsymbol{a}}_q = \boldsymbol{a}_q$. Thus the proof is complete.

For a crystalline solid body, even if all intrinsic Riemannian metrics give rise to a unique Riemannian connection, that connection still need *not* be a material connection. This situation occurs in any cubic crystal body which is not locally homogeneous. From a previous remark, the intrinsic Riemannian metrics on such body are unique to within a constant multiple. Hence the Riemannian connection is indeed unique. But this Riemannian connection cannot be a material connection, since all material connections on solid crystal bodies are always completely integrable. In particular, if the Riemannian connection is a material connection, then it must be flat, contradicting the assumption that the body is not locally homogeneous. Notice that there are many locally inhomogeneous crystal bodies, such as those considered in the theory of curvilinear aeolotropy [55].

The curvature tensors of the Riemannian connections of the intrinsic Riemannian metrics on a solid body do not vanish in general. If the curvature tensor vanishes, then in a neighborhood of any particle there exists a local coordinate system such that the components of the corresponding intrinsic Riemannian metric are δ_{ij}.

Definition 11.2. An intrinsic Riemannian metric $g_{\mathfrak{U}}$ is called *locally Euclidean*, and $\mathfrak{U}$ is said to be *regular*, if the Riemannian connection of $g_{\mathfrak{U}}$ is flat.

In general, a solid body need not have any regular atlas. However, if such atlas exists, then it can be characterized by the following.

Proposition 11.10 (NOLL). *An undistorted reference atlas $\mathfrak{U} = \{(\mathscr{U}_\alpha, \boldsymbol{r}_\alpha), \alpha \in I\}$ is regular if and only if every reference map $\boldsymbol{r}_\alpha$ on a simply connected reference neighborhood $\mathscr{U}_\alpha$ can be represented by*

$$\boldsymbol{r}_\alpha(p) = \boldsymbol{Q}_\alpha(p) \circ \boldsymbol{\psi}_{\alpha,p}, \qquad p \in \mathscr{U}_\alpha, \tag{11.23}$$

where $\boldsymbol{\psi}_\alpha$ is a configuration of $\mathscr{U}_\alpha$, and $\boldsymbol{Q}_\alpha$ is a smooth field of orthogonal tensors (in $\mathscr{Q}(3)$) on $\boldsymbol{\psi}_\alpha(\mathscr{U}_\alpha)$.

It should be noted that a solid body may possess many *distinct* intrinsic Riemannian metrics (*i.e.*, not simply differ by some constant factors) which are

[55] *Cf.* GREEN & ADKINS [*33*, §1.16]. The theory of curvilinear aeolotropy is included here as a special case.

all locally Euclidean. For instance, all intrinsic Riemannian metrics on the solid body considered in Example 11.1 are locally Euclidean. In general, the existence of a locally Euclidean intrinsic Riemannian metric on a solid body does *not* imply that all other intrinsic Riemannian metrics on that body are also locally Euclidean. This situation can be seen in the following

Example 11.2

Let $\mathscr{B}$ be a triclinic crystal body such that relative to a global coordinate system (x^1, x^2, x^3) a reference map $\boldsymbol{r}_\alpha$ corresponds to the field of orthogonal tensors $\boldsymbol{Q}$ whose components are

$$[Q_{ij}]=\begin{bmatrix}1 & 0 & 0\\ 0 & \operatorname{Sin} x^1 & \operatorname{Cos} x^1\\ 0 & -\operatorname{Cos} x^1 & \operatorname{Sin} x^1\end{bmatrix}. \tag{11.24}$$

Then the intrinsic metric $g_{\mathfrak{U}}$, where $(\mathscr{B}, \boldsymbol{r}_\alpha)\in\mathfrak{U}$, is Euclidean. Since $\mathscr{B}$ is triclinic, the metrics g with components

$$\begin{bmatrix}1 & 0 & 0\\ 0 & \operatorname{Sin} x^1 & \operatorname{Cos} x^1\\ 0 & -\operatorname{Cos} x^1 & \operatorname{Sin} x^1\end{bmatrix}\begin{bmatrix}a & 0 & 0\\ 0 & b & 0\\ 0 & 0 & c\end{bmatrix}\begin{bmatrix}1 & 0 & 0\\ 0 & \operatorname{Sin} x^1 & -\operatorname{Cos} x^1\\ 0 & \operatorname{Cos} x^1 & \operatorname{Sin} x^1\end{bmatrix} \tag{11.25}$$

relative to (x^1, x^2, x^3) are also intrinsic. However, it can be shown that their curvature tensors do not vanish unless $a=b=c$.

If we write the equation (11.20) in the form

$$\Gamma^i_{jk}=\left\{ {i \atop j\,k} \right\}+S^i_{jk}, \tag{11.26}$$

so that

$$S^i_{jk}=\tfrac{1}{2}\, g^{ia}(T_{jka}-T_{ajk}+T_{kaj}), \tag{11.27}$$

then the curvature tensor $\boldsymbol{K}$ of the Riemannian connection and the curvature tensor $\boldsymbol{R}$ of the material connection are related by[56]

$$R^i_{jkl}=K^i_{jkl}-S^i_{jk,l}+S^i_{jl,k}+S^i_{mk}\,S^m_{jl}-S^i_{ml}\,S^m_{jk}, \tag{11.28}$$

and also

$$R^i_{jkl}=K^i_{jkl}-S^i_{jk|l}+S^i_{jl|k}-S^i_{mk}\,S^m_{jl}+S^i_{ml}\,S^m_{jk}-S^i_{jm}\,T^m_{kl}, \tag{11.29}$$

where $S^i_{jk,l}$ denotes the covariant derivative of the field S^i_{jk} with respect to the Riemannian connection. Thus the curvature fields $\boldsymbol{K}$ and $\boldsymbol{R}$ are not independent.

An intrinsic metric g on $\mathscr{B}$ clearly gives rise to a volume tensor field $\boldsymbol{e}^*$ by

$$\boldsymbol{e}^*=3!\sqrt{g}\, dx'\wedge dx^2\wedge dx^3, \tag{11.30}$$

where the scalar g is the determinant of the matrix $[g_{ij}]$. Since the field $\boldsymbol{e}^*$ is intrinsic, it is related to the volume tensor $\boldsymbol{\varepsilon}^*$ on $\mathscr{B}$ by a constant multiple

$$\boldsymbol{e}^*=c\,\boldsymbol{\varepsilon}^*, \tag{11.31}$$

[56] SCHOUTEN [*30*, Ch. III, § 4].

where c is positive. For simplicity, we normalize the metric g in such a way that $e^* = \varepsilon^*$ by multiplying g the factor $c^{-\frac{2}{3}}$.

For the curvature tensor K, we introduce the field C in the same way as (10.45) by

$$C_k^{ij} = \varepsilon^{ipq} K^j_{kpq}. \tag{11.32}$$

If we lower the index j in the above expression by the metric tensor g_{jl},

$$C^i_{jk} = g_{jl} C_k^{il} = \varepsilon^{ipq} K_{jkpq}, \tag{11.33}$$

then the components C^i_{jk} satisfy the usual symmetry relation:

$$C^i_{jk} = -C^i_{kj}. \tag{11.34}$$

Therefore, as in (10.34), we introduce the field D by

$$D^{ij} = \varepsilon^{ikl} C^j_{kl} = \varepsilon^{ikl} \varepsilon^{jpq} K_{klpq}. \tag{11.35}$$

By the symmetry condition

$$K_{ijkl} = K_{klij}, \tag{11.36}$$

the tensor D is symmetric; in fact it is a constant multiple of the Einstein tensor of the metric. Of course, the relations (11.32), (11.33), and (11.35) are all invertible; we have

$$K^j_{kpq} = \tfrac{1}{2} \varepsilon_{lpq} C_k^{lj}, \tag{11.37}$$

$$C_k^{ij} = g^{jl} C^i_{lk}, \tag{11.38}$$

and

$$C^j_{kl} = \tfrac{1}{2} \varepsilon_{ikl} D^{ij}. \tag{11.39}$$

Consequently, K can be expressed in terms of D:

$$K_{jkpq} = \tfrac{1}{4} \varepsilon_{lpq} \varepsilon_{mjk} D^{lm}. \tag{11.40}$$

Using the relations (11.40), (10.46), and (10.43), we can express the identities (11.28) and (11.29) in components of the tensor fields A, B, and D. We do not give this result in detail, but note only the fact that A, B, and D are not independent. In the dislocation theory of crystals, the field D is called the *incompatibility*[57] of the metric g.

12. Equations of Motion for Elastic Bodies

A simple particle is called *elastic* if its response functional is independent of the local history (cf. Section 2). Naturally, a simple body $\mathscr{B}$ is called an *elastic body* if its particles are elastic. In this section, we shall derive the equations of motion for arbitrary elastic bodies. Our results can be generalized easily for simple

[57] BILBY [6].

bodies whose memory effects can be represented by response *functions*. In the most general case of simple bodies, whose particles are characterized by response *functionals*, our analysis could yield certain functional differential equations provided that certain smoothness conditions on the functionals are laid down. However, since there is no unique way to select such smoothness conditions, we shall consider only the simplest case, that in which $\mathscr{B}$ is elastic.

For definiteness, we choose a fixed reference atlas $\mathfrak{U}$ for $\mathscr{B}$ in such a way that $\mathbf{g}(\mathfrak{U})$ is given in Table 6.1. Relative to this fixed reference atlas $\mathfrak{U}$, the response function is denoted by $\mathbf{G}$. Thus the constitutive equation is

$$T=\mathbf{G}(F). \tag{12.1}$$

For simplicity, we assume that $\mathbf{G}$ is smooth. We can express the constitutive equation (12.1) in component form relative to the standard basis of $\mathscr{R}^3$, say

$$T^{ij}=T^i_j=\mathsf{G}^i_j([F^k_l]). \tag{12.2}$$

Then the components of the gradient of $\mathbf{G}$ are

$$\frac{\partial \mathsf{G}^i_j}{\partial F^k_l}=\mathsf{H}^{i\,l}_{j\,k}=\mathsf{H}^{i\,l}_{j\,k}([F^p_q]). \tag{12.3}$$

From (3.13), the response function $\mathbf{G}$ satisfies the condition

$$\mathbf{G}(F)=\mathbf{G}(F\,G) \quad \forall\, G\in\mathscr{G}(\mathfrak{U}), \tag{12.4}$$

or equivalently,

$$\mathsf{G}^i_j([F^p_q])=\mathsf{G}^i_j([F^p_r\,G^r_q]). \tag{12.5}$$

By virtue of these conditions, if $[H^p_q]$ belongs to the Lie algebra $\mathbf{g}(\mathfrak{U})$, then

$$\mathsf{H}^{i\,l}_{j\,k}([\delta^p_q])\,H^k_l=0. \tag{12.6}$$

More generally, if $[F^p_q]$ is an arbitrary point in $\mathscr{GL}(3)$, then

$$\mathsf{H}^{i\,l}_{j\,k}([F^p_q])\,F^k_r\,H^r_l=0, \tag{12.7}$$

for all $[H^p_q]\in\mathbf{g}(\mathfrak{U})$.

Now let $\mathscr{H}$ be a fixed material connection on $\mathscr{B}$, and suppose that φ is a configuration of $\mathscr{B}$. Then φ corresponds to a *global* coordinate system on $\mathscr{B}$, say (x^i). Relative to this coordinate system, the connection $\mathscr{H}$ can be represented by the connection symbols Γ^i_{jk}, which form smooth fields on $\varphi(\mathscr{B})$. The stress field $\boldsymbol{T}$ on $\varphi(\mathscr{B})$ can now be determined in a simple way. Suppose that $\boldsymbol{\mu}$ is a cross-section in $\mathscr{E}(\mathscr{B},\mathfrak{U})$ above some coordinate neighborhood $\mathscr{U}\subset\mathscr{B}$ as shown in (9.27). Then the stress at any point $x\in\mathscr{U}$ in the configuration φ is given by

$$T(x)=\mathbf{G}\big(F(x)\big), \tag{12.8}$$

or in component form

$$T_j^i(x) = \mathsf{G}_j^i([F_q^p(x)]), \tag{12.9}$$

where we have used the majuscule $[F_q^p]$, which is the standard notation in continuum mechanics, instead of the minuscule $[f_q^p]$ used in (9.27).

Now let the body force acting on $\mathscr{B}$ in the configuration φ be dendote by $\boldsymbol{b} = (b^i)$. Then the equation of motion is

$$\operatorname{div} \boldsymbol{T} + \rho\, \boldsymbol{b} = \rho\, \ddot{\boldsymbol{x}}, \tag{12.10}$$

or in component form

$$\frac{\partial T_j^i}{\partial x^j} + \rho\, b^i = \rho\, \ddot{x}^i, \tag{12.11}$$

where ρ denotes the mass density in φ, and $\ddot{x}^i$ denotes the acceleration field in a motion of $\mathscr{B}$, say $\varphi = \varphi(t)$. Since the representation (12.9) for the stress field holds for each instant t, we can express the first term in (12.11) by the chain rule

$$\frac{\partial T_j^i}{\partial x^j} = \mathsf{H}_{jk}^{i\,l}\, \frac{\partial F_l^k}{\partial x^j}. \tag{12.12}$$

Then (12.11) becomes

$$\mathsf{H}_{jk}^{i\,l}\, \frac{\partial F_l^k}{\partial x^j} + \rho\, b^i = \rho\, \ddot{x}^i. \tag{12.13}$$

Recall that the connection symbols Γ_{jk}^i of a material connection satisfy the condition (9.32). Hence from the condition (12.7),

$$\mathsf{H}_{jk}^{i\,l}([F_q^p]) \left(\frac{\partial F_l^k}{\partial x^m} + \Gamma_{rm}^k\, F_l^r \right) = 0. \tag{12.14}$$

In particular,

$$\mathsf{H}_{jk}^{i\,l}([F_q^p])\, \frac{\partial F_l^k}{\partial x^j} = -\, \mathsf{H}_{jk}^{i\,l}([F_q^p])\, \Gamma_{rj}^k\, F_l^r. \tag{12.15}$$

Therefore, (12.13) can be written in the form

$$-\, \mathsf{H}_{jk}^{i\,l}\, \Gamma_{rj}^k\, F_l^r + \rho\, b^i = \rho\, \ddot{x}^i. \tag{12.16}$$

These are the equations of motion for the coordinate neighborhood $\mathscr{U}$. However, as we observed in Section 9, the condition (9.32) is independent of the choice of the cross-section $\boldsymbol{\mu}$. Consequently, the form of (12.16) is independent of $\mathscr{U}$.

It should be noted that the connection symbols Γ_{jk}^i and the components F_q^p of the cross-section $\boldsymbol{\mu}$ are referred to the coordinate system (x^i) induced by the configuration φ. In a motion of $\mathscr{B}$, φ is a 1-parameter family of configurations: $\varphi = \varphi(t)$. Then Γ_{jk}^i and F_q^p also depend implicitly on t. Such implicit dependences, of course, can be rendered explicit by the transformation laws of Γ_{jk}^i and F_q^p.

Suppose that $\bar{\varphi}$ is a fixed reference configuration of $\mathscr{B}$, say it corresponds to the global coordinate system (X^1, X^2, X^3). Then a motion $\varphi(t)$ of $\mathscr{B}$ is characterized by the deformation functions

$$x^i = x^i(X^\alpha, t), \tag{12.17}$$

which represent the mappings

$$\varphi_t \circ \bar{\varphi}^{-1}: \ \bar{\varphi}(\mathscr{B}) \to \varphi_t(\mathscr{B}). \tag{12.18}$$

Let $\bar{\Gamma}^\alpha_{\beta\gamma}$ and $\bar{F}^\mu_q$ be the connection symbols and the components of the cross-section $\boldsymbol{\mu}$ in the coordinate system (X^α), and let Γ^i_{jk} and F^p_q be the corresponding quantities in the coordinate systems (x^i) for the configurations φ_t. Then from (9.18) and (9.27),

$$\Gamma^i_{jk} = \bar{\Gamma}^\alpha_{\beta\gamma} \frac{\partial x^i}{\partial X^\alpha} \frac{\partial X^\beta}{\partial x^j} \frac{\partial X^\gamma}{\partial x^k} - \frac{\partial^2 x^i}{\partial X^\alpha \partial X^\beta} \frac{\partial X^\alpha}{\partial x^j} \frac{\partial X^\beta}{\partial x^k}, \tag{12.19}$$

and

$$F^p_q = \bar{F}^\mu_q \frac{\partial x^p}{\partial X^\mu}, \tag{12.20}$$

where we have used the identity

$$\frac{\partial^2 x^j}{\partial X^\alpha \partial X^\beta} \frac{\partial X^\alpha}{\partial x^i} \frac{\partial X^\beta}{\partial x^k} = - \frac{\partial^2 X^\alpha}{\partial x^i \partial x^k} \frac{\partial x^j}{\partial X^\alpha} \tag{12.21}$$

in deriving (12.19). Here the matrices

$$\left[\frac{\partial x^i}{\partial X^\alpha}\right] \quad \text{and} \quad \left[\frac{\partial X^\alpha}{\partial x^i}\right]$$

are the inverse of each other, since

$$\frac{\partial x^i}{\partial x^j} = \frac{\partial x^i}{\partial X^\alpha} \frac{\partial X^\alpha}{\partial x^j} = \delta^i_j, \tag{12.22}$$

and

$$\frac{\partial X^\alpha}{\partial X^\beta} = \frac{\partial X^\alpha}{\partial x^i} \frac{\partial x^i}{\partial X^\beta} = \delta^\alpha_\beta. \tag{12.23}$$

Substituting the relations (12.19) and (12.20) into the equations of motion (12.16), we get

$$\mathsf{H}^{il}_{jk} \left(\frac{\partial^2 x^k}{\partial X^\mu \partial X^\beta} \frac{\partial X^\beta}{\partial x^j} - \bar{\Gamma}^\alpha_{\mu\gamma} \frac{\partial x^k}{\partial X^\alpha} \frac{\partial X^\gamma}{\partial x^j} \right) \bar{F}^\mu_l + \rho b^i = \rho \ddot{x}^i. \tag{12.24}$$

The quantities $\bar{F}^\mu_l$ and $\bar{\Gamma}^\alpha_{\mu\gamma}$ in the above equation are now certain fixed functions of (X^α), independent of t. The argument of the functions H^{il}_{jk} is the matrix $[F^p_q]$, which can be expressed in terms of the deformation gradient $[\partial x^p / \partial X^\alpha]$ and the fixed field $[\bar{F}^\alpha_q]$ by the relation (12.20).

We mention here that the usual equations of motion for homogeneous bodies can be obtained from (12.24) in the following way: We choose the reference configuration $\bar{\varphi}$ in such a way that the Euclidean connection in $\bar{\varphi}(\mathscr{B})$ correspond to a material connection (the existence of such $\bar{\varphi}$ characterizes a homogeneous body, *cf.* Section 4, Definition 4.3). In the coordinate system (X^α) induced by $\bar{\varphi}$, the connection symbols of the Euclidean connection vanish identically. Also, we choose the cross-section $\boldsymbol{\mu}$ to be the natural frame of (X^α), so that

$$\bar{F}_l^\mu = \delta_l^\mu. \tag{12.25}$$

Then (12.24) reduces to the usual equations of motion:

$$\mathsf{H}_{jk}^{i\mu} \frac{\partial^2 x^k}{\partial X^\mu \partial X^\alpha} \frac{\partial X^\alpha}{\partial x^j} + \rho b^i = \rho \ddot{x}^i, \tag{12.26}$$

where the argument of $\mathsf{H}_{jk}^{i\mu}$ is now simply the deformation gradient $[\partial x^i / \partial X^\alpha]$.

The equation (12.24), though it appears to be local, has in fact a global representation. From the symmetry condition (12.4), $\mathbf{G}$ can be identified as a function on the left quotient space $\mathscr{Y}$ of $\mathscr{GL}(3)$ over $\mathscr{G}(\mathfrak{U})$. By hypothesis, $\mathscr{G}(\mathfrak{U})$ forms a closed Lie subgroup of $\mathscr{GL}(3)$. Hence $\mathscr{Y}$ is endowed with an induced differentiable structure[58], and the canonical projection map

$$\mathbf{I}:\ \mathscr{GL}(3) \to \mathscr{Y} \tag{12.27}$$

is smooth. Clearly $\mathbf{G}$ can be factored by $\mathbf{I}$,

$$\mathbf{G} = \tilde{\mathbf{G}} \circ \mathbf{I}, \tag{12.28}$$

and $\tilde{\mathbf{G}}$ is smooth.

Now suppose that $\boldsymbol{\mu}_\alpha$ are the cross-sections of the bundle charts $(\mathscr{U}_\alpha, \varphi_\alpha)$ for $\mathscr{E}(\mathscr{B}, \mathfrak{U})$. Then relative to the global coordinate system (X^1, X^2, X^3) considered before, each $\boldsymbol{\mu}_\alpha$ gives rise to a field $[\underset{\alpha}{\bar{F}_l^\mu}]$ on $\mathscr{U}_\alpha$. Clearly, the projections of such fields in $\mathscr{Y}$:

$$\mathbf{I}([\underset{\alpha}{\bar{F}_l^\mu}(x)]) = \mathbf{Y}(x) = \mathbf{Y}_{\mathfrak{U}}(x), \tag{12.29}$$

do *not* depend on the bundle chart $(\mathscr{U}_\alpha, \varphi_\alpha)$. Consequently, there exists a *global* field $\mathbf{Y}_{\mathfrak{U}}$,

$$\mathbf{Y}_{\mathfrak{U}}:\ \mathscr{B} \to \mathscr{Y}, \tag{12.30}$$

which is smooth with respect to the induced differentiable structure on $\mathscr{Y}$.

From the factorization (12.88), we see that the functions H_{jk}^{il} satisfy the identities

$$\mathsf{H}_{jk}^{il}([F_q^p]) = \mathsf{H}_{jk}^{i\alpha}([F_r^p G_q^r])\, G_\alpha^l \tag{12.31}$$

[58] *Cf.* COHN [*22*, Ch. VI, § 6.5], CHEVALLEY [*21*, Ch. IV, § V].

for all $[F^p_q]\in\mathscr{G}\mathscr{L}(3)$ and $[G^p_q]\in\mathscr{G}(\mathfrak{U})$. Hence the combinations $\mathsf{H}^{i\,l}_{j\,k}\bar{F}^\mu_l$ in (12.24) satisfy the conditions

$$\mathsf{H}^{i\,l}_{j\,k}\left(\left[\frac{\partial x^p}{\partial X^\nu}\bar{F}^\nu_q\right]\right)\bar{F}^\mu_l=\mathsf{H}^{i\,l}_{j\,k}\left(\left[\frac{\partial x^p}{\partial X^\nu}\bar{F}^\nu_m G^m_q\right]\right)\bar{F}^\mu_r G^r_l \tag{12.32}$$

for all $[G^p_q]\in\mathscr{G}(\mathfrak{U})$. These conditions simply mean that the form of the equations (12.24) is independent of the choice of the cross-section $\boldsymbol{\mu}$. Therefore we can re-write the terms $\mathsf{H}^{i\,l}_{j\,k}\bar{F}^\mu_l$ in (12.24) in a form depending only on the field $\mathbf{Y}$:

$$\tilde{\mathsf{H}}^{i\,\mu}_{j\,k}\equiv\tilde{\mathsf{H}}^{i\,\mu}_{j\,k}\left(\left[\frac{\partial x^a}{\partial X^\alpha}\right]\right)=\mathsf{H}^{i\,l}_{j\,k}\left(\left[\frac{\partial x^a}{\partial X^\nu}\bar{F}^\nu_q\right]\right)\bar{F}^\mu_l. \tag{12.33}$$

Then the equations of motion take on the global form:

$$\tilde{\mathsf{H}}^{i\,\mu}_{j\,k}\left(\frac{\partial^2 x^k}{\partial X^\mu\,\partial X^\beta}-\bar{\Gamma}^\alpha_{\mu\beta}\frac{\partial x^k}{\partial X^\alpha}\right)\frac{\partial X^\beta}{\partial x^j}+\rho\,b^i=\rho\,\ddot{x}^i. \tag{12.34}$$

It should be noted, however, that the functions $\tilde{\mathsf{H}}^{i\,\mu}_{j\,k}$ now depend also on the co-ordinates X^α through the dependence of $\bar{F}^\mu_l$ on X^α as shown in (12.33).

In continuum mechanics, sometimes it is convenient to use the *Piola-Kirchhoff stress tensor* $\boldsymbol{T}_\mathsf{R}$ instead of the usual Cauchy stress tensor $\boldsymbol{T}$. Relative to a fixed reference configuration $\bar{\boldsymbol{\varphi}}$, $\boldsymbol{T}_\mathsf{R}$ and $\boldsymbol{T}$ are related by

$$T_{\mathsf{R}}{}^{\alpha}_{k}=J\,T^l_k\frac{\partial X^\alpha}{\partial x^l}, \tag{12.35}$$

where J denotes the determinant of the deformation gradient,

$$J=\det\left[\frac{\partial x^i}{\partial X^\alpha}\right]. \tag{12.36}$$

By Euler's theorem,

$$\frac{\partial}{\partial X^\alpha}\left(J\frac{\partial X^\alpha}{\partial x^i}\right)=0. \tag{12.37}$$

Hence from (12.35) we have

$$\frac{\partial T_{\mathsf{R}}{}^{\alpha}_{k}}{\partial X^\alpha}=J\frac{\partial T^l_k}{\partial x^l}. \tag{12.38}$$

Thus the equations of motion (12.11) can be written in the form

$$\frac{\partial T_{\mathsf{R}}{}^{\alpha}_{k}}{\partial X^\alpha}+\rho_\mathsf{R}\,b^k=\rho_\mathsf{R}\,\ddot{x}^k, \tag{12.39}$$

where ρ_R denotes the mass density in the reference configuration $\bar{\boldsymbol{\varphi}}$.

We introduce the response function $\mathbf{A}$ defined on $\mathscr{G}\mathscr{L}(3)$ by

$$\mathsf{A}^i_j([F^p_q])\equiv\det[F^p_q]\,\mathsf{G}^k_j([F^a_b])\,\overset{-1}{F^i_k}. \tag{12.40}$$

Then

$$\mathsf{G}^i_j = \frac{1}{\det \boldsymbol{F}} \mathsf{A}^s_j F^i_s. \tag{12.41}$$

Consequently,

$$\mathsf{H}^{i\,l}_{j\,h} = \frac{\partial}{\partial F^h_l}\left(\frac{F^i_s}{\det \boldsymbol{F}} \mathsf{A}^s_j\right) = \frac{F^i_s}{\det \boldsymbol{F}} \mathsf{B}^{s\,l}_{j\,k} + \frac{1}{\det \boldsymbol{F}} (\delta^i_k \delta^l_s - F^i_s \overset{-1}{F^l_k}) \mathsf{A}^s_j, \tag{12.42}$$

where

$$\mathsf{B}^{s\,l}_{j\,k} = \frac{\partial \mathsf{A}^s_j}{\partial F^k_l}. \tag{12.43}$$

Substituting (12.42) into (12.16), we see that the equations of motion can be written also in the form

$$-\frac{1}{\det \boldsymbol{F}} (\mathsf{B}^{s\,l}_{j\,k} F^i_s F^r_l + \mathsf{A}^s_j(\delta^i_k F^r_s - \delta^r_k F^i_s))\, \Gamma^k_{r\,j} + \rho\, b^i = \rho\, \ddot{x}^i. \tag{12.44}$$

Now we introduce the fixed reference configuration $\bar{\boldsymbol{\varphi}}$, so that the connection symbols Γ^k_{rj} and the cross section F^p_q are characterized by (12.19) and (12.20), respectively. Notice that we can interchange the indices i and j, since the functions G^i_j are symmetric in them. Substituting (12.19) and (12.20) into (12.44), we obtain

$$\frac{1}{\det \bar{\boldsymbol{F}}} \mathsf{B}^{s\,l}_{i\,k} \bar{F}^\mu_s \bar{F}^\nu_l \left(\frac{\partial^2 x^k}{\partial X^\nu \partial X^\mu} - \bar{\Gamma}^\alpha_{\nu\mu} \frac{\partial x^k}{\partial X^\alpha}\right) - \frac{1}{\det \bar{\boldsymbol{F}}} \mathsf{A}^s_i \bar{F}^\mu_s \bar{T}^\alpha_{\mu\alpha} + \rho_{\mathsf{R}}\, b^i = \rho_{\mathsf{R}}\, \ddot{x}^i, \tag{12.45}$$

where $\bar{T}^\alpha_{\mu\nu}$ denotes the components of the torsion tensor in the coordinate system (X^1, X^2, X^3), *viz*,

$$\bar{T}^\alpha_{\mu\nu} = \bar{\Gamma}^\alpha_{\mu\nu} - \bar{\Gamma}^\alpha_{\gamma\mu}. \tag{12.46}$$

By exactly the same argument as before, the form of equations (12.45) does *not* depend on the choice of the cross-section $\boldsymbol{\mu}$. Hence the terms

$$\frac{1}{\det \bar{F}} \mathsf{B}^{s\,l}_{i\,k} \bar{F}^\mu_s \bar{F}^\nu_l \quad \text{and} \quad \frac{1}{\det \bar{F}} \mathsf{A}^s_i \bar{F}^\mu_s$$

have global representations, say

$$\tilde{\mathsf{B}}^{\mu\,\nu}_{i\,k} = \tilde{\mathsf{B}}^{\mu\,\nu}_{i\,k}\left(\left[\frac{\partial x^i}{\partial X^\alpha}\right]\right) = \frac{1}{\det \bar{\boldsymbol{F}}} \mathsf{B}^{s\,l}_{i\,k} \bar{F}^\mu_s \bar{F}^\nu_l, \tag{12.47}$$

and

$$\tilde{\mathsf{A}}^\mu_i = \tilde{\mathsf{A}}^\mu_i\left(\left[\frac{\partial x^i}{\partial X^\alpha}\right]\right) = \frac{1}{\det \bar{\boldsymbol{F}}} \mathsf{A}^s_i \bar{F}^\mu_s. \tag{12.48}$$

Then the equations (12.45) become

$$\tilde{\mathsf{B}}^{\mu\,\nu}_{i\,k}\left(\frac{\partial^2 x^k}{\partial X^\mu \partial X^\nu} - \bar{\Gamma}^\alpha_{\nu\mu} \frac{\partial x^k}{\partial X^\alpha}\right) - \tilde{\mathsf{A}}^\mu_i \bar{T}^\alpha_{\mu\alpha} + \rho_{\mathsf{R}}\, b^i = \rho_{\mathsf{R}}\, \ddot{x}^i. \tag{12.49}$$

These are NOLL's *equations*[59] in Cartesian coordinates. As before, the global fields $\tilde{\mathsf{B}}_{ik}^{\mu\nu}$ and $\tilde{\mathsf{A}}_i^\mu$ depend implicitly on the coordinates (X^1, X^2, X^3) through the local formulae (12.47) and (12.48).

In the special case when $\mathscr{B}$ is a homogeneous body and $\bar{\varphi}$ is a homogeneous reference configuration, $\bar{\Gamma}_{\nu\mu}^\alpha$ and $\bar{T}_{\mu\alpha}^\alpha$ both vanish, and $\tilde{\mathsf{B}}_{ik}^{\mu\nu}$ are independent of the coordinates (X^α). Thus (12.49) reduces to the usual form:

$$\tilde{\mathsf{B}}_{ik}^{\mu\nu} \frac{\partial^2 x^k}{\partial X^\mu \partial X^\nu} + \rho_{\mathsf{R}} b^i = \rho_{\mathsf{R}} \ddot{x}^i. \tag{12.50}$$

Acknowledgment. I am indebted to my teachers, Professor C. TRUESDELL, who suggested this problem and provided many helpful comments and criticisms in the research, and Professor W. NOLL, who in 1963 gave me a copy of his unpublished study of this problem, on which the major concepts of this article are founded. I thank also Dr. C. DAFERMOS for his remarks on Section 6 and Mr. J. ELLIOT, Jr., for his kind help in preparing preprints of this and other works of mine for private communication. This work was supported by a grant of the U.S. National Science Foundation to the Johns Hopkins University.

References

[*1*] TRUESDELL, C., & W. NOLL, The Non-linear Field Theories of Mechanics. Handbuch der Physik, Vol. III/3. Berlin-Heidelberg-New York: Springer 1965.
[*2*] NOLL, W., Arch. Rational Mech. Anal. **2**, 197—226 (1958/59).
[*3*] KONDO, K., Memoirs of the Unifying Study of the Basic Problems in Engineering by Means of Geometry, Vol. I (1955), II (1958). Tokyo: Gakujutsu Bunken Fukyu-Kai.
[*4*] NYE, J. F., Acta Met. **1**, 153—162 (1953).
[*5*] BILBY, B. A., R. BULLOUGH, & E. SMITH, Proc. Roy. Soc. Lond. A**231**, 263—273 (1955).
[*6*] BILBY, B. A., Progress in Solid Mechanics **1**, 329—398 (1960). Ed. I. N. SNEDDON & R. HILL.
[*7*] KRÖNER, E., & A. SEEGER, Arch. Rational Mech. Anal. **3**, 97—119 (1959).
[*8*] CHERN, S. S., Differentiable Manifolds. Lecture Notes, Dept. Math., Univ. Chicago (1959).
[*9*] STERNBERG, S., Lectures on Differential Geometry. Prentice-Hall 1964.
[*10*] NOMIZU, K., Lie Groups and Differential Geometry. Math. Soc. Japan (1956).
[*11*] KOBAYASHI, S., & K. NOMIZU, Foundations of Differential Geometry. John Wiley & Sons Interscience Publishers 1963.
[*12*] LANG, S., Introduction to Differentiable Manifolds. John Wiley & Sons Interscience Publishers 1962.
[*13*] AUSLANDER, L., & R. MACKENZIE, Introduction to Differentiable Manifolds. McGraw-Hill 1963.
[*14*] NÔNO, T., Paper to appear in J. Math. Anal. Appl.
[*15*] LIE, S., & F. ENGEL, Theorie der Transformationsgruppen. Vol. 3. Leipzig: Teubner 1893.
[*16*] NOLL, W., Proc. Sym. Applied Math. Vol. XVII, 93—101 (1965).
[*17*] GURTIN, M. E., & W. C. WILLIAMS, Arch. Rational Mech. Anal. **23**, 163—172 (1966).
[*18*] HALMOS, P. R., Measure Theory. Van Nostrand 1950.
[*19*] COLEMAN, B. D., Arch. Rational Mech. Anal. **20**, 41—58 (1965).
[*20*] WANG, C.-C., Arch. Rational Mech. Anal. **20**, 1—40 (1965).
[*21*] CHEVALLEY, C., Theory of Lie Groups. Princeton University Press 1946.
[*22*] COHN, P. M., Lie Groups. Cambridge University Press 1957.
[*23*] COLEMAN, B. D., & W. NOLL, Arch. Rational Mech. Anal. **6**, 355—370 (1960).
[*24*] WANG, C.-C., Arch. Rational Mech. Anal. **18**, 343—366 (1965).
[*25*] WANG, C.-C., Arch. Rational Mech. Anal. **18**, 117—126 (1965).
[*26*] COLEMAN, B. D., & V. J. MIZEL, Arch. Rational Mech. Anal. **23**, 87—123 (1966).
[*27*] MIZEL, V. J., & C.-C. WANG, Arch. Rational Mech. Anal. **23**, 124—134 (1966).

59 *Cf.* NOLL [*16*], TRUESDELL & NOLL [*1*, Ch. D, § 44].

[28] STEENROD, N., The Topology of Fibre Bundles. Princeton University Press 1951.
[29] FRANK, F. C., Phil. Mag. **42**, 809—819 (1951).
[30] SCHOUTEN, J. A., Ricci-Calculus. Berlin-Göttingen-Heidelberg: Springer 1954.
[31] AMBROSE, W., & I. M. SINGER, Trans. Amer. Math. Soc. **75**, 428—443 (1953).
[32] COLEMAN, B. D., & W. NOLL, Arch. Rational Mech. Anal. **15**, 87—111 (1964).
[33] GREEN, A. E., & J. E. ADKINS, Large Elastic Deformations and Non-linear Continuum Mechanics. Oxford: Clarendon Press 1960.
[34] KRÖNER, E., Arch. Rational Mech. Anal. **3**, 273—334 (1959/60).

The Johns Hopkins University,
Baltimore, Maryland

(Received June 15, 1967)

Sonderdruck aus
Arch. Rational Mech. Anal., Vol. 29, S. 161–192
Springer-Verlag · Berlin · Heidelberg · New York

Universal Solutions for Incompressible Laminated Bodies

C.-C. WANG

Contents

1. Introduction

A theory of inhomogeneous simple bodies has been proposed recently by NOLL [1] and WANG [2]. In this paper, we apply that theory to a class of inhomogeneous incompressible isotropic bodies, called *laminated bodies*. We obtain certain exact static and dynamic solutions for such bodies.

In my previous paper [2], I have shown that the material geometric structure of an arbitrary isotropic body $\mathscr{B}$ can be characterized by an intrinsic Riemannian metric[1] $\boldsymbol{g}$, which is unique to within a positive scalar factor, such that the unique Riemannian connection is always a material connection[1]. I have shown also that such an intrinsic Riemannian metric $\boldsymbol{g}$ can be determined in the following way: Let

$$\mathfrak{U}=\{(\mathscr{U}_\alpha, \boldsymbol{r}_\alpha), \alpha\in \mathsf{I}\} \tag{1.1}$$

be an undistorted reference atlas[1] for $\mathscr{B}$, and suppose that

$$\varphi(\mathfrak{U})=\{(\mathscr{U}_\alpha, \varphi_\alpha), \alpha\in \mathsf{I}\} \tag{1.2}$$

is the corresponding material atlas. Then $\mathfrak{U}$ gives rise to an induced intrinsic metric

$$\boldsymbol{g}=\boldsymbol{g}_{\mathfrak{U}} \tag{1.3}$$

by the condition

$$\boldsymbol{g}(\boldsymbol{u},\boldsymbol{v})\equiv \boldsymbol{G}(\boldsymbol{r}_\alpha(\boldsymbol{u}),\boldsymbol{r}_\alpha(\boldsymbol{v}))\equiv \boldsymbol{G}(\varphi_\alpha^{-1}(\boldsymbol{u}),\varphi_\alpha^{-1}(\boldsymbol{v})) \tag{1.4}$$

for all $\boldsymbol{u}, \boldsymbol{v}\in\mathscr{B}_p$, $p\in\mathscr{U}_\alpha$. Here $\boldsymbol{G}$ denotes the standard Euclidean metric of the physical space. The curvature tensor of the Riemannian connection then characterizes the local inhomogeneity of $\mathscr{B}$.

Now let $\mathscr{B}$ be an isotropic body, and let $\boldsymbol{g}$ be a fixed intrinsic Riemannian metric on $\mathscr{B}$. Then $\mathscr{B}$ is called a *laminated body* if it is a disjoint union of a collection of two dimensional sub-manifolds $\mathscr{L}_\xi$, called the *laminae*, say

$$\mathscr{B}=\bigcup_{\xi\in \mathsf{A}}\mathscr{L}_\xi, \tag{1.5}$$

[1] These concepts have been introduced in Reference [2].

where A is an index set, such that for each fixed lamina $\mathscr{L}_\xi$, *locally*, there exist configurations ψ_ξ whose induced local configurations carry the intrinsic metric g on $\mathscr{L}_\xi$ onto the Euclidean metric G, *i.e.*,

$$g(u, v) \equiv G(\psi_{\xi *}(u), \psi_{\xi *}(v)) \tag{1.6}$$

for all $u, v \in \mathscr{B}_p$, $p \in \mathscr{L}_\xi \cap \mathscr{U}$, where $\mathscr{U}$ denotes the domain of ψ_ξ. In general, the configurations ψ_ξ depend on the lamina $\mathscr{L}_\xi$; moreover, they need not be defined for the whole lamina $\mathscr{L}_\xi$.

Physically, we can visualize a lamina $\mathscr{L}_\xi$ to be a locally homogeneous shell, which is infinitely thin, such that the configurations ψ_ξ are its (local) homogeneous configurations. For definiteness, we call the configurations ψ_ξ the *initial configurations* for $\mathscr{L}_\xi$, and all other configurations the *deformed configurations*. These terms, of course, refer strictly to the fixed lamina $\mathscr{L}_\xi$. In general, an initial configuration ψ_ξ for a lamina $\mathscr{L}_\xi$ need *not* be an initial configuration for any neighboring laminae of $\mathscr{L}_\xi$.

Now let ψ be a (global) configuration for the laminated body $\mathscr{B}$. Then generally ψ is a deformed configuration for all laminae $\mathscr{L}_\xi$ in $\mathscr{B}$. The deformations from the various initial configurations ψ_ξ to the deformed configuration ψ are given by

$$\kappa_\xi \equiv \psi \circ \psi_\xi^{-1} : \psi_\xi(\mathscr{U}) \to \psi(\mathscr{U}), \tag{1.7}$$

which need not be homogeneous. Thus we can regard $\mathscr{B}$ as being formed by patching together the laminae at their various deformed states. Such intrinsic incompatibility among the laminae, of course, cannot be released in any global deformation of $\mathscr{B}$. Hence in general $\mathscr{B}$ is an inhomogeneous body.

If the laminae of $\mathscr{B}$ are initially planes, then $\mathscr{B}$ is said to be *locally rectilinear*. Otherwise, $\mathscr{B}$ is said to be *curvilinear*. Laminated bodies of both kinds are considered in this paper. More specifically, we treat the following three classes of laminated bodies:

First, we treat a class of locally rectilinear bodies $\mathscr{B}$, which are formed by a collection of *thin plates*. We assume that there exist some reference configurations, say $\bar{\varphi}$, such that relative to a rectangular Cartesian coordinate system (X, Y, Z), the components of the intrinsic metric g form the matrix

$$[\bar{g}_{\langle\alpha\beta\rangle}] = \begin{bmatrix} \bar{g}_{\langle 11\rangle}, & 0, & 0 \\ 0, & \bar{g}_{\langle 22\rangle}, & \bar{g}_{\langle 33\rangle} \\ 0, & \bar{g}_{\langle 32\rangle}, & \bar{g}_{\langle 33\rangle} \end{bmatrix}, \tag{1.8}$$

where the $\bar{g}_{\langle\alpha\beta\rangle}$ are functions of X only. Second, we consider a class of locally rectilinear bodies $\bar{\mathscr{B}}$ which are formed by a collection of *thin cylindrical shells*. In this case, there exist reference configurations $\bar{\varphi}$ such that relative to a cylindrical coordinate system (R, Θ, Z) the components of g form the matrix in (1.8), but the $\bar{g}_{\langle\alpha\beta\rangle}$ are functions of R only. Finally, we treat a class of curvilinear laminated bodies $\tilde{\mathscr{B}}$ which are formed by a collection of *thin spherical shells*. For this class, there exist reference configurations $\bar{\varphi}$ such that relative to a spherical coordinate system (R, Θ, Φ) the components of g form the matrix

$$[\bar{g}_{\langle\alpha\beta\rangle}] = \operatorname{diag}[\bar{g}_{\langle 11\rangle}, \bar{g}_{\langle 22\rangle}, \bar{g}_{\langle 33\rangle}], \tag{1.9}$$

where the $\bar{g}\langle\alpha\beta\rangle$ are functions of R only, and

$$\bar{g}\langle 22\rangle = \bar{g}\langle 33\rangle . \tag{1.10}$$

Since we consider incompressible bodies only, naturally $\bar{\varphi}$ must be compatible with the internal constraint. Thus

$$\det[\bar{g}\langle\alpha\beta\rangle] = 1. \tag{1.11}$$

It should be noted that these three classes of laminated bodies by no means exhaust all possible laminated bodies. However, for these three classes, we show that certain known families of static and dynamic universal solutions for incompressible homogeneous isotropic bodies remain universal solutions[1]. This fact is hardly surprising, since these universal solutions, locally, are all "laminar deformations", which preserve certain families of laminae and their normal trajectories. Thus relative to an appropriate convected basis at any particle, the deformation gradients for these families all appear to be those of plane deformations. The symmetry of the particles and the laminated structure of the body then allow us to assert that the stress tensor fields in the deformed configurations of these families also form certain "laminar systems". Thus the arguments showing these families to be universal solutions for homogeneous bodies remain applicable here, for various classes of laminated bodies. In a sense, this situation is roughly parallel to that of the viscometric flows. Locally, they can be regarded as simple shearing flows relative to a suitable convected basis, so that the stress fields also form certain laminar systems. For this reason, some viscometric flows have been shown to be universal solutions for incompressible simple fluids and certain types of incompressible subfluids[2].

The idea of a laminated structure is not new. Most gun barrels, for example, are made to be laminated bodies like the bodies $\bar{\mathscr{B}}$ described here. Laminated plates, beams, and frames are also widely used in timber structures. Our theory here, of course, is mainly for academic interest. However, certain particular solutions to be considered in Section 4, such as those for pure torsion and pure bending, seem to suggest some practical applications. To mention a few interesting ones here, we show that for a circular rod of a fixed size and made up of a fixed material, we can *design* the torsional rigidity within a certain range by choosing a suitable laminated structure belonging to the class $\bar{\mathscr{B}}$. Similarly, a laminated beam, in general, has a bending rigidity different from that of a homogeneous beam having the same size and made up of the same material. In some sense, a laminated body is *pre-stressed.*

I give the general stress systems for the static universal solutions in Section 2, and those for the dynamic universal solutions in Section 3. Some particular solutions of interest are worked out in detail in Section 4.

In the last section I offer some conjectures to the effect that the three classes of laminated bodies $\mathscr{B}$, $\bar{\mathscr{B}}$, and $\tilde{\mathscr{B}}$ considered here *exhaust* the inhomogeneous incompressible isotropic bodies which can admit as universal solutions those families already known to be universal solutions for homogeneous bodies. Since

[1] Notice that homogeneous bodies are trivially included in all three classes of laminated bodies considered here.

[2] See the *Remark* below.

these conjectures depend on the general solutions of some highly overdetermined systems, I am not presently able to give complete proofs of them. However, if they turn out to be correct, then the laminated bodies $\mathscr{B}$, $\bar{\mathscr{B}}$, $\tilde{\mathscr{B}}$ occupy a rather singular position in that, unlike more general inhomogeneous bodies, these cannot be distinguished from a homogeneous body merely by the fact that certain families of deformations are *controllable*[1] by means of surface tractions applied on the boundaries of the bodies. But for any specific material, the global response of a laminated body in any specific deformation, of course, is always different from that of a homogeneous body having the same size and consisting of the same material, since the stress field depends not only on the deformation but also on the distribution of the intrinsic metric over the body.

Remark. For the static and the dynamic universal solutions considered here, the body could be made up of laminae consisting of certain transversely isotropic materials such that the axes of transverse isotropy are always normal to the laminae. While such a body must be materially uniform on each lamina, the laminae need *not* be materially isomorphic. This generalization is pertinent, for example, in laminated lumber structures, but since it may be obtained by essentially the same procedure as that used to lead from simple fluids to certain families of subfluids in viscometric flows, for simplicity we carry out the analysis here for materially uniform isotropic bodies only.

The notations in this paper are mostly the same as those used in my earlier paper [2] on the general theory.

2. Statical Universal Solutions for Laminated Bodies

It is known that[2] certain families of deformations are universal solutions for all homogeneous incompressible isotropic bodies. Here, we are interested in finding out whether these deformations remain universal solutions for certain inhomogeneous incompressible isotropic bodies. If such is the case, then the response of the particular kind of inhomogeneous body can be compared with that of a homogeneous body, while both bodies suffer exactly the same deformation. We now show that for certain laminated bodies, such comparisons are indeed possible. The tractions that have to be applied in order to produce the universal solutions are generally more complicated for the inhomogeneous body than for the homogeneous one. In particular, certain shear stresses not present at all in homogeneous bodies will have to occur in an inhomogeneous one so as to effect the same deformation. These stresses are determined below.

We treat locally rectilinear laminated bodies and curvilinear laminated bodies separately.

2a. Locally Rectilinear Laminated Bodies

First, we consider a laminated body $\mathscr{B}$ such that in some reference configuration $\bar{\varphi}$ there exists a rectangular Cartesian coordinate system (X, Y, Z)

---

[1] This term was introduced by SINGH & PIPKIN [4].

[2] *Cf.* TRUESDELL & NOLL [3, §§ 55—57]. However, our solutions here are slightly more general than those given in this reference. Recently, a new family of universal solutions for homogeneous bodies has been found by SINGH & PIPKIN [4].

relative to which the intrinsic Riemannian metric $\boldsymbol{g}$ is given in component form by

$$[\bar{g}^{\alpha\beta}]=\begin{bmatrix}\bar{g}^{11}, & 0, & 0\\ 0, & \bar{g}^{22}, & \bar{g}^{23}\\ 0, & \bar{g}^{32}, & \bar{g}^{33}\end{bmatrix}, \tag{2.1}$$

where the $\bar{g}^{\alpha\beta}$ are functions of X only. Since the coordinate system (X, Y, Z) is rectangular Cartesian, the components of the Euclidean metric $\boldsymbol{G}$ are

$$\bar{G}^{\alpha\beta}=\delta^{\alpha\beta}, \tag{2.2}$$

so that

$$\bar{g}\langle\alpha\beta\rangle=\bar{g}^{\alpha\beta}. \tag{2.3}$$

We denote the dual of $\boldsymbol{g}$ by $\boldsymbol{g}^*$. Then its components $\bar{g}^*_{\alpha\beta}$ form the matrix

$$[\bar{g}^*_{\alpha\beta}]=\begin{bmatrix}\bar{g}^*_{11}, & 0, & 0\\ 0, & \bar{g}^*_{22}, & \bar{g}^*_{23}\\ 0, & \bar{g}^*_{32}, & \bar{g}^*_{33}\end{bmatrix}, \tag{2.4}$$

where the $\bar{g}^*_{\alpha\beta}$ are functions of X only, and

$$[\bar{g}^*_{\alpha\beta}][\bar{g}^{\alpha\beta}]=[\bar{g}^{\alpha\beta}][\bar{g}^*_{\alpha\beta}]=[\delta^\alpha_\beta]. \tag{2.5}$$

Of course, the physical components of $\boldsymbol{g}^*$ are given by

$$\bar{g}^*\langle\alpha\beta\rangle=\bar{g}^*_{\alpha\beta}. \tag{2.6}$$

Since $\mathscr{B}$ is incompressible and simple, the stress field $\boldsymbol{T}$ in any static configuration φ is given by

$$\boldsymbol{T}=-p\boldsymbol{1}+\boldsymbol{S}=-p\boldsymbol{1}+\boldsymbol{H}(\varphi_*), \tag{2.7}$$

where φ_* denotes the local configuration induced by φ, and $\boldsymbol{H}$ is the response function. For an isotropic body, the response function can be expressed by the representation formula[1]

$$\boldsymbol{H}(\varphi_*)=h_1(\operatorname{tr}\boldsymbol{B}, \operatorname{tr}\boldsymbol{B}^{-1})\boldsymbol{B}+h_{-1}(\operatorname{tr}\boldsymbol{B}, \operatorname{tr}\boldsymbol{B}^{-1})\boldsymbol{B}^{-1}, \tag{2.8}$$

where $\boldsymbol{B}$ denotes the left Cauchy-Green tensor of φ_* relative to a fixed undistorted local reference configuration. We choose this undistorted local reference configuration to be the one in which the intrinsic Riemannian metric at the reference point coincides with the Euclidean metric of the physical space. Then from (2.7) and (2.8), the extra stress field $\boldsymbol{S}$ in the configuration $\bar{\varphi}$ is given by

$$\boldsymbol{S}=h_1(\overline{\operatorname{tr}}\,\boldsymbol{g}, \overline{\operatorname{tr}}\,\boldsymbol{g}^*)\,\boldsymbol{g}+h_{-1}(\overline{\operatorname{tr}}\,\boldsymbol{g}, \overline{\operatorname{tr}}\,\boldsymbol{g}^*)\,\boldsymbol{g}^*, \tag{2.9}$$

or equivalently,

$$\bar{S}\langle\alpha\beta\rangle=\bar{h}_1\,\bar{g}\langle\alpha\beta\rangle+\bar{h}_{-1}\,\bar{g}^*\langle\alpha\beta\rangle. \tag{2.10}$$

Here, the trace operation $\overline{\operatorname{tr}}$ is taken relative to the Euclidean metric $\boldsymbol{G}$ in $\bar{\varphi}$.

From (2.1), (2.3), (2.4), (2.6), and (2.10), we see that the matrix $[\bar{S}\langle\alpha\beta\rangle]$ is of the form

$$[\bar{S}\langle\alpha\beta\rangle]=\begin{bmatrix}\bar{S}\langle 11\rangle, & 0, & 0\\ 0, & \bar{S}\langle 22\rangle, & \bar{S}\langle 23\rangle\\ 0, & \bar{S}\langle 32\rangle, & \bar{S}\langle 33\rangle\end{bmatrix}, \tag{2.11}$$

---

[1] *Cf.* TRUESDELL & NOLL [3, § 49].

where the $\bar{S}_{\langle\alpha\beta\rangle}$ are functions of X only. Thus S forms a "laminar system" in $\bar{\varphi}(\mathscr{B})$. This result implies that $\bar{\varphi}$ is an equilibrium configuration for $\mathscr{B}$ provided that the pressure field $\bar{p}$ is given by

$$\bar{p}=\bar{p}(X)=\bar{p}_0+\bar{S}_{\langle 11\rangle}, \tag{2.12}$$

where $\bar{p}_0$ is an arbitrary constant. Consequently, the stress field T in equilibrium also forms a laminar system, and

$$[\bar{T}_{\langle\alpha\beta\rangle}]=\begin{bmatrix}\bar{T}_{\langle 11\rangle}, & 0, & 0\\ 0, & \bar{T}_{\langle 22\rangle}, & \bar{T}_{\langle 23\rangle}\\ 0, & \bar{T}_{\langle 32\rangle}, & \bar{T}_{\langle 33\rangle}\end{bmatrix}, \tag{2.13}$$

where the $\bar{T}_{\langle\alpha\beta\rangle}$ are functions of X only.

We now give some universal solutions for the laminated body $\mathscr{B}$ relative to the reference configuration $\bar{\varphi}$.

Family 0. Homogeneous plane deformations. We claim that all deformations given by

$$x=AX, \quad y=BY+CZ, \quad z=DY+EZ \tag{2.14}$$

are universal solutions for $\mathscr{B}$. Here $A, \ldots, E$ are constants such that

$$A(BE-CD)=1, \tag{2.15}$$

and (x, y, z) is a rectangular Cartesian coordinate system in the deformed configuration. Special cases of this family include simple extensions in the (X, Y, Z) directions (characterized by the condition $C=D=0$), and simple shears in the Y-Z plane (characterized by the conditions $A=B=E=1$, $D=0$ or $A=B=E=1$, $C=0$).

The fact that the plane deformations given by (2.14) are universal solutions for $\mathscr{B}$ is obvious. We can verify easily that in the deformed configurations the component form of the intrinsic metric g relative to the coordinate system (x, y, z) is the same as before, namely,

$$[g^{ij}]=\begin{bmatrix}g^{11}, & 0, & 0\\ 0, & g^{22}, & g^{23}\\ 0, & g^{32}, & g^{33}\end{bmatrix}, \tag{2.16}$$

where the g^{ij} are functions of x only and are given by

$$\begin{aligned}
g^{11}(x)&=A^2\,\bar{g}^{11}\left(\frac{x}{A}\right),\\
g^{22}(x)&=B^2\,\bar{g}^{22}\left(\frac{x}{A}\right)+2BC\,\bar{g}^{23}\left(\frac{x}{A}\right)+C^2\,\bar{g}^{33}\left(\frac{x}{A}\right),\\
g^{23}(x)&=BD\,\bar{g}^{22}\left(\frac{x}{A}\right)+(CD+BE)\,\bar{g}^{23}\left(\frac{x}{A}\right)+CE\,\bar{g}^{33}\left(\frac{x}{A}\right),
\end{aligned} \tag{2.17}$$

and

$$g^{33}(x)=D^2\,\bar{g}^{22}\left(\frac{x}{A}\right)+2DE\,\bar{g}^{23}\left(\frac{x}{A}\right)+E^2\,\bar{g}^{33}\left(\frac{x}{A}\right).$$

Similarly, the components g^*_{ij} of $\boldsymbol{g}^*$ relative to (x, y, z) form the matrix

$$[g^*_{ij}] = \begin{bmatrix} g^*_{11}, & 0, & 0 \\ 0, & g^*_{22}, & g^*_{23} \\ 0, & g^*_{32}, & g^*_{33} \end{bmatrix}, \tag{2.18}$$

where the g^*_{ij} are functions of x only and are given by

$$\begin{aligned} g^*_{11} &= \frac{1}{A^2}\, \bar{g}^*_{11}, \\ g^*_{22} &= A^2[E^2\, \bar{g}^*_{22} - 2DE\, \bar{g}^*_{23} + D^2\, \bar{g}^*_{33}], \\ g^*_{23} &= A^2[-CE\, \bar{g}^*_{22} + (BE+CD)\, \bar{g}^*_{23} - BD\, \bar{g}^*_{33}], \\ \text{and} \quad g^*_{33} &= A^2[C^2\, \bar{g}^*_{22} - 2BC\, \bar{g}^*_{23} + B^2\, \bar{g}^*_{33}]. \end{aligned} \tag{2.19}$$

Here the arguments of g^*_{ij} and $\bar{g}^*_{\alpha\beta}$ are again x and x/A, respectively.

From the constitutive equation, the extra stress field in the deformed state is given by

$$S\langle ij\rangle = h_1\, g\langle ij\rangle + h_{-1}\, g^*\langle ij\rangle, \tag{2.20}$$

where the arguments of h_1 and h_{-1} are $\operatorname{tr} \boldsymbol{g}$ and $\operatorname{tr} \boldsymbol{g}^*$. Here the trace operation tr is taken with respect to the Euclidean metric in the deformed state, and

$$g\langle ij\rangle = g^{ij}, \qquad g^*\langle ij\rangle = g^*_{ij}. \tag{2.21}$$

It should be noted that in the deformed configuration the shear stress $S\langle 23\rangle$, in general, does not vanish even if the deformation is a simple extension. Also, not every homogeneous deformation is a universal solution for a laminated body such as $\mathscr{B}$. For instance, a simple shear given by

$$x = X, \qquad y = Y + KX, \qquad z = Z \tag{2.22}$$

is *not* a universal solution for $\mathscr{B}$ unless $\mathscr{B}$ happens to be homogeneous. Such departure of the global response of $\mathscr{B}$ from that of a homogeneous body is typical of inhomogeneous bodies. Globally, the response of $\mathscr{B}$ appears to be "anisotropic", although locally, at each particle, the stress is indeed described by an "isotropic" response function, namely, equation (2.8). This apparent paradox cannot be resolved by considering the responses of arbitrarily small but finite subbodies of $\mathscr{B}$, since the inhomogeneity field, which is characterized by the curvature tensor of the intrinsic Riemannian metric on $\mathscr{B}$ in this case, may be non-vanishing everywhere.

Next, we give some inhomogeneous universal solutions for $\mathscr{B}$.

Family 1. We use cylindrical coordinate systems in the deformed configuration. The deformation functions are

$$r = \sqrt{2AX}, \qquad \theta = BY + CZ, \qquad z = DY + EZ, \tag{2.23}$$

where the constants $A, \ldots, E$, again satisfy the condition (2.15).

For this family, the component form of the intrinsic metric $\boldsymbol{g}$ relative to the coordinate system (r, θ, z) in the deformed configuration, again, is given by

(2.16). However, the g^{ij} are now functions of r and are given by

$$
\begin{aligned}
g^{11} &= \frac{A^2}{r^2}\,\bar{g}^{11}, \\
g^{22} &= B^2\,\bar{g}^{22} + 2BC\,\bar{g}^{23} + C^2\,\bar{g}^{33}, \\
g^{23} &= BD\,\bar{g}^{22} + (CD+BE)\,\bar{g}^{23} + CE\,\bar{g}^{33}, \\
\text{and}\quad g^{33} &= D^2\,\bar{g}^{22} + 2DE\,\bar{g}^{23} + E^2\,\bar{g}^{33},
\end{aligned}
\tag{2.24}
$$

where the arguments of g^{ij} and $\bar{g}^{\alpha\beta}$, of course, are r and $r^2/2A$, respectively. Similarly, the covariant components g^*_{ij} of the dual metric g^* relative to (r, θ, z) form the matrix (2.18) and are given by

$$
\begin{aligned}
g^*_{11} &= \frac{r^2}{A^2}\,\bar{g}^*_{11}, \\
g^*_{22} &= A^2\left[E^2\,\bar{g}^*_{22} - 2DE\,\bar{g}^*_{23} + D^2\,\bar{g}^*_{33}\right], \\
g^*_{23} &= A^2\left[-CE\,\bar{g}^*_{22} + (BE+CD)\,\bar{g}^*_{23} - BD\,\bar{g}^*_{33}\right], \\
\text{and}\quad g^*_{33} &= A^2\left[C^2\,\bar{g}^*_{22} - 2BC\,\bar{g}^*_{23} + B^2\,\bar{g}^*_{33}\right].
\end{aligned}
\tag{2.25}
$$

Here, again, the arguments of g^*_{ij} and $\bar{g}^*_{\alpha\beta}$ are r and $r^2/2A$, respectively. Since the component form of the Euclidean metric G relative to (r, θ, z) is

$$
[G^{ij}] = \operatorname{diag}\left[1, \frac{1}{r^2}, 1\right],
\tag{2.26}
$$

the physical components $g\langle ij\rangle$ and $g^*\langle ij\rangle$ are now given by

$$
\begin{aligned}
g\langle 11\rangle &= g^{11}, \\
g\langle 22\rangle &= r^2\,g^{22}, \\
g\langle 23\rangle &= r\,g^{23}, \\
\text{and}\quad g\langle 33\rangle &= g^{33}, \\
\text{and also}\quad g^*\langle 11\rangle &= g^*_{11}, \\
g^*\langle 22\rangle &= \frac{1}{r^2}\,g^*_{22}, \\
g^*\langle 23\rangle &= \frac{1}{r}\,g^*_{23}, \\
\text{and}\quad g^*\langle 33\rangle &= g^*_{33}.
\end{aligned}
\tag{2.27}
$$

Thus the $g\langle ij\rangle$ and $g^*\langle ij\rangle$ are functions of r only; moreover,

$$
g\langle 12\rangle = g\langle 13\rangle = g^*\langle 12\rangle = g^*\langle 13\rangle = 0.
\tag{2.28}
$$

Substituting these results into the constitutive equation (2.20), we see that in the deformed configuration the physical components $S\langle ij\rangle$ of the extra stress are functions of r only. Also,

$$S\langle 12\rangle = S\langle 13\rangle = 0. \tag{2.29}$$

Consequently, the deformed configuration is in equilibrium under a pressure field p given by

$$p = p(r) = S\langle 11\rangle + \int \frac{S\langle 11\rangle - S\langle 22\rangle}{r}\, dr. \tag{2.30}$$

Thus we have shown that the deformations belonging to this family are static universal solutions for $\mathscr{B}$.

As before, the global response of $\mathscr{B}$ in these deformations is different from that of a homogeneous body. For instance, in a pure bending (characterized by the condition $C=D=0$), the shear stress $S\langle 23\rangle$ does not vanish unless $\bar{g}^{23}=0$; moreover, the relation between the bending moment and the curvature now depends also on the metric $\boldsymbol{g}$. The explicit form of such global response relations can be determined by integrating the surface traction over the boundary of $\mathscr{B}$ in the deformed configuration. We shall work out some examples in detail in Section 4.

The above family of deformations shows that we can visualize the laminated body $\mathscr{B}$ as being formed locally by laminae which are initially circular cylinders also. This fact reflects the condition that a cylinder is a developable surface.

Now let $\overline{\mathscr{B}}$ be a laminated body whose intrinsic metric relative to a cylindrical coordinate system (R, Θ, Z) in a (global) reference configuration $\bar{\varphi}$ is given in component form by (2.1), where the $\bar{g}^{\alpha\beta}$ are functions of R only. Then the deformations belonging to the following family, which are the inverses of those of the preceding family, are universal solutions.

Family 2. We use a rectangular Cartesian coordinate system (x, y, z) in the deformed configuration. Then the deformation functions are

$$x = \tfrac{1}{2} A R^2, \qquad y = B\Theta + CZ, \qquad z = D\Theta + EZ, \tag{2.31}$$

where the constants $A, \ldots, E$ satisfy the condition (2.15).

Clearly, in the deformed configuration, the components g^{ij} of $\boldsymbol{g}$ relative to (x, y, z) again form the matrix shown in (2.16); moreover, they are given by

$$\begin{aligned} g_{11} &= 2Ax\,\bar{g}^{11}, \\ g^{22} &= B^2\,\bar{g}^{22} + 2BC\,\bar{g}^{23} + C^2\,\bar{g}^{33}, \\ g^{23} &= BD\,\bar{g}^{22} + (CD+BE)\,\bar{g}^{23} + CE\,\bar{g}^{33}, \\ \text{and} \quad g^{33} &= D^2\,\bar{g}^{22} + 2DE\,\bar{g}^{23} + E^2\,\bar{g}^{33}, \end{aligned} \tag{2.32}$$

where the arguments of g^{ij} and $\bar{g}^{\alpha\beta}$ are x and $\sqrt{2x/A}$, respectively. The components g^*_{ij} of $\boldsymbol{g}^*$ relative to (x, y, z) can be determined in a similar way, and they are functions of x only such that

$$g^*_{12} = g^*_{13} = 0. \tag{2.33}$$

Substituting these results into the constitutive equation, we see that the extra stress components $S\langle ij\rangle$ are functions of x only, and the condition (2.29) again holds. Thus we have shown that the deformations belonging to this family are universal solutions for $\bar{\mathscr{B}}$.

Now it is clear that the deformations belonging to the following family are also universal solutions for $\bar{\mathscr{B}}$:

Family 3. We use a cylindrical coordinate system (r, θ, z) in the deformed configuration. The deformation functions are

$$r=\sqrt{A R^2+B}, \quad \theta=C\Theta+DZ, \quad z=E\Theta+FZ, \tag{2.34}$$

where $A, \ldots, F$ are constants satisfying the condition

$$A(CF-DE)=1. \tag{2.35}$$

Locally, we can visualize this family of deformations to be the compositions of a deformation belonging to Family 2, a rigid translation in the laminar planes, and a deformation belonging to Family 1. Since each step is a universal solution, so is their composition. Thus we can expect the deformations given by (2.34) to be universal solutions.

The components of $\boldsymbol{g}$ and $\boldsymbol{g}^*$ relative to (r, θ, z) in the deformed configuration can be computed in exactly the same way as before. For example, we now have

$$\begin{aligned} g^{11} &= \frac{A(r^2-B)}{r^2}\,\bar{g}^{11}, \\ g^{22} &= C^2\,\bar{g}^{22}+2CD\,\bar{g}^{23}+D^2\,\bar{g}^{33}, \end{aligned} \tag{2.36}$$

and so forth, where the g^{ij} and g^*_{ij} are functions of r only. The physical components can then be determined by using the Euclidean metric $\boldsymbol{G}$ as shown in (2.27). Finally, the stress field is given by (2.20), which shows that the deformed configuration is in equilibrium under the pressure field given by (2.30).

For the various special cases in this family, such as pure torsions (characterized by the conditions $A=C=F=1$ and $B=E=0$), torsions with extensions (characterized by the conditions $C=1$ and $B=E=0$), eversions (characterized by the conditions $C=1$, $D=E=0$, and $A<0$), and pure bendings (characterized by the conditions $F=1$ and $D=E=0$), the global responses of $\bar{\mathscr{B}}$ depend not only on the functions h_1 and h_{-1} but also explicitly on the intrinsic metric $\boldsymbol{g}$. We shall see some examples in Section 4.

2b. Curvilinear Laminated Bodies

We consider a laminated body $\tilde{\mathscr{B}}$ whose intrinsic metric relative to a spherical coordinate system (R, Θ, Φ) in a reference configuration $\bar{\varphi}$ is given by

$$[\bar{g}_{\langle\alpha\beta\rangle}]=\operatorname{diag}[\bar{g}_{\langle 11\rangle}, \bar{g}_{\langle 22\rangle}, \bar{g}_{\langle 33\rangle}], \tag{2.37}$$

where the $\bar{g}_{\langle\alpha\beta\rangle}$ are functions of R only, and where

$$\bar{g}_{\langle 22\rangle}=\bar{g}_{\langle 33\rangle}. \tag{2.38}$$

Thus the laminae are concentric spherical shells, whose initial radii, in general, are different from those in the configuration $\bar{\varphi}$.

Of course, the component form of $\boldsymbol{G}$ relative to (R, Θ, Φ) is

$$[\bar{G}^{\alpha\beta}]=\operatorname{diag}\left[1, \frac{1}{R^2}, \frac{1}{R^2 \sin^2\Phi}\right]. \tag{2.39}$$

Thus

$$[\bar{g}^{\alpha\beta}]=\operatorname{diag}[\bar{g}^{11}, \bar{g}^{22}, \bar{g}^{33}], \tag{2.40}$$

where

$$\bar{g}^{11}=\bar{g}_{\langle 11\rangle},$$

$$\bar{g}^{22}=\frac{1}{R^2}\bar{g}_{\langle 22\rangle}, \tag{2.41}$$

and

$$\bar{g}^{33}=\frac{1}{R^2 \sin^2\Phi}\bar{g}_{\langle 33\rangle}.$$

Relative to the same spherical coordinate system (R, Θ, Φ), the components of the dual metric g^* are given by

$$[\bar{g}^*_{\langle\alpha\beta\rangle}]=\operatorname{diag}[\bar{g}^*_{\langle 11\rangle}, \bar{g}^*_{\langle 22\rangle}, \bar{g}^*_{\langle 33\rangle}]= \\ =\operatorname{diag}\left[\frac{1}{\bar{g}_{\langle 11\rangle}}, \frac{1}{\bar{g}_{\langle 22\rangle}}, \frac{1}{\bar{g}_{\langle 33\rangle}}\right]. \tag{2.42}$$

Hence the $\bar{g}^*_{\langle\alpha\beta\rangle}$ are functions of R only, and

$$\bar{g}^*_{\langle 22\rangle}=\bar{g}^*_{\langle 33\rangle}. \tag{2.43}$$

Also,

$$[\bar{g}^*_{\alpha\beta}]=\operatorname{diag}[\bar{g}^*_{11}, \bar{g}^*_{22}, \bar{g}^*_{33}], \tag{2.44}$$

where

$$\bar{g}^*_{11}=\bar{g}^*_{\langle 11\rangle}=\frac{1}{\bar{g}_{\langle 11\rangle}},$$

$$\bar{g}^*_{22}=R^2\bar{g}^*_{\langle 22\rangle}=\frac{R^2}{\bar{g}_{\langle 22\rangle}}, \tag{2.45}$$

and

$$\bar{g}^*_{33}=R^2\sin^2\Phi\,\bar{g}^*_{\langle 33\rangle}=\frac{R^2\sin^2\Phi}{\bar{g}_{\langle 33\rangle}}.$$

From (2.37), (2.42), and (2.20), we see that in the reference configuration $\bar{\varphi}$,

$$[\bar{S}_{\langle\alpha\beta\rangle}]=\operatorname{diag}[\bar{S}_{\langle 11\rangle}, \bar{S}_{\langle 22\rangle}, \bar{S}_{\langle 33\rangle}], \tag{2.46}$$

where the $\bar{S}_{\langle\alpha\beta\rangle}$ are functions of R only, and

$$\bar{S}_{\langle 22\rangle}=\bar{S}_{\langle 33\rangle}. \tag{2.47}$$

Then $\bar{\varphi}(\tilde{\mathscr{B}})$ is in equilibrium under the pressure field

$$\bar{p}=\bar{p}(R)=\bar{S}_{\langle 11\rangle}+\int\frac{2\bar{S}_{\langle 11\rangle}-\bar{S}_{\langle 22\rangle}-\bar{S}_{\langle 33\rangle}}{R}\,dR. \tag{2.48}$$

We shall see that the deformations belonging to the following family are universal solutions for $\tilde{\mathscr{B}}$:

Family 4. We use a spherical coordinate system (r, θ, φ) in the deformed configuration. The deformation functions are

$$r=(\pm R^3+A)^{\frac{1}{3}}, \qquad \theta=\pm\Theta, \qquad \varphi=\Phi, \tag{2.49}$$

where A is an arbitrary constant. Here, if the sign is positive, then the deformation is an inflation, while if the sign is negative, the deformation is an eversion.

It is easily verfied that the components of the intrinsic metric relative to the coordinate system (r, θ, φ) in the deformed configuration are

$$[g_{\langle ij\rangle}]=\operatorname{diag}[g_{\langle 11\rangle}, g_{\langle 22\rangle}, g_{\langle 33\rangle}], \tag{2.50}$$

where

$$g_{\langle 11\rangle}=\frac{(r^3-A)^{\frac{4}{3}}}{r^4}\,\bar{g}_{\langle 11\rangle},$$

$$g_{\langle 22\rangle}=\frac{r^2}{(r^3-A)^{\frac{2}{3}}}\,\bar{g}_{\langle 22\rangle}, \tag{2.51}$$

and

$$g_{\langle 33\rangle}=\frac{r^2}{(r^3-A)^{\frac{2}{3}}}\,\bar{g}_{\langle 33\rangle}=g_{\langle 22\rangle},$$

and also,

$$[g^*_{\langle ij\rangle}]=\operatorname{diag}[g^*_{\langle 11\rangle}, g^*_{\langle 22\rangle}, g^*_{\langle 33\rangle}], \tag{2.52}$$

where

$$g^*_{\langle 11\rangle}=\frac{r^4}{(r^3-A)^{\frac{4}{3}}}\,\bar{g}^*_{\langle 11\rangle},$$

$$g^*_{\langle 22\rangle}=\frac{(r^3-A)^{\frac{2}{3}}}{r^2}\,\bar{g}^*_{\langle 22\rangle}, \tag{2.53}$$

and

$$g^*_{\langle 33\rangle}=\frac{(r^3-A)^{\frac{2}{3}}}{r^2}\,\bar{g}^*_{\langle 33\rangle}=g^*_{\langle 22\rangle}.$$

Here the arguments of $g_{\langle ij\rangle}$, $g^*_{\langle ij\rangle}$, and $\bar{g}_{\langle\alpha\beta\rangle}$, $\bar{g}^*_{\langle\alpha\beta\rangle}$ are r and $\pm(r^3-A)^{\frac{1}{3}}$, respectively.

Substituting these results into the constitutive equation (2.20), we see that in the deformed configuration

$$[S_{\langle ij\rangle}]=\operatorname{diag}[S_{\langle 11\rangle}, S_{\langle 22\rangle}, S_{\langle 33\rangle}], \tag{2.54}$$

where $S_{\langle ij\rangle}$ are functions of r only, and

$$S_{\langle 22\rangle}=S_{\langle 33\rangle}. \tag{2.55}$$

Thus the deformed states are all in equilibrium under the pressure field

$$p=p(r)=S_{\langle 11\rangle}+\int\frac{2S_{\langle 11\rangle}-S_{\langle 22\rangle}-S_{\langle 33\rangle}}{r}\,dr. \tag{2.56}$$

Hence we have shown that the deformations belonging to this family are universal solutions for $\tilde{\mathscr{B}}$. As before, the global response of $\tilde{\mathscr{B}}$ in this family of deformations now depends on h_1, h_{-1}, and g.

It should be noted that the laminated body $\tilde{\mathscr{B}}$ is not locally rectilinear, since a sphere is not a developable surface.

So far, we have considered five families of universal solutions for homogeneous bodies which turn out to be universal solutions for certain laminated bodies also. There is, however, one more family of universal solutions for homogeneous bodies found recently by SINGH & PIPKIN [4]:

Family 5. We use cylindrical coordinate systems (R, Θ, Z) and (r, θ, z) in the reference and the deformed configurations, respectively. The deformation functions are

$$r = A R, \quad \theta = B \log R + C \Theta, \quad z = D Z, \tag{2.57}$$

where $A, \ldots, D$, are constants such that

$$A^2 C D = 1. \tag{2.58}$$

I have not found any class of simple inhomogeneous laminated bodies for which this family yields universal solutions [1].

3. Dynamical Universal Solutions for Laminated Bodies

In this section, we give a class of exact solutions of the dynamical equations for the various kinds of incompressible laminated bodies considered in the previous section. These dynamical solutions are required to satisfy the following condition: At each instant t, the instantaneous configuration is a possible configuration for equilibrium if an appropriate pressure field is applied. Such solutions are called *quasi-equilibrated motions* [2]. As before, we require these solutions to be controllable, so that the body is not subject to any body force. The fact that these motions are dynamic universal solutions for all incompressible isotropic homogeneous simple bodies (not necessarily elastic) has been remarked recently by CARROLL [5] and FOSDICK [6]. Here we show that these motions remain universal solutions for the three classes of laminated (simple) bodies.

Let $\bar{\varphi}$ be a reference configuration, and suppose that χ_t, $t \in (-\infty, \infty)$ is a 1-parameter family of deformations from $\bar{\varphi}$. Then $\chi_t \circ \bar{\varphi}$ is a dynamically possible motion if and only if the dynamical equation

$$\operatorname{div} \boldsymbol{T} = \rho \ddot{\boldsymbol{x}} \tag{3.1}$$

is satisfied at each instant t. Since the body is incompressible and simple, the stress field $\boldsymbol{T}$ is given by

$$\boldsymbol{T} = -p\boldsymbol{1} + \boldsymbol{S} = -p\boldsymbol{1} + \underset{s=0}{\overset{\infty}{\mathfrak{H}}} (\chi_{t-s} \circ \bar{\varphi}_*), \tag{3.2}$$

where $\mathfrak{H}$ is the response functional for the extra stress $\boldsymbol{S}$. The argument of $\mathfrak{H}$ is the local history of the particles. Thus a necessary and sufficient condition for a dynamically possible motion is that at each instant t there exists a (single-valued) pressure field p such that

$$\operatorname{grad} p = \operatorname{div} \boldsymbol{S} - \rho \ddot{\boldsymbol{x}}. \tag{3.3}$$

---

[1] See *Conjecture* 4 in Section 5.

[2] *Cf.* TRUESDELL & NOLL [3, § 61]. However, our solutions here are more complete than those given in this reference.

Now suppose that the motion $\chi_t \circ \bar{\varphi}$ be quasi-equilibrated. Then, by definition, we can determine a pressure field p_0 such that the stress field

$$T_0 = -p_0 \mathbf{1} + S = -p_0 \mathbf{1} + \underset{s=0}{\overset{\infty}{\mathfrak{H}}} (\chi_{t-s} \circ \bar{\varphi}_*) \tag{3.4}$$

satisfies the equation of equilibrium:

$$0 = \operatorname{div} T_0 = -\operatorname{grad} p_0 + \operatorname{div} S. \tag{3.5}$$

Comparing (3.3) and (3.5), we see that[1] a quasi-equilibrated motion is dynamically possible if and only if the acceleration field is conservative with a single-valued potential function ζ:

$$\ddot{x} = -\operatorname{grad} \zeta. \tag{3.6}$$

In this case, we then have

$$p = \rho \zeta + p_0. \tag{3.7}$$

It should be noted that, from classical hydrodynamics, motions satisfying the condition (3.6) are circulation-preserving.

Now we proceed to determine the dynamically possible quasi-equilibrated motions whose instantaneous deformations from a fixed reference configuration $\bar{\varphi}$ are the various statical universal solutions considered in the previous section. Again, we classify the motions into five families according to those for the statical solutions, so that the previous scalar parameters now become functions of t, which are subject to the restrictions due to the condition (3.6), of course.

Remark. A more general class of solutions, which differ from the ones just described by arbitrary rigid motions, can be considered also. For this general class, six more functions of t are needed to characterize the motions. However, since the explicit representation of this general class and the analysis for the restrictions on the scalar functions due to the condition (3.6) are quite complex, we treat here the simple class[2] only.

3a. Locally Rectilinear Laminated Bodies

We consider first the laminated body $\mathscr{B}$ introduced in the preceding section.

Family 0. Homogeneous plane motions. The deformation functions are given by (2.14) such that $A, \ldots, E$ are functions of t, which satisfy the condition (2.15).

Using the usual objectivity argument[3], we can verify that the extra stress field in the motion is given in component form by

$$[S\langle ij\rangle] = \begin{bmatrix} S\langle 11\rangle, & 0, & 0 \\ 0, & S\langle 22\rangle, & S\langle 23\rangle \\ 0, & S\langle 32\rangle, & S\langle 33\rangle \end{bmatrix}, \tag{3.8}$$

and

$$A = \frac{1}{BE - CD}.$$

[1] This result is due to TRUESDELL. *Cf.* TRUESDELL & NOLL [3, § 61].

[2] TRUESDELL, in his original paper, included quasi-equilibrated motions which differ from the simple class by certain special rigid motions, but his solutions for the restrictions due to the condition (3.6) are not complete.

[3] *Cf.* FOSDICK [6]. This argument was introduced by COLEMAN & NOLL in their papers on viscometric flows for incompressible simple fluids and was used by WANG for subfluids.

where the $S\langle ij\rangle$ are functions of x only. Thus the instantaneous configuration is a possible equilibrium state under the pressure field p_0 given by (2.12). Hence this family consists of quasi-equilibrated motions for $\mathcal{B}$. We proceed to determine the restrictions on the functions $A, \ldots, E$ due to the condition (3.6), which characterizes those of these motions that are dynamic universal solutions for $\mathcal{B}$.

From (2.14) and (2.15), the acceleration field $\boldsymbol{a}$ in the motions is given in component form by

$$a_1 = \frac{\ddot{A}}{A} x,$$

$$a_2 = A(E\ddot{B} - D\ddot{C})\, y + A(B\ddot{C} - C\ddot{B})\, z, \tag{3.9}$$

and

$$a_3 = A(E\ddot{D} - D\ddot{E})\, y + A(B\ddot{E} - C\ddot{D})\, z.$$

Then the condition (3.6) requires that

$$\frac{\partial a_i}{\partial x^j} - \frac{\partial a_j}{\partial x^i} = 0, \qquad \forall\, i, j = 1, 2, 3. \tag{3.10}$$

Substituting (3.9) into these expressions, we obtain the following differential equation:

$$B\ddot{C} - C\ddot{B} = E\ddot{D} - D\ddot{E}. \tag{3.11}$$

Thus (2.15) and (3.11) are the governing equations for the functions $A, \ldots, E$.

Integrating (3.11) once, we obtain

$$B\dot{C} - C\dot{B} = E\dot{D} - D\dot{E} + k, \tag{3.12}$$

where k is a constant. Now we consider the following two possibilities:

Case 1. $B(t_0) \neq 0$ for some t_0. Then the general solution for t near t_0 is given by

$$B, D, E = \text{arbitrary functions of } t, \text{ but } B \neq 0;$$

$$C = B\left[k' + \int_{t_0}^{t} \frac{1}{B^2} (E\dot{D} - D\dot{E} + k)\, dt\right],$$

where k' is a constant such that $E(t_0) - k' D(t_0) \neq 0$; and

$$A = \frac{1}{BE - CD}.$$

Case 2. $B(t_0) = 0$ for some t_0. In this case, we must have

$$D(t_0)\, C(t_0) \neq 0. \tag{3.13}$$

Then the general solution for t near t_0 is given by

$$C, D, E = \text{arbitrary functions of } t, \text{ but } CD \neq 0,$$

$$B = -C \int_{t_0}^{t} \frac{1}{C^2} [E\dot{D} - D\dot{E} + k]\, dt;$$

Any particular solution for $t\in(-\infty, \infty)$ is then obtained by piecing together solutions belonging to one of the above two cases, on various intervals. It should be noted that a zero for the function B need *not* be an isolated one. It may happen that B is identically zero on an interval (a, b) without being identically zero for all t.

If the condition (3.6) is satisfied, then the acceleration potential ζ for this family is given by

$$\begin{aligned} -\zeta=&\frac{1}{2}\frac{\ddot{A}}{A}x^2+\frac{1}{2}A(E\ddot{B}-D\ddot{C})y^2+\frac{1}{2}A(B\ddot{E}-C\ddot{D})z^2\\ &+A(B\ddot{C}-C\ddot{B})yz+\Phi(t), \end{aligned} \tag{3.14}$$

where $\Phi(t)$ is an arbitrary function of t.

Next, we consider quasi-equilibrated motions belonging to the following

Family 1. The deformation functions are

$$r^2=2AX+B, \qquad \theta=CY+DZ, \qquad z=EY+FZ, \tag{3.15}$$

where $A, \ldots, F$ are functions of t satisfying the condition (2.35). It should be noted that the function B here does *not* correspond to a superimposed rigid motion. Again, it can be shown that the extra stress at any instant t is given in component form by (3.8), where the $S\langle ij\rangle$ are functions of r only. Thus the instantaneous configuration is a possible equilibrium state under the pressure field p_0 given by (2.30), so that this family consists of quasi-equilibrated motions for $\mathscr{B}$. We wish to determine the restrictions on the functions $A, \ldots, F$ due to the condition (3.6).

From (3.15), the velocity field and the acceleration field can be characterized by

$$\dot{r}=\frac{\dot{A}}{2A}r+\left(\frac{\dot{B}}{2}-\frac{B\dot{A}}{2A}\right)\frac{1}{r},$$

$$\dot{\theta}=A(F\dot{C}-E\dot{D})\theta+A(C\dot{D}-D\dot{C})z,$$

$$\dot{z}=A(F\dot{E}-E\dot{F})\theta+A(C\dot{F}-D\dot{E})z,$$

$$\begin{aligned} \ddot{r}=&\left[\frac{\ddot{A}}{2A}-\left(\frac{\dot{A}}{2A}\right)^2\right]r+\left[\frac{\dot{A}}{A}\left(\frac{\dot{A}B}{A}-\dot{B}\right)-\left(\frac{\ddot{A}B}{A}-\ddot{B}\right)\right]\frac{1}{2r}\\ &-\left(\frac{\dot{A}B}{A}-\dot{B}\right)^2\frac{1}{4r^3}, \end{aligned} \tag{3.16}$$

$$\ddot{\theta}=A(F\ddot{C}-E\ddot{D})\theta+A(C\ddot{D}-D\ddot{C})z,$$

and

$$\ddot{z}=A(F\ddot{E}-E\ddot{F})\theta+A(C\ddot{F}-D\ddot{E})z.$$

Then the covariant components a_r, a_θ, and a_z of $\boldsymbol{a}$ are given by

$$a_r = \ddot{r} - r\dot{\theta}^2,$$
$$a_\theta = r^2\ddot{\theta} - 2r\dot{r}\dot{\theta}, \tag{3.17}$$

and

$$a_z = \ddot{z}.$$

The restrictions on $A, \ldots, F$ due to the condition (3.6) again take on the form (3.10).

Substituting (3.17) into (3.10), we obtain the following system of differential equations:

$$C\dot{D} - D\dot{C} = 0, \tag{3.18}$$

$$F\ddot{E} - E\ddot{F} = 0, \tag{3.19}$$

and

$$F\ddot{C} - E\ddot{D} - A(F\dot{C} - E\dot{D})\left(F\dot{C} - E\dot{D} + \frac{\dot{A}}{A^2}\right) = 0. \tag{3.20}$$

These and the previous condition (2.35) then form the governing equations for the functions $A, \ldots, F$. We wish to determine their general solutions.

First, we notice that the function B does not enter into any one of these equations, so that it is arbitrary. Second, we claim that the function C must either be identical to zero for all t or never equal to zero for any t. To see this fact, suppose that $C \neq 0$. Then the integral of (3.18) is

$$D = kC, \tag{3.21}$$

where k is a constant. Hence if $C(t_0) = 0$ for some t_0 while $C(t) \neq 0$ for all $t \in (t_0, t_0 + \varepsilon)$ or for all $t \in (t_0 - \varepsilon, t_0)$, then from (3.21) we must have

$$D(t_0) = 0. \tag{3.22}$$

But since from the condition (2.35), C and D cannot vanish simultaneously, we see that if C vanishes at any one instant t_0, then it must be identical to zero for all t. Hence there are two possibilities:

Case 1. $C \neq 0$ for any t. In this case, the equation (3.21) must be satisfied for all t. Then the equation (2.35) reduces to

$$AC(F - kE) = 1. \tag{3.23}$$

Also, equation (3.20) reduces to

$$\ddot{C} + \dot{C}\left[\frac{\dot{C}}{C} + \frac{\dot{A}}{A}\right] = 0, \tag{3.24}$$

which can be integrated, yielding

$$\dot{C} = \frac{k'}{AC}, \tag{3.25}$$

where k' is a constant. As before, equation (3.19) can be integrated once also, and we get

$$F\dot{E} - E\dot{F} = k'', \tag{3.26}$$

where k'' is a constant. Now we consider the following three possibilities:

1a). $F \equiv 0$ for all t. In this subcase, $k''=0$, and the equation (3.26) is satisfied for arbitrary E. Further, equation (3.23) becomes

$$\frac{1}{AC} = -kE. \tag{3.27}$$

Then the general solution is given by

$$B, E = \text{arbitrary functions of } t, \text{ but } E \neq 0;$$

$$F \equiv 0;$$

$$C = k_1 \left[\int_0^t E\,dt + k_2\right] \neq 0;$$

$$D = -k_3 \left[\int_0^t E\,dt + k_2\right] \neq 0;$$

and

$$A = -\frac{1}{kEC} = \frac{1}{k_3 E \left[\int_0^t E\,dt + k_2\right]}.$$

1b). $F(t_0) \neq 0$ for some t_0. We consider the general solution for t near t_0. Integrating equation (3.26), we get

$$E = F\left[k''' + k'' \int_{t_0}^t \frac{1}{F^2}\,dt\right]. \tag{3.28}$$

Then we put

$$\xi = \frac{1}{AC} = F\left[1 - k\left(k''' + k'' \int_{t_0}^t \frac{1}{F^2}\,dt\right]\right. \neq 0, \tag{3.29}$$

and the general solution is given by

$$B, F = \text{arbitrary functions of } t, \text{ but } F \neq 0;$$

$$E = F\left[k''' + k'' \int_{t_0}^t \frac{1}{F^2}\,dt\right];$$

$$C = k_1 \left[\int_{t_0}^t \xi\,dt + k_2\right] \neq 0;$$

$$D = kC = k k_1 \left[\int_{t_0}^t \xi\,dt + k_2\right];$$

and

$$A = \frac{1}{C\xi}.$$

1c). $F(t_0)=0$ for some t_0, but $F \not\equiv 0$. We claim that t_0 must be an isolated root of F. Indeed, since F does not vanish identically, without loss of generality we can choose a t_0 in such a way that either

$$F(t) \neq 0, \qquad t \in (t_0, t_0 + \varepsilon), \tag{3.30}$$

or

$$F(t) \neq 0, \qquad t \in (t_0 - \varepsilon, t_0). \tag{3.31}$$

Then the constant k'' in (3.26) must be non-zero, since otherwise $E(t_0)$ would vanish also, contradicting the condition (3.23). Now if $k'' \neq 0$, then from (3.26), $\dot{F}(t_0) \neq 0$ for any t_0 such that $F(t_0)=0$. Thus t_0 must be an isolated root of F.

We now determine the general solution of the system for t near t_0. From (3.23), we see that $k E(t_0) \neq 0$. Integrating (3.26), we obtain

$$F = -k'' E \int_{t_0}^{t} \frac{dt}{E^2}. \tag{3.32}$$

Then we put

$$\xi = \frac{1}{AC} = -E\left(k + k'' \int_{t_0}^{t} \frac{dt}{E^2}\right) \neq 0, \tag{3.33}$$

and the general solution is as follows:

$B, E =$ arbitrary functions of t, but $E \neq 0$;

$$F = -k'' E \int_{t_0}^{t} \frac{dt}{E^2} \neq 0;$$

$$C = k_1 \left[\int_{t_0}^{t} \xi\, dt + k_2\right] \neq 0;$$

$$D = k C = k k_1 \left[\int_{t_0}^{t} \xi\, dt + k_2\right] \neq 0;$$

and

$$A = \frac{1}{C \xi}.$$

Next, we consider

Case 2. $C \equiv 0$ for all t. In this case, the governing equations reduce to

$$A D E = -1, \tag{3.34}$$

$$F \dot{E} - E \dot{F} = k'', \tag{3.26}$$

and

$$\ddot{D} + \dot{D}\left(\frac{\dot{A}}{A} + \frac{\dot{D}}{D}\right) = 0. \tag{3.35}$$

We can integrate the last equation immediately and obtain

$$\dot{D} = \frac{k'''}{A D}. \tag{3.36}$$

Again, there are three possibilities:

2a). $F \equiv 0$ for all t. For this subcase the general solution is exactly the same as that given in the previous subcase 1a), except that now $C \equiv 0$.

2b). $F(t_0) \neq 0$ for some t_0. The general solution for t near t_0 is

B E, F, same as those given in subcase 1b) except that now $E \neq 0$;

$C \equiv 0$; and

A, D, same as those given in subcase 1a).

2c). $F(t_0)=0$ for some t_0, but $F \not\equiv 0$. Again, t_0 must be an isolated root for F, and the general solution for t near t_0 is

B, E, F, same as those given in subcase 1c);

$C \equiv 0$; and

A, D, same as those given in subcase 1a).

Finally, if we piece together solutions belonging to the various cases on different intervals, then we can obtain all particular solutions for the system. In any case, since the condition (3.6) is satisfied, the acceleration potential exists and is given by

$$\begin{aligned} -\zeta = & \frac{1}{2}\left[\frac{\ddot{A}}{2A} - \left(\frac{\dot{A}}{2A}\right)\right]^2 r^2 + \frac{1}{2}\left[\frac{\dot{A}}{A}\left(\frac{\dot{A}B}{A} - \dot{B}\right) - \left(\frac{A\ddot{B}}{A} - \ddot{B}\right)\right] \log r + \\ & + \frac{1}{8}\left(\frac{\dot{A}B}{A} - \dot{B}\right)^2 \frac{1}{r^2} - \frac{1}{2} A (F\dot{C} - E\dot{D})\, r^2 \theta^2 - \\ & - \frac{1}{2} A \left(\frac{\dot{A}B}{A} - \dot{B}\right)(F\dot{C} - E\dot{D})\,\theta^2 + \frac{1}{2} A (C\ddot{F} - D\ddot{E})\, z^2 + \Phi(t), \end{aligned} \tag{3.37}$$

where $\Phi(t)$ is arbitrary.

Next, we consider the laminated body $\bar{\mathscr{B}}$ introduced in the preceding section. There are two families of quasi-equilibrated motions:

Family 2. The deformation functions are given by (2.31), but now $A, \ldots, E$ are functions of t satisfying the condition (2.15).

For this family, it can be shown that the extra stress field at any instant t is given by (3.8), and the acceleration field is given by (3.9). Thus the general solution is identical with that given in Family 0. This fact is not surprising, since for any motion belonging to this family, the deformation from the configuration at any instant t to that at a later instant $t+\Delta t$ can be identified as a special case belonging to Family 0.

Family 3. The deformation functions are given by (2.34), but $A, \ldots, F$ are now functions of t. By the same argument as the one given above for Family 2, the result for this family is identical with that for Family 1.

Remark. Relative to suitable convected bases, the components of the local histories for all four families considered so far are indistinguishable. This situation is parallel to that of the viscometric flows.

3b. Curvilinear Laminated Bodies

We consider quasi-equilibrated motions for the laminated body $\tilde{\mathscr{B}}$ introduced in the previous section.

Family 4. The deformation functions are given by (2.47), where A is a function of t. The extra stress at any instant is again given by (2.52) and (2.53), so that the motions are quasi-equilibrated for the curvilinear laminated body $\tilde{\mathscr{B}}$.

The velocity field and the acceleration field can be determined in the same way as before. Here we have

$$\dot{r}=\frac{\dot{A}}{3r^2},$$
$$\dot{\theta}=\dot{\varphi}=0,$$
$$\ddot{r}=\frac{\ddot{A}}{3r^2}-\frac{2\dot{A}^2}{9r^5},\tag{3.38}$$

and

$$\ddot{\theta}=\ddot{\varphi}=0.$$

Thus the covariant components of the acceleration field are

$$a_r=\frac{\ddot{A}}{3r^2}-\frac{2\dot{A}^2}{9r^5}$$

and

$$a_\theta=a_\varphi=0.\tag{3.39}$$

In this case, it can be shown easily that the condition (3.6), or equivalently the condition (3.10), is always satisfied for arbitrary A, and the acceleration potential is given by

$$-\zeta=-\frac{\ddot{A}}{3r}+\frac{\dot{A}^2}{18r^4}+\Phi(t).\tag{3.40}$$

In any fixed motion belonging to the above families of dynamical universal solutions, the global response of a laminated body, of course, is different from that of a homogeneous body. We shall see some examples in the next section.

4. Particular Solutions

In the preceding two sections, we have seen certain families of exact static and dynamic universal solutions for the various laminated bodies $\mathscr{B}$, $\bar{\mathscr{B}}$, and $\tilde{\mathscr{B}}$. Now we work out some special cases of these universal solutions in detail. For simplicity, we assume that the particles of these bodies are elastic and are characterized by the constitutive equation (2.8).

First, we consider a simple shear, which is characterized by

$$x=X,\qquad y=Y+DZ,\qquad z=Z.\tag{4.1}$$

Clearly, this deformation is a special case of Family 0 given before by (2.14). For the moment, we focus our attention on the stress tensor at a fixed reference point $p\in\mathscr{B}$. We assume that the components of the intrinsic metric $\boldsymbol{g}$ at p has the simple form

$$[\bar{g}\langle\alpha\beta\rangle]=\operatorname{diag}[\bar{g}\langle 11\rangle,\ \bar{g}\langle 22\rangle,\ \bar{g}\langle 33\rangle],\tag{4.2}$$

in the (deformed) reference configuration $\bar{\boldsymbol{\varphi}}$.

From the general stress system considered in Section 2, the shear stress in the deformed configuration described by (4.1) is

$$T\langle 23\rangle = (h_1\,\bar{g}\langle 33\rangle - h_{-1}\,\bar{g}^*\langle 22\rangle)\,D$$
$$= \left(h_1\,\bar{g}\langle 33\rangle - h_{-1}\,\frac{1}{\bar{g}\langle 22\rangle}\right) D, \tag{4.3}$$

where the arguments of the functions h_1 and h_{-1} are

$$\operatorname{tr} g = \bar{g}\langle 11\rangle + \bar{g}\langle 22\rangle + \bar{g}\langle 33\rangle + D^2\,\bar{g}\langle 33\rangle,$$

and

$$\operatorname{tr} g^* = \frac{1}{\bar{g}\langle 11\rangle} + \frac{1}{\bar{g}\langle 22\rangle} + \frac{1}{\bar{g}\langle 33\rangle} + \frac{D^2}{\bar{g}\langle 22\rangle}. \tag{4.4}$$

Thus the shear modulus at p in the reference configuration $\bar{\varphi}$ is

$$\bar{\mu} = h_1\,\bar{g}\langle 33\rangle - h_{-1}\,\frac{1}{\bar{g}\langle 22\rangle}, \tag{4.5}$$

which is different from the shear modulus μ at the (initial) undistorted natural state, namely

$$\mu = h_1 - h_{-1}. \tag{4.6}$$

However, since we have assumed that $\bar{g}\langle 23\rangle = 0$ in $\bar{\varphi}$, the shear modulus $\bar{\mu}$ at p is still an even function of D:

$$\bar{\mu} = \bar{\mu}(D^2). \tag{4.7}$$

Consequently, if $\bar{\mu}$ is a smooth function, then for sufficiently small D

$$\bar{\mu} = \bar{\mu}_0 + \bar{\mu}_2\,D^2 + o(D^2), \tag{4.8}$$

where $\bar{\mu}_0$ and $\bar{\mu}_2$ are constants given by

$$\bar{\mu}_0 = \bar{\mu}(0) = \lim_{D\to 0} \frac{T\langle 23\rangle}{D},$$

and

$$\bar{\mu}_2 = \bar{\mu}'(0). \tag{4.9}$$

In order to see the effect of the initial deformation as characterized by (4.2) on the shear modulus more explicitly, let us choose the material to be of the Mooney-Rivlin type. Then the constitutive equation reduces to

$$\boldsymbol{T} = -p\,\mathbf{1} + \mu\left[\left(\tfrac{1}{2}+\beta\right)\boldsymbol{B} - \left(\tfrac{1}{2}-\beta\right)\boldsymbol{B}^{-1}\right], \tag{4.10}$$

where μ and β are constants. It is known that such a material is always hyperelastic, the stored-energy function being

$$\sigma = \sigma(\boldsymbol{B}) = \frac{\mu}{2\rho}\left[\left(\tfrac{1}{2}+\beta\right)(\mathbf{I}-3) + \left(\tfrac{1}{2}-\beta\right)(\mathbf{II}-3)\right], \tag{4.11}$$

where $\mathbf{I}$ and $\mathbf{II}$ denote the first and the second principal invariants of $\boldsymbol{B}$. A necessary and sufficient condition for this stored-energy function to be positive-definite for all unimodular $\boldsymbol{B}$ is that

$$\mu>0, \quad -\tfrac{1}{2}\leqq\beta\leqq\tfrac{1}{2}. \tag{4.12}$$

We assume that these conditions are satisfied here. It should be noted that the constant μ is the shear modulus as given by (4.6) for this material at the undistorted natural state and is independent of D.

From (4.5), the shear modulus $\bar{\mu}$ at the (deformed) reference state $\bar{\varphi}$ is given by

$$\bar{\mu}=\mu\left[(\tfrac{1}{2}+\beta)\,\bar{g}_{\langle 33\rangle}+(\tfrac{1}{2}-\beta)\,\frac{1}{\bar{g}_{\langle 22\rangle}}\right], \tag{4.13}$$

which is again a constant independent of D, but in general, $\bar{\mu}\neq\mu$. In fact, we can adjust the values of $\bar{g}_{\langle 22\rangle}$ and $\bar{g}_{\langle 33\rangle}$ in such a way that $\bar{\mu}$ can take on any positive value we like. In particular, $\bar{\mu}>\mu$ if and only if

$$(\tfrac{1}{2}+\beta)\,\bar{g}_{\langle 33\rangle}+(\tfrac{1}{2}-\beta)\,\frac{1}{\bar{g}_{\langle 22\rangle}}>1. \tag{4.14}$$

Now we go back to the general situation where the functions h_1 and h_{-1} are arbitrary. The normal stresses in the simple shear deformation can be written down immediately also:

$$\begin{aligned} T_{\langle 11\rangle}&=-p+h_1\,\bar{g}_{\langle 11\rangle}+h_{-1}\,\frac{1}{\bar{g}_{\langle 11\rangle}},\\ T_{\langle 22\rangle}&=-p+h_1\big(\bar{g}_{\langle 22\rangle}+D^2\,\bar{g}_{\langle 33\rangle}\big)+h_{-1}\,\frac{1}{\bar{g}_{\langle 22\rangle}},\\ T_{\langle 33\rangle}&=-p+h_1\,\bar{g}_{\langle 33\rangle}+h_{-1}\left(\frac{1}{\bar{g}_{\langle 33\rangle}}+\frac{D^2}{\bar{g}_{\langle 22\rangle}}\right). \end{aligned} \tag{4.15}$$

Thus if we hold the hydrostatic pressure fixed in the deformation, then the increments of the normal stresses are given by

$$\begin{aligned} \Delta T_{\langle 11\rangle}&=0,\\ \Delta T_{\langle 22\rangle}&=h_1\,D^2\,\bar{g}_{\langle 33\rangle}, \end{aligned} \tag{4.16}$$

and

$$\Delta T_{\langle 33\rangle}=h_{-1}\,D^2\,\frac{1}{\bar{g}_{\langle 22\rangle}}.$$

Clearly, these expressions are even functions of D and are of at least second order in D. Also, the universal relation[1]

$$\frac{\Delta T_{\langle 22\rangle}-\Delta T_{\langle 33\rangle}}{\Delta T_{\langle 23\rangle}}=D \tag{4.17}$$

holds.

For Mooney-Rivlin materials, the relations (4.16) reduce to

$$\Delta T_{\langle 22\rangle}=\mu(\tfrac{1}{2}+\beta)\,D^2\,\bar{g}_{\langle 33\rangle},$$

and

$$\Delta T_{\langle 33\rangle}=-\mu(\tfrac{1}{2}-\beta)\,D^2\,\frac{1}{\bar{g}_{\langle 22\rangle}}. \tag{4.18}$$

[1] This relation no longer holds if $\bar{g}_{\langle 23\rangle}\neq 0$.

Thus in this case we can adjust the magnitudes of $\Delta T\langle 22\rangle$ and $\Delta T\langle 33\rangle$ in any way we like by choosing suitable values for $\bar{g}\langle 22\rangle$ and $\bar{g}\langle 33\rangle$. We *cannot* adjust the values of $\Delta T\langle 22\rangle$, $\Delta T\langle 33\rangle$, and $\Delta T\langle 23\rangle$ independently, however, since the universal relation (4.17) must hold in any case.

We now consider pure torsion for the laminated body $\bar{\mathscr{B}}$. For simplicity, we choose $\bar{\mathscr{B}}$ to be a circular cylinder. Also, we assume that the intrinsic metric has the simple form (4.2) at *all* particles. Then we can define the shear modulus $\bar{\mu}$ for each fixed lamina, so that $\bar{\mu}$ now depends explicitly on R.

For a pure torsion, the deformation functions are given by

$$r=R, \qquad \theta=\Theta+DZ, \qquad z=Z, \tag{4.19}$$

which are special cases of (2.34). From the general stress system found in the Section 2, the shear stress in the deformed configuration is given by

$$T\langle 23\rangle=\Delta T\langle 23\rangle=D\,r\left(h_1\,\bar{g}\langle 33\rangle-h_{-1}\frac{1}{\bar{g}\langle 22\rangle}\right)=D\,r\,\bar{\mu}. \tag{4.20}$$

Thus the laminated body $\bar{\mathscr{B}}$ shares with the homogeneous body the property that in a pure torsion the shear stress can be determined by the shear modulus alone[1].

From (4.20), the resultant axial torque in the deformed configuration is given by

$$\tau=\int 2\pi r^2\,T\langle 23\rangle\,dr=2\pi D\int r^3\,\bar{\mu}(D^2 r^2, r)\,dr. \tag{4.21}$$

Here, we have made explicit the dependence of $\bar{\mu}$ on the radius r. The limiting value of the torsional rigidity as $D\to 0$ is then given by

$$\lim_{D\to 0}\frac{\tau}{D}=2\pi\int r^3\,\bar{\mu}_0(r)\,dr, \tag{4.22}$$

where $\bar{\mu}_0$ denotes the limiting shear moduli of the laminae as shown in equation (4.9). This result might have some practical applications, since it shows that we can design the torsional rigidity of a circular rod *having a fixed radius* by an appropriate lamination within the rod. The range for the possible torsional rigidities, of course, depends on the total variation of $\bar{\mu}$ on the initial deformations, as characterized by the relation (4.5). For a Mooney-Rivlin material, in principle, we can choose the torsional rigidity in any way we like, since from (4.13), the range for $\bar{\mu}$ is infinite.

In pure torsion, the resultant axial force does not vanish in general. In fact, even in the reference configuration $\bar{\varphi}$ the resultant axial force need not vanish, since the laminae are not necessarily in their initial states, in $\bar{\varphi}$. From the general stress system found before, we know that in the configuration $\bar{\varphi}$ the extra stress components are given by

$$\bar{S}\langle 11\rangle=\bar{h}_1\,\bar{g}\langle 11\rangle+\bar{h}_{-1}\frac{1}{\bar{g}\langle 11\rangle}, \tag{4.23}$$

---

[1] This property does not depend on the fact that the intrinsic metric has the special form as given by (4.2). If $\bar{g}\langle 23\rangle$ is not equal to zero, then the shear stress in the reference configuration $\bar{\varphi}$ does not vanish, but the *increment* of the shear stress in a pure torsion still can be determined by the shear moduli of the laminae.

etc. Then from (2.30),

$$\bar{T}_{\langle 11\rangle} = -\int \frac{1}{R}\left[\bar{h}_1(\bar{g}_{\langle 11\rangle} - \bar{g}_{\langle 22\rangle}) + \bar{h}_{-1}\left(\frac{1}{\bar{g}_{\langle 11\rangle}} - \frac{1}{\bar{g}_{\langle 22\rangle}}\right)\right] dR. \tag{4.24}$$

Consequently,

$$\begin{aligned} \bar{T}_{\langle 33\rangle} &= \bar{h}_1(\bar{g}_{\langle 33\rangle} - \bar{g}_{\langle 11\rangle}) + \bar{h}_{-1}\left(\frac{1}{\bar{g}_{\langle 33\rangle}} - \frac{1}{\bar{g}_{\langle 11\rangle}}\right) - \\ &\quad - \int \frac{1}{R}\left[\bar{h}_1(\bar{g}_{\langle 11\rangle} - \bar{g}_{\langle 22\rangle}) + \bar{h}_{-1}\left(\frac{1}{\bar{g}_{\langle 11\rangle}} - \frac{1}{\bar{g}_{\langle 22\rangle}}\right)\right] dR, \end{aligned} \tag{4.25}$$

and the axial force is

$$\bar{N} = 2\pi \int R\,\bar{T}_{\langle 33\rangle}\, dR. \tag{4.26}$$

For simplicity, we assume that the body $\bar{\mathscr{B}}$ is a solid circular rod with outer radius R_0, and the boundary condition is

$$\bar{T}_{\langle 11\rangle}|_{R=R_0} = 0. \tag{4.27}$$

Then the axial force $\bar{N}$ as given by (4.26) has a fixed value:

$$\bar{N} = \pi \int_0^{R_0} R\left[\bar{h}_1(2\bar{g}_{\langle 33\rangle} - \bar{g}_{\langle 11\rangle} - \bar{g}_{\langle 22\rangle}) + \bar{h}_{-1}\left(\frac{2}{\bar{g}_{\langle 33\rangle}} - \frac{1}{\bar{g}_{\langle 11\rangle}} - \frac{1}{\bar{g}_{\langle 22\rangle}}\right)\right] dR. \tag{4.28}$$

Of course, if we wish, we could choose the fields $\bar{g}_{\langle\alpha\beta\rangle}$ in such a way that $\bar{N}$ is equal to zero.

We can compute the *increment* of the axial force due to the pure torsion in the following way: First, the increments of the extra-stress components are given by

$$\begin{aligned} \Delta S_{\langle 11\rangle} &= 0, \\ \Delta S_{\langle 22\rangle} &= h_1 D^2 r^2 \bar{g}_{\langle 33\rangle}, \\ \text{and}\quad \Delta S_{\langle 33\rangle} &= h_{-1} D^2 r^2 \frac{1}{\bar{g}_{\langle 22\rangle}}. \end{aligned} \tag{4.29}$$

Thus an increment of the pressure field is necessary and is given by

$$\Delta p = -D^2 \int r\, h_1\, \bar{g}_{\langle 33\rangle}\, dr, \tag{4.30}$$

so that the increments of the normal stresses are

$$\begin{aligned} \Delta T_{\langle 11\rangle} &= D^2 \int r\, h_1\, \bar{g}_{\langle 33\rangle}\, dr, \\ \Delta T_{\langle 22\rangle} &= \Delta T_{\langle 11\rangle} + h_1 D^2 r^2 \bar{g}_{\langle 33\rangle}, \\ \text{and}\quad \Delta T_{\langle 33\rangle} &= \Delta T_{\langle 11\rangle} + h_{-1} D^2 r^2 \frac{1}{\bar{g}_{\langle 22\rangle}}, \end{aligned} \tag{4.31}$$

where the integration limits again can be set by the boundary condition (4.27), now for the field $\Delta T_{\langle 11\rangle}$.

The increment of the resultant axial force can now be determined:

$$\begin{aligned} \Delta N &= 2\pi \int_0^{r_0} r\, \Delta T_{\langle 33\rangle}\, dr = \\ &= -\pi D^2 \int_0^{r_0} r^3 \left(h_1\, \bar{g}_{\langle 33\rangle} - 2h_{-1} \frac{1}{\bar{g}_{\langle 22\rangle}}\right) dr, \end{aligned} \tag{4.32}$$

where r_0 is equal to R_0. From this expression, we see that for sufficiently small D, ΔN is of second or higher order in D. Also, if h_1 and h_{-1} obey the *E-inequalities*,

$$h_1 > 0, \quad h_{-1} \leqq 0, \tag{4.33}$$

then ΔN is always negative. These results are extensions of a version of the classical Poynting effect in pure torsion for the laminated body $\bar{\mathscr{B}}$. It should be noted that any Mooney-Rivlin material obeying the condition (4.12) satisfies also the E-inequality (4.33).

The exact Poynting effect, however, is characterized by the axial elongation in a torsion free of the increment ΔN. Thus we now consider the problem of torsion with extension and inflation. The deformation functions are

$$r = \sqrt{A}\, R, \quad \theta = \Theta + D Z, \quad z = F Z, \tag{4.34}$$

which are special cases of (2.34). From (2.35),

$$A F = 1. \tag{4.35}$$

First, we determine the torsional rigidity when the axial elongation F is held constant. From the general stress system for Family 3 determined in Section 2, the shear stress $T\langle 23\rangle$ is given by

$$T\langle 23\rangle = \left(h_1 F \bar{g}\langle 33\rangle - h_{-1} \frac{1}{\bar{g}\langle 22\rangle}\right) D r, \tag{4.36}$$

where the arguments of h_1 and h_{-1} are

$$\operatorname{tr} \boldsymbol{g} = A \bar{g}\langle 11\rangle + A \bar{g}\langle 22\rangle + F^2 \bar{g}\langle 33\rangle + D^2 r^2 \bar{g}\langle 33\rangle,$$

and

$$\operatorname{tr} \boldsymbol{g}^* = \frac{1}{A \bar{g}\langle 11\rangle} + \frac{1}{A \bar{g}\langle 22\rangle} + \frac{1}{F^2 \bar{g}\langle 33\rangle} + A D^2 r^2 \frac{1}{\bar{g}\langle 22\rangle}, \tag{4.37}$$

while the argument of $\bar{g}\langle\alpha\alpha\rangle$ is $r/\sqrt{A}$, of course. It should be noted that the expression (4.36) can be put into the standard form

$$T\langle 23\rangle = \frac{D}{F} r \left(h_1 \tilde{g}\langle 33\rangle - h_{-1} \frac{1}{\tilde{g}\langle 22\rangle}\right) = \frac{D}{F} r \tilde{\mu}, \tag{4.38}$$

so that it becomes a special case of (4.20). Here $\tilde{g}\langle\alpha\alpha\rangle$, of course, denotes the components of the intrinsic metric in the configuration $\tilde{\varphi}$ which differs from the reference configuration $\bar{\varphi}$ by a pure axial stretch of amount F, and $\tilde{\mu}$ is the field of shear moduli in $\tilde{\varphi}$. Thus

$$\tilde{g}\langle 11\rangle = A \bar{g}\langle 11\rangle,$$

and

$$\tilde{g}\langle 22\rangle = A \bar{g}\langle 22\rangle, \tag{4.39}$$

$$\tilde{g}\langle 33\rangle = F^2 \bar{g}\langle 33\rangle.$$

The axial torque can now be determined either from (4.21) or directly from (4.38),

$$\tau = 2\pi \frac{D}{F} \int_0^{r_0} r^3 \tilde{\mu}\left(\frac{D^2}{F^2} r^2, r\right) dr. \tag{4.40}$$

Likewise, the limiting value of the axial rigidity in the state $\tilde{\varphi}$ is given by

$$\lim_{D\to 0} \frac{\tau}{D/F} = 2\pi \int_0^{r_0} r^3 \tilde{\mu}_0(r)\, dr. \tag{4.41}$$

By exactly the same argument, we can determine also the increment of the axial force for the pure torsion applied to the configuration $\tilde{\varphi}$,

$$\begin{aligned} \widetilde{\Delta N} &= -\pi \left(\frac{D}{F}\right)^2 \int_0^{r_0} r^3 \left(h_1\, \tilde{g}_{\langle 33\rangle} - 2h_{-1} \frac{1}{\tilde{g}_{\langle 22\rangle}}\right) dr = \\ &= -\pi \left(\frac{D}{F}\right)^2 \int_0^{R_0} \left(h_1\, \bar{g}_{\langle 33\rangle} - 2h_{-1} \frac{1}{\bar{g}_{\langle 22\rangle}}\right) R^3\, dR. \end{aligned} \tag{4.42}$$

But since the axial force in the configuration $\tilde{\varphi}$, in general, is different from that in the reference configuration $\bar{\varphi}$, the quantity $\widetilde{\Delta N}$ alone does *not* represent the total increment of the axial force for the deformation given by (4.35).

To determine the increment of the axial force from the configuration $\bar{\varphi}$ to the configuration $\tilde{\varphi}$, we must determine the resultant axial force in each of these two configurations. Of course, $\bar{N}$ has been found by (4.28). By the same argument,

$$\begin{aligned} \tilde{N} &= \pi \int_0^{r_0} r \left[\tilde{h}_1 (2\tilde{g}_{\langle 33\rangle} - \tilde{g}_{\langle 11\rangle} - \tilde{g}_{\langle 22\rangle}) + \tilde{h}_{-1} \left(\frac{2}{\tilde{g}_{\langle 33\rangle}} - \frac{1}{\tilde{g}_{\langle 11\rangle}} - \frac{1}{\tilde{g}_{\langle 22\rangle}}\right)\right] dr \\ &= \pi \int_0^{R_0} R \left[\tilde{h}_1 (2F\, \bar{g}_{\langle 33\rangle} - A^2\, \bar{g}_{\langle 11\rangle} - A^2\, \bar{g}_{\langle 22\rangle}) + \right. \\ &\quad \left. + \tilde{h}_{-1} \left(\frac{2}{F^3\, \bar{g}_{\langle 33\rangle}} - \frac{1}{\bar{g}_{\langle 11\rangle}} - \frac{1}{\bar{g}_{\langle 22\rangle}}\right)\right] dR, \end{aligned} \tag{4.43}$$

where the arguments of $\tilde{h}_1$, and h_{-1} are

$$\widetilde{\operatorname{tr}}\, \mathbf{g} = A\, \bar{g}_{\langle 11\rangle} + A\, \bar{g}_{\langle 22\rangle} + F^2\, \bar{g}_{\langle 33\rangle},$$

and

$$\widetilde{\operatorname{tr}}\, \mathbf{g}^* = \frac{1}{A\, \bar{g}_{\langle 11\rangle}} + \frac{1}{A\, \bar{g}_{\langle 22\rangle}} + \frac{1}{F^2\, \bar{g}_{\langle 33\rangle}}. \tag{4.44}$$

Thus the increment is

$$\begin{aligned} \tilde{N} - \bar{N} &= \pi(F-1) \int_0^{R_0} R \left[\bar{h}_1 \left(2\bar{g}_{\langle 33\rangle} + \frac{F+1}{F^2} (\bar{g}_{\langle 11\rangle} + \bar{g}_{\langle 22\rangle})\right) - \right. \\ &\quad \left. - 2\bar{h}_{-1} \frac{F^2+F+1}{F^3}\, \bar{g}_{\langle 33\rangle}\right] dR + \\ &\quad + \pi \int_0^{R_0} R \left[(\tilde{h}_1 - \bar{h}_1)(2F\, \bar{g}_{\langle 33\rangle} - A^2\, \bar{g}_{\langle 11\rangle} - A^2\, \bar{g}_{\langle 22\rangle}) + \right. \\ &\quad \left. + (\tilde{h}_{-1} - \bar{h}_{-1}) \left(\frac{2}{F^3\, \bar{g}_{\langle 33\rangle}} - \frac{1}{\bar{g}_{\langle 11\rangle}} - \frac{1}{\bar{g}_{\langle 22\rangle}}\right)\right] dR. \end{aligned} \tag{4.45}$$

Assuming that h_1 and h_{-1} are smooth functions, we can express the differences $\tilde{h}_1-\bar{h}_1$ and $\tilde{h}_{-1}-\bar{h}_{-1}$ approximately by

$$\begin{aligned}\tilde{h}_1-\bar{h}_1=(F-1)\Bigg[&\overline{\frac{\partial h_1}{\partial \mathbf{I}}}(-\bar{g}\langle 11\rangle-\bar{g}\langle 22\rangle+2\bar{g}\langle 33\rangle)+\\&+\overline{\frac{\partial h_1}{\partial \mathbf{II}}}\left(\frac{1}{\bar{g}\langle 11\rangle}+\frac{1}{\bar{g}\langle 22\rangle}-\frac{2}{\bar{g}\langle 33\rangle}\right)\Bigg]+o(F-1),\end{aligned} \tag{4.46}$$

and

$$\begin{aligned}\tilde{h}_{-1}-\bar{h}_{-1}=(F-1)\Bigg[&\overline{\frac{\partial h_{-1}}{\partial \mathbf{I}}}(-\bar{g}\langle 11\rangle-\bar{g}\langle 22\rangle+2\bar{g}\langle 33\rangle)+\\&+\overline{\frac{\partial h_{-1}}{\partial \mathbf{II}}}\left(\frac{1}{\bar{g}\langle 11\rangle}+\frac{1}{\bar{g}\langle 22\rangle}-\frac{2}{\bar{g}\langle 33\rangle}\right)\Bigg]+o(F-1).\end{aligned} \tag{4.47}$$

From these expressions, we see that the quantity $\tilde{N}-\bar{N}$ is of first order in $F-1$.

We now lay down the condition

$$\Delta N=\tilde{N}-N+\widetilde{\Delta N}=0 \tag{4.48}$$

for the exact Poynting problem. From (4.42) and (4.45), this condition determines a fixed relation between the twist D and the elongation F. Since $\tilde{N}-\bar{N}$ is of first order in $(F-1)$, and $\widetilde{\Delta N}$ is of second order in D/F, for sufficiently small D the condition (4.48) implies that

$$(F-1)\propto\left(\frac{D}{F}\right)^2. \tag{4.49}$$

This is the desired answer to the exact Poynting problem for small twist.

The limiting value for the ratio of $F-1$ to $(D/F)^2$ as $D\to 0$, $F\to 1$, is called the *Poynting modulus*, denoted by Π. For the laminated body $\mathscr{B}$ considered here, the Poynting modulus can be determined immediately by the expressions (4.43), (4.45), (4.46), and (4.47):

$$\Pi=\lim_{\substack{D\to 0\\ F\to 1}}\frac{F-1}{(D/F)^2}=\frac{\Pi_1}{\Pi_2}, \tag{4.50}$$

where

$$\Pi_1=\int_0^{R_0}\left(\bar{h}_1\,\bar{g}\langle 33\rangle-2\bar{h}_{-1}\frac{1}{\bar{g}\langle 22\rangle}\right)R^3\,dR, \tag{4.51}$$

and

$$\begin{aligned}\Pi_2=&\int_0^{R_0}2[\bar{h}_1(\bar{g}\langle 11\rangle+\bar{g}\langle 22\rangle+\bar{g}\langle 33\rangle)-3\bar{h}_{-1}\,\bar{g}\langle 33\rangle]\,R\,dR+\\&+\int_0^{R_0}\Bigg[\overline{\frac{\partial h_1}{\partial \mathbf{I}}}(2\bar{g}\langle 33\rangle-\bar{g}\langle 11\rangle-\bar{g}\langle 22\rangle)^2+\\&+\left(\overline{\frac{\partial h_1}{\partial \mathbf{II}}}-\overline{\frac{\partial h_{-1}}{\partial \mathbf{I}}}\right)(2\bar{g}\langle 33\rangle-\bar{g}\langle 11\rangle-\bar{g}\langle 22\rangle)\\&\cdot\left(\frac{1}{\bar{g}\langle 11\rangle}+\frac{1}{\bar{g}\langle 22\rangle}-\frac{2}{\bar{g}\langle 33\rangle}\right)-\overline{\frac{\partial h_{-1}}{\partial \mathbf{II}}}\left(\frac{1}{\bar{g}\langle 11\rangle}+\frac{1}{\bar{g}\langle 22\rangle}-\frac{2}{\bar{g}\langle 33\rangle}\right)^2\Bigg]R\,dR.\end{aligned} \tag{4.52}$$

Of course, we can obtain the usual Poynting modulus for a homogeneous circular rod by setting $\bar{g}_{\langle\alpha\alpha\rangle}=1$ in the above expressions. Then

$$\hat{\Pi}_1=\tfrac{1}{4}(\hat{h}_1-2\hat{h}_{-1})R_0^4,$$

and (4.53)

$$\hat{\Pi}_2=3(\hat{h}_1-\hat{h}_{-1})R_0^2,$$

so that[1]

$$\hat{\Pi}=\frac{1}{12}\left(\frac{\hat{h}_1-2\hat{h}_{-1}}{\hat{h}_1-\hat{h}_{-1}}\right)R_0^2. \tag{4.54}$$

Here the arguments of h_1 and $\hat{h}_{-1}$ are the constants (3.3).

Now we consider the problem of radial oscillation of a laminated hollow circular cylinder. The solution for this problem is a special case of the dynamical universal solutions belonging to Family 3 found in the Section 3. For simplicity, we assume that the metric g again has the special component form (4.2) at all particles relative to a cylindrical coordinate system (R, Θ, Z) in the reference configuration $\bar{\varphi}$. Also, we assume that the cylinder occupies the region

$$0<R_1<R<R_2. \tag{4.55}$$

In the deformed configuration, we use the cylindrical coordinate system (r, θ, z). Then the deformation functions are

$$r^2=R^2-R_1^2+r_1^2, \qquad \theta=\Theta, \qquad z=Z, \tag{4.56}$$

where $r_1=r_1(t)$ is the inner radius at time t.

From the general solution for Family 3 in Section 3, we know that (4.56) defines a dynamically possible quasi-equilibrated motion for the laminated body, the acceleration potential being

$$-\zeta=(r_1\dot{r}_1)\log r+\frac{1}{2}\frac{r_1^2\dot{r}_1^2}{r^2}. \tag{4.57}$$

Then from the general result (3.7), the radial stress in the motion is given by

$$\begin{aligned} T_{\langle rr\rangle}=\rho\left[(r_1\ddot{r}_1+\dot{r}_1^2)\log r+\frac{1}{2}\frac{r_1^2\dot{r}_1^2}{r^2}\right]+\Phi(t)-\\ -\int\frac{1}{r}\left(\frac{R^2}{r^2}\bar{g}_{\langle 11\rangle}-\frac{r^2}{R^2}\bar{g}_{\langle 22\rangle}\right)(h_1-h_{-1}\bar{g}_{\langle 33\rangle})\,dr, \end{aligned} \tag{4.58}$$

where the arguments of h_1 and h_{-1} are

$$\operatorname{tr} g=\frac{R^2}{r^2}\bar{g}_{\langle 11\rangle}+\frac{r^2}{R^2}\bar{g}_{\langle 22\rangle}+\bar{g}_{\langle 33\rangle},$$

and (4.59)

$$\operatorname{tr} g^*=\frac{r^2}{R^2}\frac{1}{\bar{g}_{\langle 11\rangle}}+\frac{R^2}{r^2}\frac{1}{\bar{g}_{\langle 22\rangle}}+\frac{1}{\bar{g}_{\langle 33\rangle}},$$

---

[1] *Cf.* TRUESDELL & NOLL [3, Eq. (57.24)].

while the argument of $\bar{g}\langle\alpha\alpha\rangle$ is R. Of course,

$$R=\sqrt{r^2-r_1^2+R_1^2}. \tag{4.60}$$

Now let $P_1(t)$ and $P_2(t)$ be the pressure on the inner and the outer faces of the cylinder. Then

$$P_1(t)-P_2(t)=T\langle rr\rangle|_{r_2}-T\langle rr\rangle|_{r_1}. \tag{4.61}$$

The right-hand side of this equation can be determined in terms of the function $r_1(t)$ from the expression (4.58). If we set

$$x=\frac{r_1}{R_1}, \qquad \delta=\frac{R_2^2}{R_1^2}-1>0,$$

$$u=\frac{r^2}{R^2}, \tag{4.62}$$

and

$$f(x,\delta)=\frac{1}{\rho R_1^2}\int_{\frac{\delta+x^2}{\delta+1}}^{x^2}\left(\frac{1}{u}\,\bar{g}\langle 11\rangle-u\,\bar{g}\langle 22\rangle\right)(h_1-h_{-1}\,\bar{g}\langle 33\rangle)\,\frac{du}{u(1-u)},$$

where the argument for $\bar{g}\langle\alpha\alpha\rangle$ is now given by

$$R=R_1\sqrt{\frac{x^2-1}{u-1}}, \tag{4.63}$$

then the governing equation for x is

$$x\log\left(1+\frac{\delta}{x^2}\right)\ddot{x}+\left[\log\left(1+\frac{\delta}{x^2}\right)-\frac{\delta}{\delta+x^2}\right]\dot{x}^2+f(x,\delta)=\frac{P_1(t)-P_2(t)}{\frac{1}{2}\rho R_1^2}. \tag{4.64}$$

This equation has exactly the same form[1] as that for the radial oscillations of a homogeneous hollow cylinder, the only difference being the new representation for the function f. For a homogeneous cylinder, $\bar{g}\langle\alpha\alpha\rangle=1$. Then[2] $(4.62)_4$ reduces to

$$\hat{f}(x,\delta)=\frac{1}{\rho R_1^2}\int_{\frac{\delta+x^2}{\delta+1}}^{x^2}(1+u)(h_1-h_{-1})\,\frac{du}{u}, \tag{4.65}$$

where both arguments of h_1 and h_{-1} are

$$1+u+\frac{1}{u}.$$

Thus the effect of lamination on the radial oscillations is entirely characterized by the form of the function $f(x,\delta)$, as given by $(4.62)_4$.

For a laminated cylinder, the value of $f(1,\delta)$, in general, need *not* be equal to zero, because it is the difference of pressures on the inner and the outer surfaces when the cylinder is in the reference state $\bar{\varphi}$. If $f(1,\delta)\neq 0$, the integrand in $(4.62)_4$ has a singularity at the point $u=1$. For a homogeneous cylinder, of course, $\hat{f}(1,\delta)=0$, and the singularity at $u=1$ is a removable one, as shown by (4.65).

---

[1] TRUESDELL & NOLL [3, Eq. (62.5)]. This equation was obtained originally by KNOWLES.
[2] TRUESDELL & NOLL [3, Eq. $(62.4)_4$].

A neutral position x_0 for a laminated cylinder is characterized by the equation

$$f(x_0, \delta) = 0. \tag{4.66}$$

Clearly,

$$x(t) \equiv x_0 \qquad \forall t \tag{4.67}$$

is a solution of the governing equation (4.64) when $P_1(t) - P_2(t)$ is held at the value zero for all t. This neutral position is a (locally) stable one if

$$f'(x_0, \delta) > 0. \tag{4.68}$$

In this case, free oscillation about x_0 is characterized by the energy equation

$$\tfrac{1}{2} \dot{x}^2 x^2 \log\left(1 + \frac{\delta}{x^2}\right) + F(x, \delta) = C, \tag{4.69}$$

where

$$\begin{aligned} F(x, \delta) &= \int_{x_0}^{x} \xi f(\xi, \delta)\, d\xi = \\ &= \frac{1}{\rho R_1^2} \int_{x_0}^{x} \xi\, d\xi \int_{\frac{\xi^2 + \delta}{1 + \delta}}^{\xi^2} \left(\frac{1}{u} \bar{g}_{\langle 11 \rangle} - u\, \bar{g}_{\langle 22 \rangle}\right) (h_1 - h_{-1}\, \bar{g}_{\langle 33 \rangle}) \frac{du}{u(1-u)}. \end{aligned} \tag{4.70}$$

The constant of integration, C, can be determined by a set of initial conditions, say

$$x(0) = \bar{x}, \qquad \dot{x}(0) = \dot{\bar{x}}. \tag{4.71}$$

Then an oscillatory motion is possible if and only if the equation

$$F(x, \delta) = C \tag{4.72}$$

has at least two roots, one on each side of the neutral position x_0. In such motion, of course, $F(x, \delta) < C$ except for those x such that $\dot{x} = 0$, corresponding to the extremal positions in the oscillation. For a stable neutral position x_0, oscillatory motion is always possible for sufficiently small values of C.

Additional analysis concerning free or forced oscillation for a laminated cylinder can be carried out in exactly the same way as those for a homogeneous cylinder. We pass over the details. Clearly, a similar analysis for the problem of radial oscillations for a laminated spherical shell can be worked out also. Further, there are many interesting static problems other than the torsion problem, such as bending and eversion, which are also special cases of the universal solutions given in Section 2. As a general rule, the effects of lamination on the global response all can be expressed by certain integrals depending on the initial fields $\bar{g}_{\langle \alpha\beta \rangle}$, in the reference state $\bar{\varphi}$. I leave the details of these additional applications of our general solutions to the reader.

5. Some Conjectures Concerning Laminated Bodies

Suppose that a certain family of deformations is given. Then we can consider the problem of determining the most general form of the field g such that the given deformations can be identified as universal solutions. In general, this problem need not have any solution, since the governing equations for tghe field, clearly, form a highly overdetermined system. However, for certain special families of deformations which already allow

$$g=G \tag{5.1}$$

to be a solution, we have observed above some other non-trivial solutions. These are the metrics defining the various laminated bodies treated in the preceding sections. But in view of the complexity of the governing equations, I have not been able to show whether the observed solutions are complete. I offer here the following conjectures:

Conjecture 1. In order that the deformations given by (2.14), (2.23), (2.31), or (2.34) be universal solutions, it is necessary and sufficient that the body be a locally rectilinear laminated body of the class $\mathscr{B}$ or $\overline{\mathscr{B}}$.

Of course, if this conjecture is correct, then we have also the following one:

Conjecture 2. In order that *all* homogeneous deformations be universal solutions, it is necessary and sufficient that the body be homogeneous (and the reference configuration be its homogeneous state).

I propose also

Conjecture 3. In order that the deformations given by (2.49) be universal solutions, it is necessary and sufficient that the body be a curvilinear laminated body of the class $\tilde{\mathscr{B}}$.

Conjecture 4. In order that the deformations given by (2.57) be universal solutions, it is necessary and sufficient that the body be homogeneous.

If these conjectures turn out to be correct, then the analysis in this paper solves the problem set in the introduction: to compare the global response of inhomogeneous incompressible isotropic bodies with that of homogeneous bodies while both suffer exactly the same deformation. It should be noted that the preceding conjectures do *not* rule out the possibility of having universal solutions for certain inhomogeneous bodies which are *not* universal solutions for homogeneous bodies. These solutions, however, are still wanting.

Acknowledgement. I am indebted to C. TRUESDELL for criticism of a previous draft of this paper. The work reported here was supported by a Grant from the U. S. National Science Foundation to the Johns Hopkins University.

References

1. NOLL, W., Arch. Rational Mech. Anal. **27**, 1—32 (1967).
2. WANG, C.-C., Arch. Rational Mech. Anal. **27**, 33—94 (1967).
3. TRUESDELL, C., & W. NOLL, The Non-Linear Field Theories of Mechanics. In: Flügge's Handbuch der Physik, Bd. III/3. Berlin-Heidelberg-New York: Springer 1965.
4. SINGH, M., & A. C. PIPKIN, ZAMP **16**, 706—709 (1965).
5. CARROLL, M. M., Int. J. Engng. Sci. **15**, 515—525 (1967).
6. FOSDICK, R. L., Dynamically possible motions of incompressible, isotropic, simple materials. Arch. Rational Mech. Anal. forthcoming.

The Johns Hopkins University
Baltimore, Maryland

(Received January 20, 1968)